The A to Z of Mathematics

The A to Z of Mathematics

A Basic Guide

Thomas H. Sidebotham
St. Bede's College
Christchurch, New Zealand

A JOHN WILEY & SONS, INC., PUBLICATION

This book is printed on acid-free paper. ⊗

Copyright © 2002 by John Wiley & Sons, Inc., New York. All rights reserved.

Published simultaneously in Canada.

For ordering and customer service, call 1-800-CALL-WILEY.

Library of Congress Cataloging-in-Publication Data is available

The A to Z of Mathematics : A Basic Guide—Thomas H. Sidebotham

ISBN 0-471-15045-2

Printed in the United States of America.

10 9 8 7 6 5 4 3 2 1

To my wife, Patricia

*Who persuaded me to get started and
supported me until it was finished*

Contents

Preface

Throughout the world many people suffer from the same problem: math anxiety. No other area of skill seems to polarize people so readily into two contrasting groups, those who can do math and those who cannot. Of the two groups, the second one is by far the larger. To succeed in mathematics you need to understand the basics, and only then can you learn with confidence. Many people fall at this first hurdle and then struggle later. My aim in writing this book is to guide you through the basics so that you can develop an understanding of mathematical processes. As you study this book you will become aware of how mathematics relates to everyday life and situations with which you are familiar. Study this book in depth, simply browse, or search for the meaning of a word, and learn your math again. Why should you go to this trouble? Whatever your age, mathematics is one of the basic requirements of life. This study of mathematics will make a difference.

This book is written in an appropriate language for explaining basic mathematics to the general reader, and uses examples drawn from everyday life. There are many worked examples with detailed steps of working. Each step of working is accompanied by an explanation. It is this process of showing HOW and explaining WHY that gives this book its unique style. Those mathematical abbreviations that often frustrate readers are written in full and the text is "user-friendly." For quick reference the format of the book is alphabetical, and it covers topics in basic mathematics. They are linked together with cross-references so that a theme can be followed through. This book is a great deal more than a dictionary. Under each entry there is a straightforward explanation of the term, followed in many cases by carefully worked examples, showing the relevance of mathematics in the world around us. At the end of some entries the reader is directed to other references in the book if some prior knowledge is needed. The mathematics is reliable and up to date, and encompasses a wide range of topics so that everyone will find something of interest.

The material in the book falls into three categories.

1. There are processes that explain specific skills; a typical example is the entry Algebra.
2. There are straightforward definitions of words with applications in the world around us, as in the entry Proportion.
3. There is a variety of enrichment material that has good entertainment value, like Hexomino.

I believe there is something of interest for everyone. If you are curious about mathematics and it intrigues you, now may be the time to take the initiative and discover that you indeed have skills in this area of knowledge. Some people need the maturity of a few more years before they achieve success. If you are making a career change and need to revise your mathematical knowledge, then this book is for you. The book will appeal to everyone, even students, who may be interested in, or need to catch up on, basic mathematics. If you are a parent who desires to help your son or daughter and lack the expertise, then this book is for you also. The style and presentation of the book are chosen specifically to suit the lay reader. It is a useful resource for home schooling situations. I hope it is a rich source of ideas for mathematics teachers and also those who are in teacher training, whatever subject in which they are specializing.

You will need a scientific calculator to follow through the steps of working in some of the examples. In statistical topics the reader is referred to the calculator handbook for its use, because brands of calculators can vary so much. The whole book is cross-referenced. If readers are not familiar with the explanations given in a specific entry, they are advised to first read the references at the end of the entry to prepare the groundwork. This book contains an abundance of diagrams, equations, tables, graphs, and worked examples. An emphasis is placed on SI units throughout.

If you are keen to acquire the basic skills of mathematics in this book, I offer the following advice. Do not read it like you may read a novel in which you can skim and still enjoy the book and have a good grasp of the story and plot. To grasp mathematics you must examine the detail of every word and symbol. Have a pen, a calculator, and paper at hand to try out the processes and verify them for yourself.

It is my hope that this book will be on a bookshelf in every home, and will be used by family members as a reference and guide. I am sure you will find it useful, interesting, and entertaining.

TOM SIDEBOTHAM

Acknowledgments

I take this opportunity to thank the following persons, without whose help and guidance this book would not have been possible: Steve Quigley, for encouraging, enthusing, and redirecting me, and for sharing the vision; Heather Haselkorn, for all her efforts on my behalf to keep the project afloat and for maintaining the lines of communication; David Byatt, for scrutinizing the text and offering judicious alternatives (the accuracy is entirely the responsibility of the author); Dr. David Sidebotham, for his computer skills in enabling me to transplant the book in New York; Stephanie Lentz and her team at TechBooks for transforming my manuscript so wonderfully well; and lastly, the Angel at my elbow throughout the writing.

$$A$$

ABSOLUTE VALUE

The absolute value of a number is the distance of the number from the origin 0, measured along the number line. In the accompanying figure the absolute value of 3 at point B is the distance of point B from 0, which is 3. The absolute value of -2 at point A is the distance of point A from 0, which is 2.

The symbol for the absolute value of a number is two vertical parallel lines placed around the number. At the point B, the absolute value of 3 is written as

$$|\text{point } B - \text{point } 0| = |3 - 0|$$

$$= |3|$$

$$= 3$$

If the subtraction had taken place in the opposite way, the answer would still be 3, but the working would be

$$\text{Absolute value of } 3 = |\text{point } 0 - \text{point } B|$$

$$= |0 - 3|$$

$$= |-3|$$

$$= 3 \qquad \text{as before}$$

Similarly for the point A:

$$\text{Absolute value of } -2 = |\text{point } A - \text{point } 0|$$

$$= |-2 - 0|$$

$$= |-2|$$

$$= 2$$

The absolute value of the line segment AB is obtained in the following way:

$$\text{Absolute value of } AB = |\text{point } A - \text{point } B| \quad \text{or} \quad |\text{point } B - \text{point } A|$$
$$= |-2 - 3|$$
$$= |-5|$$
$$= 5$$

When an absolute value is applied to a subtraction, it is safe to subtract the two numbers on the number line in either direction; one gets the same answer for the length of the line segment that joins them.

References: Integers, Number Line.

ABSTRACT

"Abstract" means separated from practical problems in the real world. Mathematics is an abstract subject and is practiced using symbols, but these symbols can represent real-life things. Real problems can be solved using abstract mathematics, and then the symbols used can be related back to solve the practical problem. In the example that follows the practical problem is about finding the width of a lawn. In the abstract, the symbol representing this width is x. The value of x is found using abstract mathematics, and so the width of the lawn is known and the problem solved.

Example. Andrew has 24 square meters of ready-made lawn and plans to lay it in his garden. His wife, Jo, suggests that a good shape is a rectangle that is twice as long as it is wide. What width should Andrew make the lawn?

Solution. This is a real-life problem, but it can be solved using abstract symbols in the following way: Let the width of the lawn be x meters; its length will be $2x$, because its length is twice its width (see the figure): The calculation goes as follows:

Area of rectangle = length × width

$24 = 2x \times x$ This is an abstract equation

$24 = 2x^2$ Multiplying the terms on the right-hand side

$12 = x^2$ Dividing both sides of the equation by 2

$$\sqrt{12} = x \qquad \text{Taking the square root of both sides of the equation}$$

$$x = 3.464 \qquad \text{Using a calculator for the square root}$$

Now relate the abstract symbol x back to the real-life situation. The width of the lawn is 3.46 meters (three significant figures).

References: Accuracy, Konigsberg Bridge Problem, Square Root.

ACCELERATION

If a car is increasing its velocity, for example, upon changing lanes on a freeway, its acceleration is the rate at which its velocity is changing with respect to time. The SI unit of acceleration is meters per second per second, which is abbreviated m s^{-2} or m/s^2. Another unit of acceleration is centimeters per second per second, which is abbreviated cm s^{-2} or cm/s^2.

As a body falls to earth it has an acceleration due to gravity that is approximately $10\,\text{m s}^{-2}$, and this acceleration is a fixed value at different places on the earth's surface. This means that when a stone is falling through the air, its velocity increases by $10\,\text{m s}^{-1}$ for every second it is falling. In this example we ignore the air resistance. Suppose the velocity of the stone is measured every second and recorded in a table as follows:

Time t in s	0	1	2	3
Velocity v in m s^{-1}	0	10	20	30

The values from the table are written as ordered pairs and then plotted to draw a velocity–time graph (see figure a). In this case the ordered pairs are (0, 0), (1, 10), (2, 20), and (3, 30).

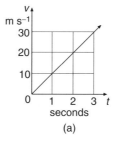

(a)

$$\text{The slope of this graph} = \frac{\text{rise}}{\text{run}} \qquad \text{Formula for slope}$$

$$= \frac{30\ \text{m/s}}{3\ \text{s}}$$

$$= 10\,\text{m s}^{-2}$$

The acceleration of the stone is $10\,\text{m s}^{-2}$.

The slope of a velocity–time graph gives the acceleration. If the stone is thrown upward, the velocity is decreasing, and the negative acceleration is called retardation. When the acceleration is not a fixed quantity, but varies, the velocity–time graph is a curve. This is the situation, for example, for a motorcyclist who accelerates from a standing position.

Example. A motorcyclist is accelerating from a standing position, and the velocities in m s^{-1} are recorded every 10 seconds. This information is shown in the table. The velocity of the cyclist can be expressed by the formula $v = 0.03\,t^2$. Find the acceleration of the cyclist when the time is 20 seconds.

Time in seconds	0	10	20	30
Velocity in m s^{-1}	0	3	12	27

Solution. Using the data in the table, we draw the velocity–time graph (see figure b). The acceleration of the motorcyclist is changing, which is indicated by the curved graph. To find the acceleration at $t = 20$ seconds, we find the slope of the curve at the point on the graph where t is 20 seconds, and this slope gives the acceleration at that instant. The slope of the curve is the slope of the tangent to the curve at that point.

Using a ruler, we draw a tangent as accurately as possible at the point on the curve where $t = 20$ seconds. The tangent should just touch the curve at this point. The length of the tangent is not critical, but should not be too short, otherwise accuracy will be lost. The slope of this tangent will give the acceleration of the cyclist at 20 seconds.

Complete the right-angled triangle, and estimate the rise and the run, but take care with the units:

$$\text{Slope of tangent} = \frac{\text{rise}}{\text{run}}$$

$$= \frac{22\ \text{m/s}}{20\ \text{seconds}}$$

$$= 1.1\ \text{m s}^{-2}$$

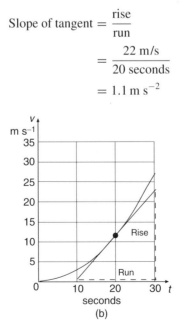

(b)

The acceleration of the cyclist at the time when $t = 20$ seconds is estimated to be 1.1 m s^{-2}.

References: Ordered Pairs, Retardation, SI Units, Slope, Tangent, Velocity, Velocity–Time Graph.

ACCURACY

When the dimensions of a quantity are measured, such as the height of a building, the resulting measurement cannot be exact, because of the limitations of the measuring instrument. The measurement is approximate, and is given to a degree of accuracy in the form of decimal places (dp), significant figures (sf), or the nearest whole unit. Whenever measurements are made or calculations done using measured quantities, the degree of accuracy should always be stated with the answer. This process of giving an answer in approximate form is called rounding.

One day David was listening to the radio when he heard the news of an earthquake. He heard that its center was 50 kilometers below the surface of the earth. Later, he was discussing the earthquake at home; Jane had heard on the radio that the center was 48 kilometers below ground and William said it was 47.4 kilometers underground. All three measurements are correct, but differ because were rounded to different degrees of accuracy by the three different radio stations.

- David's measurement of 50 km had been rounded to one significant figure.
- Jane's measurement of 48 km had been rounded to two significant figures.
- William's measurement of 47.4 km had been rounded to three significant figures.

Numbers are rounded in order to supply different people with the kind of information they require. For example, suppose a journalist was told that the attendance for the final of the 100 meters race at the Olympic Games was 95,287 people. In his report he would probably round the figure to the nearest thousand, because readers would not be interested in the exact figure. In his article he would write up the attendance as 95,000. The organizers of the games, who are interested in the receipts, would prefer a more exact figure of 95,290, which is rounded to the nearest 10 people.

Suppose the length of a small room is measured to be 3.472 meters. This measurement can be rounded to different degrees of accuracy:

3.47 m	(2 dp)	Showing two decimal places
3.5 m	(1 dp)	Showing one decimal place
3.0 m	(2 sf)	Showing two figures

The zero must be inserted, otherwise one figure would be showing:

3 m	(1 sf)	Showing one figure

The method used in rounding numbers to a given number of decimal places is illustrated in the following examples.

Example 1. Suppose we are rounding the number 3.47 to 1 dp, which means that only one number is written after the decimal point. We round up to 3.5 and not down to 3.4, because 3.47 is closer to 3.5 than it is to 3.4 (see figure a).

(a)

Example 2. Round the number 3.423 to 2 dp. In this example we write the answer with two numbers written after the decimal point. The number 3.423 rounds down to 3.42, because it is closer to 3.42 than it is to 3.43 (see figure b).

(b)

Example 3. Write 3.45 to 1 dp. For a number like 3.45, which is exactly halfway between 3.4 and 3.5, we always round up to the higher value, which in this example means we round to 3.5 (see figure c).

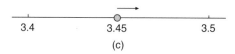

(c)

The examples above illustrate rounding numbers to a specific number of decimal places. Rounding numbers to significant figures works in a similar way. The difference is that figures are counted irrespective of the position of the decimal point. Sometimes zero is not included in the count of significant figures. This will be explained in some of the following examples.

Example 4. Round 20.8 to two significant figures. The number 20.8 rounds up to 21, because it is closer to 21 than it is to 20 (see figure d).

(d)

Example 5. Round 0.35 to one significant figure. The number 0.35 rounds up to 0.4, because 0.35 is exactly halfway between 0.3 and 0.4 (see figure e). Note that we write a zero to show there is no whole number, but this zero does not count as a significant figure.

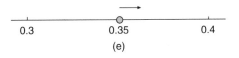

(e)

Example 6. Round 8.032 to 3 sf. The number 8.032 rounds down to 8.03, because it is closer to 8.03 than it is to 8.04 (see figure f). Note that zero counts as a significant figure when it acts as a placeholder between other figures.

(f)

Example 7. Write 0.6049 to 2 sf. The number 0.6049 rounds down to 0.60, because it is closer to 0.60 than it is to 0.61 (see figure g). This example shows that when rounding to 2 sf we only examine the third figure, which is 4, and ignore the 9, which does not affect the second significant figure. Note also that the zero in the answer must be inserted to show the required number of two figures.

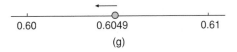

(g)

Example 8. Round off 208 to 2 sf. The number 208 rounds up to 210, because it is closer to 210 than it is to 200 (see figure h). This is an interesting example, because we cannot write 21 as the answer. It has the required two figures, but 21 is not roughly the same size as 208. So a zero must be inserted to make sure the answer of 210 is approximately the same size as 208. Alternatively, the answer can be expressed in standard form as 2.1×10^2, which is rounded to 2 sf.

(h)

References: Approximation, Decimal, Limits of Accuracy, Standard Form.

ACRE

The acre is a unit of area for measuring the size of a piece of land, and is an imperial unit. One acre is 4840 square yards, or 43,560 square feet. Historically, the acre was defined to be the amount of land that a team of two oxen could plough in 1 day. One acre is approximately 0.40 hectare, or 1 hectare is about 2.47 acres.

For practical purposes, 1 hectare is roughly $2\frac{1}{2}$ acres, or 5 acres is approximately 2 hectares.

Reference: Hectare, Imperial System of Units, SI Units.

ACUTE ANGLE

This entry also discusses *right angle, straight angle, obtuse angle,* and *reflex angle.* In order to define *acute angles* it is first necessary to explain a right angle. An angle of 90 degrees, which is written as 90°, is called a *right angle.* A right angle is indicated in the following diagrams by a box, and represents a quarter turn. A flagpole makes a right angle with the ground (see figure a).

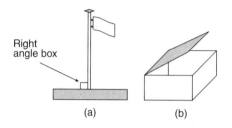

(a)   (b)

Angles that are less than one right angle, that is, less than 90°, are called *acute angles.* If the lid of a box is opened through an acute angle and then let go, the lid will fall back onto the top of the box (see figure b).

Every triangle must have at least two acute angles (see figure c).

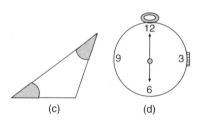

(c)   (d)

An angle of 180° is called a *straight angle* and corresponds to a half turn.

On a clock face when the time is 6 o'clock, the angle between the two hands is a straight angle (see figure d).

An *obtuse angle* is greater than 90°, but less than 180°. If the lid on a box is opened through an obtuse angle and let go, it will fall open and not fall back onto the top of the box (see figure e).

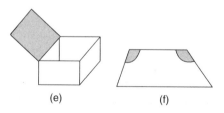

(e)   (f)

The trapezium shown in figure f has two obtuse angles.

A *reflex angle* is greater than 180°, but less than 360°. An example of a reflex angle is a three-quarters turn. When the lid of a box is opened fully, the lid has turned through a reflex angle (see figure g).

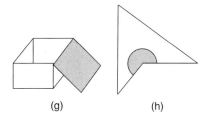

(g)   (h)

When a quadrilateral contains a reflex angle, it is called a reentrant quadrilateral (see figure h).

ADDEND

Numbers that are added together are called addends. The answer obtained from their addition is called the sum:

$$\text{Addend} + \text{addend} = \text{sum}$$

More than two addends might be involved. For example, to get the sum of $3 + 9 + 6$; we add together the pair of numbers 3 and 9 to get 12, and then add the 12 to 6 to get 18. Or, in adding numbers in our head, we might develop the skill of first arranging them into pairs that add to make 10, because it is easy to add numbers onto 10. For example,

$$7 + 8 = 7 + (3 + 5) \qquad \text{Splitting 8 up into } 3 + 5$$
$$= (7 + 3) + 5 \qquad \text{Combining the 7 with the 3 to make 10}$$
$$= 15$$

Reference: Associative Law.

ADJACENT ANGLES

Angles such as x and y in figure a that lie side by side are called *adjacent angles*.

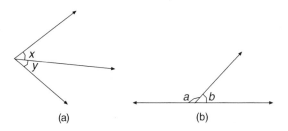

(a)   (b)

Any number of adjacent angles that form a straight line add up to 180°. When two adjacent angles form a straight line they add up to 180°. In figure b, angles *a* and *b* are *supplementary angles:*

$$a + b = 180°$$

Example 1. If a spade makes an acute angle of 76° with the horizontal ground, find the obtuse angle that the spade makes with the ground.

Solution. Let this obtuse angle be *x* (see figure c):

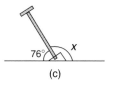

(c)

$x + 76° = 180°$ Sum of adjacent angles = 180°

$x = 104°$ Subtracting 76° from both sides of the equation to solve it

The spade makes an obtuse angle of 104° with the ground.

Example 2. Light rays are reflected from a mirror as shown in figure d. Find the angle *x* between the two rays.

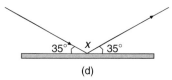

(d)

Solution. The calculation goes as follows:

$35° + x + 35° = 180°$ Sum of adjacent angles = 180°

$x + 70° = 180°$ Simplifying

$x = 110°$ Subtracting 70° from both sides of the equation to solve it

Reference: Geometry Theorems.

ALGEBRA

Algebra is the abstract study of the properties of numbers, using letters to stand for the numbers; these letters are called variables. Variables stand for unknown quantities and we use the operations of arithmetic to try to find their value. This entry explains how to simplify, or rewrite, expressions by adding, subtracting, multiplying, and dividing terms.

Example 1. Bill is doing a preliminary sketch of the ground floor of a house he is designing. He is not sure of some of the dimensions and uses variables x and y to represent them. All measurements are in meters. His sketch is shown in figure a, and is not to scale. Find the perimeter of the house.

Solution. The perimeter is the distance all the way around the house:

Perimeter $= x + 3y + 8 + y + 18 + 3x$

Perimeter $= x + 3x + 3y + y + 8 + 18$ Grouping similar terms together

Perimeter $= 4x + 4y + 26$ meters   Adding similar terms together.

This expression for the perimeter cannot be simplified further.

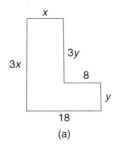

(a)

When terms are similar and can be simplified by adding or subtracting, we say they are "like terms." The process of adding and subtracting like terms is called "collecting terms." Terms are like terms if they are exactly the same except for the number in front of them. This number written in front of the term is called the coefficient of the term. For example, the coefficient of y in the term $-3y$ is -3.

Examples of sets of like terms are (i) $2a, 4a, -6a, 24a$; (ii) $xy, 3xy, 5xy, -14xy$; (iii) $4x^2, x^2, 2x^2, -16x^2$.

Example 2. Simplify these expressions: (i) $-2xy + 4xy - 3xy$, (ii) $a + 2b - 2a + ab$.

Solution. For part (i)

$-2xy + 4xy - 3xy = -1xy$ The terms are all like terms, and calculating $-2 + 4 - 3 = -1$ gives the coefficient of xy

$= -xy$ When the coefficient of a term is 1 or -1, it is not written with the term

For part (ii)

$$a + 2b - 2a + ab = a - 2a + 2b + ab \quad \text{Grouping the like terms together}$$

$$= -a + 2b + ab \quad \text{There are no more like terms}$$

Example 3. Now find the area of the house.

Solution. To find the area of the ground floor of the house, we need to divide the shape up into two rectangles, find the area of each, and add them together (see figure b):

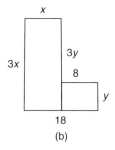

(b)

$$\text{Area of larger rectangle} = \text{length} \times \text{width}$$

$$= 3x \times x$$

$$= 3x^2 \text{ square meters} \quad x^2 \text{ is shorthand for } x \times x$$

$$\text{Area of the smaller rectangle} = \text{length} \times \text{width}$$

$$= 8 \times y$$

$$= 8y \text{ square meters} \quad 8y \text{ is shorthand for } 8 \times y$$

$$\text{Total area of the house} = \text{sum of areas of the two rectangles}$$

$$= 3x^2 + 8y$$

This expression cannot be simplified, because the two terms are not like terms.

In the example of finding the area of the house we multiplied terms together. We now study some more examples that explain how to simplify expressions when terms are multiplied.

If the length and width of a square are x, then

$$\text{Area of a square} = \text{length} \times \text{width}$$

$$A = x \cdot x \quad \text{The dot may be used for "times"}$$

$$A = x^2 \quad \text{In words we say "}x\text{ squared"}$$

If the length, width, and height of a cube are all x, then

$$\text{Volume of a cube} = \text{length} \times \text{width} \times \text{height}$$

$$V = x \cdot x \cdot x$$

$$V = x^3 \qquad \text{In words we say ``}x\text{ cubed''}$$

Example 4. Simplify these expressions: (i) $2 \times 3x$, (ii) $2ab \times 4b$

Solution. For part (i)

$$2 \times 3x = 2 \times 3 \times x \qquad \text{Inserting the } \times \text{ sign between 3 and } x$$

$$= 6 \times x \qquad \text{Multiplying the numbers 2 and 3 first}$$

$$= 6x \qquad \text{Writing without the } \times \text{ sign}$$

For part (ii)

$$2ab \times 4b = 2 \cdot a \cdot b \cdot 4 \cdot b \quad \text{Inserting dots for the } \times \text{ signs}$$

$$= 2 \cdot 4 \cdot a \cdot b \cdot b \quad \text{Grouping terms in alphabetical order with numbers first.}$$

$$= 8 \cdot a \cdot b^2 \qquad b \cdot b = b^2$$

$$= 8ab^2$$

The process of dividing algebraic terms and simplifying the answer, is explained under the entry Canceling.

References: Abstract, Algebraic Equations, Coefficient, Equations, Factor, Indices, Substitution, Variable.

ALGEBRAIC EQUATIONS

References: Balancing an Equation, Equations, Linear Equation, Quadratic Equations, Simultaneous Equations, Solving an Equation.

ALGEBRAIC FRACTIONS

Algebraic fractions, like arithmetic fractions, can be canceled, added, subtracted, multiplied, and divided. Canceling fractions is explained under the entry Canceling. It may be helpful to realize that if two fractions are equivalent, they can be written differently, as demonstrated here.

When the variable y is multiplied by the fraction $\frac{1}{5}$ the result is $\frac{1}{5}y$. Also, when 5 divides y, the result is written as $\frac{y}{5}$.

Since multiplying by $\frac{1}{5}$ achieves the same result as dividing by 5, it follows that $\frac{1}{5}y$ and $\frac{y}{5}$ are equivalent. Similarly, $\frac{2}{3}x$ and $\frac{2x}{3}$ are equivalent.

In order to add and subtract fractions they must have the same denominator. For example,

$$\frac{x}{7} + \frac{y}{7} = \frac{x+y}{7} \qquad \text{The two fractions have the same denominator, 7}$$

If the denominators are not the same, each fraction has to be converted to an equivalent fraction so that the denominators are the same size.

Example. Simplify

$$\frac{x}{5} + \frac{2x}{3}$$

Solution. The two denominators are 5 and 3. The lowest common multiple of 5 and 3 is the lowest positive number they both divide into exactly, which is 15. When working with fractions the lowest common multiple is called the lowest common denominator. Each fraction is then written with a denominator of 15, using equivalent fractions, as set out here:

$$\frac{x}{5} \times \frac{3}{3} = \frac{3x}{15} \qquad \text{and} \qquad \frac{2x}{3} \times \frac{5}{5} = \frac{10x}{15}$$

Now,

$$\frac{x}{5} + \frac{2x}{3} = \frac{3x}{15} + \frac{10x}{15} \qquad \text{Replacing each fraction in the sum by its equivalent fraction}$$

$$= \frac{3x + 10x}{15} \qquad \text{The denominators are the same and numerators are added}$$

$$= \frac{13x}{15} \qquad \text{Which is a single fraction}$$

Example. Simplify this expression

$$\frac{2}{x^2} - \frac{3}{xy}$$

by subtracting the two fractions.

Solution. The denominators are x^2 and xy. The lowest common multiple of x^2 and xy is x^2y. Thus

$$\frac{2}{x^2} - \frac{3}{xy} = \frac{2}{x^2} \times \frac{y}{y} - \frac{3}{xy} \times \frac{x}{x}$$ Replacing each fraction by its equivalent fraction

$$= \frac{2y}{x^2y} - \frac{3x}{x^2y}$$ Simplifying each fraction

$$= \frac{2y - 3x}{x^2y}$$ The denominators are the same and fractions are subtracted

Two fractions can be multiplied to obtain a single fraction. The numerators are multiplied together, and the denominators are multiplied together.

Example. Simplify

$$\frac{2x}{3} \times \frac{5x}{4}$$

Solution. Write

$$\frac{2x}{3} \times \frac{5x}{4} = \frac{2x \times 5x}{3 \times 4}$$

$$= \frac{10x^2}{12}$$

$$= \frac{5x^2}{6} \times \frac{2}{2}$$ 2 is a common factor of numerator and denominator

$$= \frac{5x^2}{6}$$ Canceling the 2's

Two fractions can be divided to obtain a single fraction. The method is explained in the following example.

Example. Simplify

$$\frac{3x}{2} \div \frac{1}{2y}$$

by dividing the fractions.

Solution. Write

$$\frac{3x}{2} \div \frac{1}{2y} = \frac{3x}{2} \times \frac{2y}{1}$$ Instead of dividing, multiply by the reciprocal of the second fraction

$$= \frac{6xy}{2}$$ Multiplying the numerators and denominators

$$= \frac{3xy}{1} \times \frac{2}{2}$$ 2 is a common factor

$$= 3xy$$ Canceling the 2's

Example. On his way to work, in the car, David reckons his speed through the traffic is 30 kilometers per hour, abbreviated km h^{-1}, whereas for the return journey it is 40 km h^{-1}. Calculate his average speed.

Solution.

$$\text{Average speed} = \frac{\text{total distance traveled}}{\text{total time taken}}$$ Formula for average speed

Let the distance to work be x kilometers. Therefore the total distance traveled is $2x$ kilometers. On the outward journey the time taken is T_1 hours and for the return journey the time is T_2 hours. Write

$$T_1 = \frac{x}{30} \quad \text{and} \quad T_2 = \frac{x}{40}$$ The formula used is time = distance ÷ speed.

Now write

$$\text{Total time} = T_1 + T_2$$

$$= \frac{x}{30} + \frac{x}{40}$$

$$= \frac{x}{30} \times \frac{4}{4} + \frac{x}{40} \times \frac{3}{3}$$ The lowest common denominator is 120

$$= \frac{4x}{120} + \frac{3x}{120}$$ Simplifying each fraction

$$= \frac{7x}{120}$$ Adding fractions with equal denominators

Now,

$$\text{Average speed} = \frac{2x}{(7x/120)}$$ Substituting distance and time into average speed formula

$$= 2x \times \frac{120}{7x}$$ Multiplying by the reciprocal of $\frac{7x}{120}$

$$= \frac{240}{7} \times \frac{x}{x}$$ x is a common factor of numerator and denominator

$$= \frac{240}{7}$$ Canceling the x's

The average speed is 34.3 km h^{-1}, to one decimal place.

References: Average Speed, Canceling, Equivalent Fractions, Rational Expression, Reciprocal.

ALTERNATE ANGLES

Figure a shows two parallel lines indicated by arrows. The line cutting across them is called a transversal. A pair of angles such as a and b that are on alternate sides of the transversal and lie between the parallel lines are equal in size and are called alternate angles.

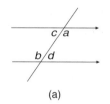

(a)

Alternate angles form a shape like a reversed letter z, as shown in figure b.

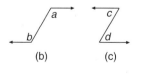

(b)   (c)

Angle a = angle b

The pair of alternate angles shown in figure c forms a letter z:

Angle c = angle d

Example. Figure d shows a concrete pillar set in the seabed as a support for a bridge. If a laser beam focused on the foot of the pillar makes an angle of 47° with the sea level, find the angle, marked x, that the laser beam makes with the line of the seabed. The seabed is parallel to the sea level.

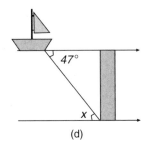

(d)

Solution. Write

$$x = 47° \qquad \text{Alternate angles are equal}$$

The laser beam makes an angle of 47° with the seabed.

There are two more geometry theorems that relate to two parallel lines and their transversal. They are as follows:

1. Corresponding angles are equal in size.
2. Cointerior angles are supplementary, which means they add together to equal 180°.

With reference to figure e, a pair of angles that are on the same side of the transversal and occupy a similar position are equal in size and are called corresponding angles. Angles a and b are corresponding angles, and so are angles c and d:

$$\text{Angle } a = \text{angle } b$$
$$\text{Angle } c = \text{angle } d$$

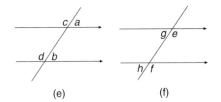

(e)   (f)

Figure f shows two more pairs of corresponding angles that are equal in size:

$$\text{Angle } e = \text{angle } f$$

$$\text{Angle } g = \text{angle } h$$

Example. A ladder leans against a vertical wall, and the angle the ladder makes with the horizontal ground is 65°. Find the angle the ladder makes with the top of the wall (figure g).

Solution. Write

$$x = 65° \qquad \text{Corresponding angles are equal}$$

The ladder makes an angle of 65° with the top of the wall.

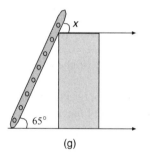

(g)

Cointerior angles are pairs of angles that lie between two parallel lines and are situated on the same side of the transversal:

$$\text{The sum of cointerior angles} = 180°$$

In figure h, angles *a* and *b* are cointerior, and so are *c* and *d*:

$$\text{Angle } a + \text{angle } b = 180°$$

$$\text{Angle } c + \text{angle } d = 180°$$

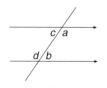

(h)

Example. Figure i shows a chimney on the roof of a house. One side of the chimney makes an angle of 110° with the roof. Find the size of the angle the other side of the chimney makes with the roof, marked x in the figure.

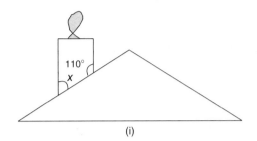

(i)

Solution. Write

$$x + 110° = 180° \qquad \text{Sum of cointerior angles} = 180°$$

$$x = 70° \qquad \text{Subtracting } 110° \text{ from both sides of the equation}$$

The chimney makes an angle of 70° with the roof.

Example. In figure j, the two angles of a triangle are 60° and 70°, and the arrows indicate parallel lines. Find the sizes of the angles x, y, and z.

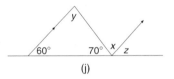

(j)

Solution. Write

$$y + 60° + 70° = 180° \qquad \text{Sum of the angles of a triangle} = 180°$$

$$y + 130° = 180°$$

$$y = 50° \qquad \text{Subtracting } 130° \text{ from both sides of the equation}$$

$$x = 50° \qquad \text{Alternate angles are equal}$$

$$z = 60° \qquad \text{Corresponding angles are equal}$$

From the results of the above calculations it can be seen that the sum of the three angles of the triangle is the same total as the sum of the three angles on the straight line, and that the sum is 180°.

References: Angle Sum of a Triangle, Geometry Theorems.

ALTITUDE

This word is used in two different ways. The altitude of an object is the distance of the object above the surface of the earth, and is often called its vertical height. In geometry, altitude takes on a slightly different meaning, and refers to the altitude of a polygon or a polyhedron. In this context, the term altitude is explained under the entry Base (geometry).

References: Base (geometry), Concurrent, Polygon, Polyhedron.

AMPLITUDE

Amplitude is a feature of periodic curves, like the sine or the cosine curves. The amplitude of the sine curve is the greatest distance of a point on the curve from the x-axis, and is indicated by a in the figure. For the sine curve $y = \sin x$, the amplitude is $a = 1$. For the curve $y = 2 \cos x$, the amplitude is $a = 2$.

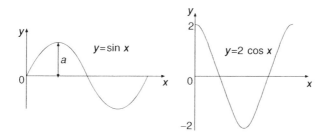

References: Cycle, Frequency.

ANALYTICAL GEOMETRY

Reference: Cartesian Coordinates.

ANGLE

This is a measure of turning or rotation; the units of angle measurement are degrees or radians. The angle x in the figure is the amount of turning between two rays $\overrightarrow{OA}$ and $\overrightarrow{OB}$. The symbol for angle is $\angle$. The angle in the figure can be represented by the three capital letters, $\angle AOB$, or by a single small letter, $\angle x$. It is also common in trigonometry to represent an angle by the letters of the Greek alphabet, for example, $\angle\theta$, $\angle\alpha$, or $\angle\beta$.

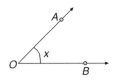

References: Acute Angle, Degree, Radian, Ray.

ANGLE BETWEEN A LINE AND A PLANE

This topic is part of three-dimensional trigonometry. To find the angle between a line and a plane, the line is projected onto the plane, and then the angle between the projected line and the original line is calculated. This angle is the angle between the line and the plane.

Suppose a straight nail *ON* is hammered into a piece of wood at an angle so that the nail is not upright (see figure a). The projection of the nail *ON* onto the plane of the wood is *OW*, as shown in figure a. The projection of *ON* onto the plane can be considered to be the shadow cast by the nail *ON* when parallel rays of light shine at right angles to the plane.

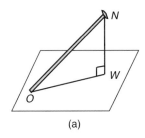

(a)

The angle *NOW* is the angle between the line *ON* and the plane of the wood. The method of calculating the angle between a line and a plane is explained in the following example.

Example. The longest diagonal of a cuboid, which is a box, is the line *DF* (see figure b). Calculate the angle between the line *DF* and the base *EFGH*.

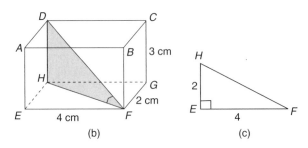

(b)   (c)

Solution. The projection of the line *DF* onto the base is the line *HF*. So the angle the line *DF* makes with the base *EFGH* is the angle *DFH*.

In the process of finding this angle, we use two right-angled triangles, *HEF* and *DHF*. The first step is to calculate the length of *HF*, using triangle *HEF*, which is positioned in the base of the cuboid (figure c):

$HF^2 = HE^2 + EF^2$	Pythagoras' Theorem
$HF^2 = 2^2 + 4^2$	Substituting $HE = 2$ meters and $EF = 4$ meters
$HF^2 = 4 + 16$	Squaring 2 and 4
$HF = \sqrt{20}$	Taking the square root

In triangle *DHF*, using $\sqrt{20}$ as the length of *HF* gives

$$\text{Tangent (angle } DFH\,) = \frac{3}{\sqrt{20}} \qquad \text{Using tangent of an angle}$$
$$= \frac{\text{opposite side}}{\text{adjacent side}}$$

$$\text{Angle } DFH = \tan^{-1}\left(\frac{3}{\sqrt{20}}\right) \qquad \text{If } \tan a = b, \text{ then } a = \tan^{-1} b$$

$$\text{Angle } DFH = 33.9° \quad \text{to 1 dp} \qquad \text{Using calculator}$$

The angle between the line *DF* and the plane base *EFGH* is 33.9°.

References: Angle between Two Planes, Plane, Pythagoras' Theorem, Trigonometry.

ANGLE BETWEEN TWO PLANES

This topic is part of three-dimensional trigonometry. Suppose two planes which are inclined to each other intersect in the straight line *XY* (see figure a). The two planes can be thought of as hinged together, rather like an opening trapdoor. If the sloping plane swings down onto the other plane, then the angle through which it turns is the angle between the two planes. This angle is also called the dihedral angle with respect to the two planes.

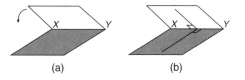

(a)   (b)

To find the angle between the two planes, we select two lines, one in each plane, which intersect at the hinge *XY* (see figure b). Both lines are at right angles to

XY. The angle between these two lines is the angle between the two planes. This concept is demonstrated in the following example, in which one of the planes is a triangle.

Example. Figure c shows an upright, square-based pyramid. Calculate the angle between the plane *ABCD*, which is the base of the pyramid, and the sloping plane *VBC*.

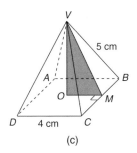

(c)

Solution. The two lines we select in the planes are *VM* and *OM*, as both lines are at right angles to their line of intersection *BC*. These lines are selected because they are two sides of a right-angled triangle *VOM*, whose lengths we can find. The required angle between the two planes is angle *VMO*.

In triangle *VOM*, the length of *OM* is 2 cm, and the length of *VM* is found from the right-angled triangle *VMB* in the following way (see figure d):

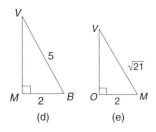

(d)   (e)

$$VM^2 + MB^2 = VB^2$$ Pythagoras' Theorem

$$VM^2 + 2^2 = 5^2$$ Substituting lengths $MB = 2$ and $VB = 5$

$$VM^2 = 25 - 4$$ Squaring 5 and 2, and rearranging equation

$$VM = \sqrt{25 - 4}$$ Taking square roots

$$VM = \sqrt{21}$$ Subtracting the numbers

The length of *VM* is now used in triangle *VOM* (see figure e):

cosine (angle *VMO*) = $\dfrac{2}{\sqrt{21}}$ Using cosine of an angle = $\dfrac{\text{adjacent side}}{\text{hypotenuse}}$

Angle *VMO* = $\cos^{-1}\left(\dfrac{2}{\sqrt{21}}\right)$ If $\cos a = b$, then $a = \cos^{-1} b$

Angle *VMO* = 64.1°   to 1 dp   Using calculator

The angle between the plane *ABCD* and the sloping plane *VBC* is 64.1°.

References: Angle, Angle between a Line and a Plane, Dihedral Angle, Pyramid, Pythagoras' Theorem, Square Root, Trigonometry.

ANGLE BISECTOR

This is a line that divides an angle into two equal angles. The angle bisector has a ruler-and-compass construction, which is described in the example below.

Example. A piece of pizza has been left for Jacob and Nathan's supper. Doing a drawing of the pizza, show how to construct a line, using ruler and compasses only, that equally divides the piece of pizza into two.

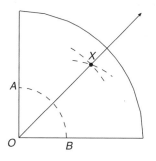

Solution. The drawing of the pizza is shown in the figure with the construction lines drawn dashed. Open the compasses and insert the point at *O*, and draw an arc *AB*. Using the same radius, or longer if you wish, insert the point of the compass at *A* and draw an arc. Then insert the compass point at *B*, and using the same radius, draw another arc to meet the arc from *A* at *X*. The line *OX* is the angle bisector of angle *AOB*, and divides the pizza equally. The point *X* is equidistant from the lines *OA* and *OB*.

References: Arc, Bisect, Equidistant, Radius.

ANGLE IN A SEMICIRCLE

This is a geometry theorem about a triangle drawn in a semicircle. This theorem is a special case of another theorem given under the entry Angle at the Center and Circumference of a Circle.

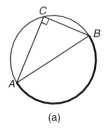

(a)

In this special case the arc that subtends the angles is the semicircle *AB*. (See figure a), in which case the angle at the center of the circle is 180° and the angle at the circumference is half of 180°, which is 90°. The theorem is stated as follows:

In figure a, if *AB* is a diameter of the circle and *C* is a point on the circumference, then angle *ACB* is a right angle.

Example. Suppose you are presented with a drawing of a circle and asked to find its diameter.

Solution. This method makes use of the theorem that the angle in a semicircle is a right angle. Tear off a corner of a piece of paper and place it on the circle with the corner of the paper containing the right angle just touching a point on the inside circumference of the circle, as shown in figure b. The sides of the piece of paper will cross over the circle at two points which are labeled *A* and *B* in figure b. The dashed line *AB* is the diameter of the circle.

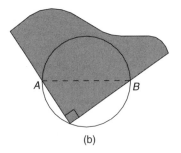

(b)

Example. Joanne loves skating and regularly practices her skills at the Big Apple skating rink, which is in the shape of a circle. One of her activities is to start at a point *S* on the edge of the circle, skate through the center of the rink, and pick up a ball at *O* (see figure c). Having collected the ball, she carries on in a straight line until she meets the edge of the rink again at the point *A*. On her way back to her starting point at *S* she touches another point *B*, which is also on the edge of the rink. If the size of angle *ASB* = 47°, find the size of angle *BAS*.

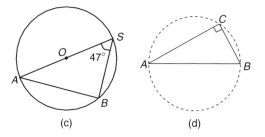

(c)   (d)

Solution. Write

Angle $ABS = 90°$ Angle in a semicircle is a right angle

Angle $BAS = 43°$ Angle sum of triangle $ABC = 180°$

The theorem that the angle in a semicircle is a right angle has a converse theorem, and is stated here:

If there is a right-angled triangle ABC which is right-angled at C (see figure d), and the circumcircle of the triangle is drawn, then line AB is the diameter of the circumcircle.

References: Altitude, Angle Sum of a Triangle, Angles at the Center and Circumference of a Circle, Circumcircle, Converse of a Theorem, Cyclic Quadrilateral, Orthocenter, Semicircle, Subtend.

ANGLE IN THE ALTERNATE SEGMENT

This theorem is a circle geometry theorem. A circle passes through the three vertices of a triangle ABC and at one of these vertices, say C, a tangent is drawn to the circle (see figure a). The chord BC divides the circle into two segments. If one of the segments, say the smaller one, is shaded, then the other segment is referred to as the alternate

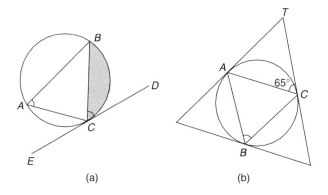

(a)   (b)

segment. This theorem states that the angle between the tangent *CD* and the chord *BC* is equal to the angle at *A* in the alternate segment, or

$$\text{Angle } BCD = \text{Angle } BAC$$

Similarly,

$$\text{Angle } ACE = \text{Angle } ABC$$

Example. Three straight footpaths touch a circular playground at the points *A*, *B*, and *C*, and angle *TCA* $= 65°$ (see figure b).

    (a) What is the size of angle *ABC*?

    (b) Name another angle the same size as $65°$.

Solution. (a) Write

        Angle *ABC* = Angle *ACT*   Angle in the alternate segment

        Angle *ABC* $= 65°$

    (b) Angle *ABC* is also equal to angle *TAC*, because of the theorem Angle in the Alternate Segment:

$$\text{Angle } TAC = 65°$$

References: Angle Sum of a Triangle, Chord, Circle Geometry Theorems, Segment of a Circle, Tangent.

ANGLE OF DEPRESSION

Suppose a surveyor is at the point *A* on the top of a wall looking horizontally out to sea (see figure). The angle *x* through which she lowers her gaze to look at a buoy *B* out at sea is the angle of depression of the buoy from her position *A*.

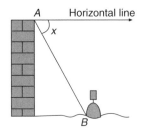

Reference: Angle of Elevation.

ANGLE OF ELEVATION

Suppose Darren has climbed to the top of his house and is looking horizontally into the distance (see figure). The angle *y* through which he raises his eyes to gaze up at the top of a flagpole is the angle of elevation of the top of the flagpole from his position *A*.

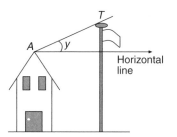

ANGLE OF INCLINATION

This is the angle a certain line makes with another line, or the angle it makes with a plane. Suppose Helen is abseiling down the wall of a building and her rope makes an angle of 33° with the wall (see figure). The angle of inclination of the rope to the wall is 33°. In other words, the rope is inclined at 33° to the wall.

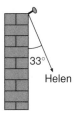

The phrase "angle of inclination" is also used to describe the angle between two planes.

Reference: Angle of Depression, Angle of Elevation, Plane.

ANGLE SUM OF A TRIANGLE

This is a geometry theorem, which states that the three angles of a triangle add up to 180°. Alternatively, this can be expressed as follows: The three angles of a triangle are supplementary.

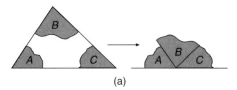

(a)

This geometry theorem can be demonstrated by the following experiment: Draw a triangle on a sheet of paper and using scissors, carefully cut it out. Tear off each of the three angles *A*, *B*, and *C* and rearrange them as shown in figure a and you will discover that they form a straight line, which is an angle of 180°.

Example.   Figure b shows the gable end of a house with an angle at the vertex of 136°. Find the angle that the roof makes with the horizontal, marked by *x* in figure b.

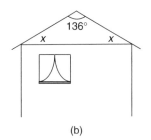

(b)

Solution.   The rooflines are symmetrical, and the triangle in figure b is therefore isosceles. The two equal base angles are marked as *x*. Write

$$2x + 136° = 180°$$ Sum of angles of a triangle is $180°$

$$2x = 180° - 136°$$ Subtracting $136°$ from both sides of equation

$$2x = 44°$$

$$x = 22°$$ Dividing both sides of equation by 2

The roof makes an angle of 22° with the horizontal.

Example.   In figure c, find the size of angle *y* if angle *ABC* is 90°.

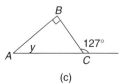

(c)

Solution. Write

Angle $ACB = 53°$ Sum of adjacent angles is $180°$

$y + 53° + 90° = 180°$ Sum of angles of triangle $ABC = 180°$

$y = 37°$ Subtracting $53°$ and $90°$ from both sides of equation

References: Adjacent Angles, Equations, Geometry Theorems, Supplementary Angles, Vertex.

ANGLES AT A POINT

When two or more angles meet at a point and together make up a full turn, we say they are angles at a point. In figure a, the three angles $100°$, $95°$, and x are angles at a point. When two or more angles meet at a point and together make up a full turn, the sum of the angles is $360°$, since there are $360°$ in a full turn. Angles that add up to $360°$ are called conjugate angles. With reference to figure a, we write

$$x + 100° + 95° = 360° \qquad \text{Sum of angles at a point} = 360°$$

$$x = 165°$$

(a)

Example. A family share a birthday cake, so that Jo has a slice with an angle of $121°$, Sarah's has an angle of $37°$, and Andy's slice has an angle of $162°$. What angle does Luke's slice have?

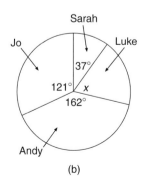

(b)

Solution. A view of the top of the cake is drawn in figure b, with Luke's slice having an angle x. The figure is not drawn to scale. Write

$$121° + 37° + 162° + x = 360° \qquad \text{Sum of angles at a point} = 360°$$

$$320° + x = 360°$$

$$x = 40°$$

Luke's slice has an angle of 40°.

Reference: Geometry Theorems.

ANGLES AT THE CENTER AND CIRCUMFERENCE OF A CIRCLE

This statement refers to a circle geometry theorem. The angle subtended at the center O of a circle by the arc AB is equal to twice the angle subtended by the same arc at any point C on the circumference of the same circle.

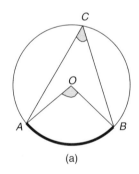

(a)

This means that angle AOB = twice angle ACB.

Example. If O is the center of the circle and angle $AOB = 104°$, find the size of angle ACB (figure b).

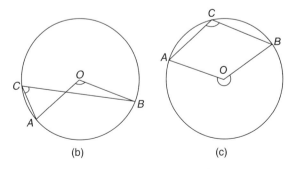

(b)   (c)

Solution. Write

Angle $ACB = 104° ÷ 2$ Angle at center is twice angle at circumference

Angle $ACB = 52°$

A special case of this theorem occurs when the arc AB is greater than a semicircle; the diagram looks quite different, as shown in figure c. Nevertheless, the theorem is still true, but the angles are larger:

Reflex angle AOB = twice obtuse angle ACB

Example.   In figure d, angle $ADC = 64°$ and O is the center of the circle. Find the sizes of angle AOC and angle ABC, denoted by x and y, respectively.

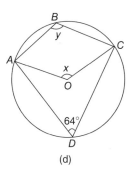

(d)

Solution. Write

x = twice $64°$ Angle at center is twice angle at circumference

$x = 128°$

and

Reflex angle $AOC = 360° − 128$ Sum of angles at a point $= 360°$

$= 232°$

$y = 232° ÷ 2$ Angle at center is twice angle at circumference

$y = 116°$

References: Geometry Theorems, Circle Geometry Theorems, Obtuse Angle, Reflex Angle, Semicircle, Subtend.

ANGLES ON THE SAME ARC

This entry refers to a circle geometry theorem, and states that all angles subtended at the circumference of a circle by the arc *AB* are equal in size. In figure a, there are two angles on the circumference of the circle, one at the point *R* and one at *Q*, that are subtended by the arc *AB*. Therefore, this theorem states the following:

The two angles *ARB* and *AQB* are equal in size.

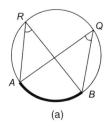

(a)

The theorem is not limited to just two angles, but is true for any number of angles at the circumference of the circle, provided they are all subtended by the arc *AB*.

Example. William is having shooting practice at soccer training and his coach has drawn a large circle on the gymnasium floor. The straight line *AB* in figure b represents the goal, and players are shooting from anywhere on the major arc of the circle trying to score. The coach is at point *C* and William is at point *W*. If angle *AWB* = 34°, find angle *ACB*, which is the coach's shooting angle.

Solution. The angle required is angle *x* in figure b:

$$x = 34° \qquad \text{Angles on the same arc are equal}$$

The coach's shooting angle is 34°, the same as William's.

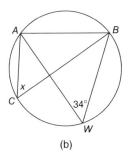

(b)

Example. In figure c, *AB* is parallel to *CD*, indicated by the arrows. The straight lines *AC* and *BD* intersect at point *E*. Prove that triangle *ABE* is an isosceles triangle.

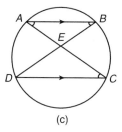

(c)

Solution. Write

      Angle *ACD* = angle *ABD*   Angles on the same arc are equal

      Angle *ACD* = angle *BAC*   Alternate angles are equal

      Angle *ABD* = angle *BAC*   Both angles are equal to angle *ACD*

Therefore, triangle *ABE* is isosceles, because its base angles are equal.

References: Alternate Angles, Circle Geometry Theorems, Geometry Theorems, Isosceles Triangle, Subtend.

ANNULUS

This is the region between two concentric circles, and is drawn shaded in figure a. Another name for an annulus is a ring. If the radii of the circles are *R* and *r*, then the area of the annulus is given as follows:

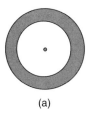

(a)

$A = \pi R^2 - \pi r^2$ Formula for the area of a circle is $A = \pi R^2$

$A = \pi(R^2 - r^2)$ Factoring, since π is a common factor

$A = \pi(R - r)(R + r)$ Factoring, using the difference of two squares

Example. Find the volume of plastic required to make a plastic pipe (figure b) which is 20 cm long and whose inner and outer radii of the circular ends are 3 and 4 centimeters, respectively.

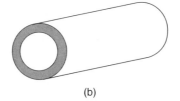

(b)

Solution. Write

Volume = area of annulus × length	Volume of a prism = area of cross section × length
Volume = $\pi(R - r)(R + r) \times$ length	Using formula for area of annulus
Volume = $\pi(4 - 3)(4 + 3) \times 20$	Substituting $R = 4, r = 3,$ and length = 20
Volume = $439.8 \, \text{cm}^3$ (1 dp)	Using calculator

References: Concentric Circles, Factoring, Prism.

APPROXIMATION

Reference: Accuracy.

ARC

An arc is a section, or part, of a curve. It can also be a section of a line graph. In the figure the arc *AB* is part of a circle.

When the arc of a circle is less than a semicircle, it is called a minor arc, and when the arc is greater than a semicircle, it is called a major arc. In the figure, the arc *AB* of the circle is a minor arc because it is less than a semicircle.

References: Line Graph, Networks.

ARC LENGTH

The length of an arc of a circle can be calculated as in the following example.

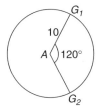

Example. Amanda is schooling up her horse George for a show. He is trotting in a circle at the end of a rope held by Amanda, who is at point A in the figure. The length of the rope is 10 meters.

(a) As Amanda turns through an angle of $120°$ find the distance trotted by George along the arc of the circle from G_1 to G_2.

(b) If George trots a distance of 20 meters, through what angle has the rope turned?

Solution. (a) The circumference C of the whole circle is given by the formula

$C = \pi \times$ diameter

$C = \pi \times 20$ \qquad\qquad Substituting 20 meters for the diameter of the circle

$C = 20\pi$

The arc length trotted by George is a fraction of C, and this fraction is $120°/360°$, where $120°$ is the angle turned through by the rope, and a full turn is $360°$.

$$\text{Arc length} = \frac{120}{360} \times 20\pi \qquad \text{Multiplying the fraction of the full turn by the circumference}$$

$$= 20.9 \quad (1 \text{ dp})$$

The distance trotted along the arc of the circle by George is 20.9 meters.

(b) Suppose the angle turned through by the rope is x. The equation is

$$\frac{x}{360} \times C = 20$$

$$x = \frac{360 \times 20}{C} \qquad \text{Making } x \text{ the subject of the formula}$$

$$x = \frac{360 \times 20}{20\pi} \qquad \text{Substituting } C = 20\pi \text{ from earlier}$$

$$x = 114.6° \quad \text{(1 dp)} \qquad \text{Using the calculator}$$

The rope turns through an angle of $114.6°$. For an alternative method using radians refer to the entry Radian.

References: Arc, Circumference, Radian.

AREA

The area of a surface is a measure of the two-dimensional space it occupies. It is measured in square units, which is written as units². We can find the area of a shape, say a rectangle, by counting the number of squares that its surface occupies. The area of the rectangle in figure a is found by counting the number of square centimeters it occupies. Square centimeters are abbreviated cm^2. The area of this rectangle is 12 cm^2. Alternatively, the length of the rectangle, which is 4 cm, and the width, which is 3 cm, can be measured and the area calculated using the formula

$$\text{Area} = \text{length} \times \text{width}$$

$$A = 4 \times 3$$

$$= 12$$

The area of this rectangle is 12 cm^2.

1	2	3	4
5	6	7	8
9	10	11	12

(a)

For some shapes, like circles, it is difficult to find the area accurately by counting squares, because the process would involve piecing together small parts of the circle to make up whole squares, like a jigsaw. The only satisfactory method is to use a formula to find its area.

Example. Calculate the area of the circle in figure b.

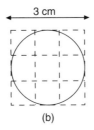

3 cm

(b)

Solution. We need to know the radius of the circle, which is $R = 1.5$ cm. We write

Area $= \pi \times 1.5^2$ Formula for the area of a circle is Area $= \pi \times R^2$

Area $= 7.07$ (to 2 dp)   Using a calculator

The area of the circle is 7.07 cm ^{2}.

The units commonly used for area are as follows:

- Square millimeters, mm^2.
- Square centimeters, cm^2.
- Square meters, m^2.
- Hectares, ha.
- Square kilometers, km^2.

The relationships between these units of area are as follows:

$$100 \, mm^2 = 1 \, cm^2$$
$$1,000,000 \, mm^2 = 1 \, m^2$$
$$10,000 \, cm^2 = 1 \, m^2$$
$$10,000 \, m^2 = 1 \, ha$$
$$1,000,000 \, m^2 = 1 \, km^2$$
$$100 \, ha = 1 \, km^2$$

Large numbers may be written in index form. For example $10,000 = 10^4$. Another unit of area, in the Imperial system of units, is the acre. One hectare is roughly 2.5 acres.

Methods for finding the areas of other shapes, such as the parallelogram, the trapezium, the triangle, etc., are given in the relevant entries.

References: Acre, Circle, Index, Metric Units, Survey, Radius, Rectangle.

ARITHMETIC

This is the branch of mathematics that uses numbers in processes such as addition, subtraction, multiplication, division, and taking square roots.

ARITHMETIC MEAN

Arithmetic mean is often abbreviated to just the one word mean. It is one of three averages which are used in statistics, the other two being median and mode. The symbol for mean is $\bar{x}$. Mean is often incorrectly referred to as "average." The mean is a single quantity used to represent a set of quantities or a group. The method of finding the mean of a set of quantities is to add up all the quantities and divide this total sum by the number of quantities.

Example. Sarah has just started at the local child care center. She is in a small group of six children, whose ages are given in the table. Find the mean age of the children.

Name	Sarah	Luke	Samuel	Simon	Francis
Age in years	2	3	4	1	4

Solution. Write

$$\bar{x} = \frac{2 + 3 + 4 + 1 + 4}{5}$$ Sum the ages and then divide by the number of children

$$\bar{x} = \frac{14}{5}$$

$$\bar{x} = 2.8$$ Using calculator.

The mean age of the children is 2.8 years.

Sometimes the mean is calculated and then an extra quantity is added to the group as illustrated in the following example.

Example. The mean mass of a rowing eight at Plato College is 87.5 kilograms (abbreviated kg). The mass of the cox is 49.8 kg. What is the mean mass of the whole crew?

Solution. Write

The eight rowers weigh $8 \times 87.5 = 700$ kg

The whole crew weighs $700 + 49.8 = 749.8$ kg   Mass of original crew + cox

$$\text{Mean mass of the crew} = \frac{749.8}{9}$$ Total mass ÷ 9 crewmembers

The mean mass of the whole crew is 83.3 kg (to 1dp).

In the following example the mean is calculated when the data are contained in a frequency table.

Example. The frequency table shows the marks, out of 10, gained by 29 students who took a mathematics test. Find the mean.

Mark	0	1	2	3	4	5	6	7	8	9	10
Frequency	2	1	3	3	4	6	5	2	2	1	0

Solution. The frequency row tells us, for example, that 6 students scored a mark of 5 out of 10. In order to find the sum of all the marks, we multiply the marks by the frequencies. Then, to obtain the mean, we divide this sum by the total number of students, which is 29. The symbol for mean is $\bar{x}$:

$$\bar{x} = \frac{(2 \times 0) + (1 \times 1) + (3 \times 2) + (3 \times 3) + (4 \times 4) + (6 \times 5) + (5 \times 6) + (2 \times 7) + (2 \times 8) + (1 \times 9) + (0 \times 10)}{29}$$

$$\bar{x} = \frac{131}{29}$$

$$\bar{x} = 4.5 \quad \text{(to 1 dp)}$$

The mean mark in the test is 4.5.

Data that have a large range can be handled more easily if the data are grouped into class intervals. For example, if the students' marks in a test have a large range, say, out of 100 instead of out of 10, the marks can grouped into class intervals of, say, 10 marks. When data are grouped into class intervals the individual score of each student is lost, but the mean can still be calculated. The middle of each class interval is used to represent the mark in the test obtained by the students, as illustrated in the example below.

Example. The grouped frequency table shows the percentage marks gained by 100 students who took a mathematics examination. Find the mean mark.

Mark %	0–9	10–19	20–29	30–39	40–49
Frequency	1	3	12	15	20
Mark %	50–59	60–69	70–79	80–89	90–100
Frequency	23	12	8	4	2

Solution. For more complicated statistical calculations, such as this example, it is convenient to use a vertical table. It is necessary to incorporate two extra columns, one for the middle mark and one for the product of the middle mark and the frequency. To obtain the sum of the marks of all the students, it is necessary to multiply the middle marks in each class interval by the corresponding frequencies, since the individual marks are not known.

Mark (x)	Frequency (f)	Middle mark (m)	f × m
0–9	1	4.5	4.5
10–19	3	14.5	43.5
20–29	12	24.5	294
30–39	15	34.5	517.5
40–49	20	44.5	890
50–59	23	54.5	1253.5
60–69	12	64.5	774
70–79	8	74.5	596
80–89	4	84.5	338
90–100	2	94.5	189
Totals	100		4900

The formula for the mean is

$$\bar{x} = \frac{\text{Sum of } f \times m}{\text{Sum of } f}$$

Which may be abbreviated to $\bar{x} = \frac{\Sigma f \times m}{\Sigma f}$, where Σ means "the sum of"

$$\bar{x} = \frac{4900}{100}$$

The mean score in the examination is 49%.

References: Class Interval, Frequency Table, Median of a Set of Data, Mode.

ARRANGEMENTS

These are also called permutations.

Reference: Combinations.

ARROW GRAPH

This is sometimes called an arrow diagram. Suppose there is a relation "is the capital of" between a set of cities and a set of countries. One pairing off could be "London is the capital of England," and on the graph we would draw an arrow from London to England to show that this relation exists between them. Figure a is an arrow graph showing four possible pairings for the relation "is the capital of." If the directions of the arrows were reversed, the result would be the inverse relation: "has as its capital"

Example.   Draw an arrow graph for the relation "is greater than" between the set of numbers {2, 4, 6} and another set {2, 4, 5}, write down the ordered pairs for this relation, and draw a Cartesian graph.

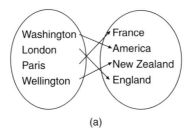

(a)

Solution. The arrow graph is drawn in figure b.

The ordered pairs are {(4, 2) (6, 2) (6, 4) (6, 5)}.

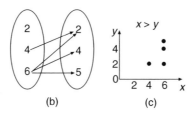

(b)   (c)

The Cartesian graph is shown in figure c.

The set of numbers {4, 6}, which includes the first numbers in the ordered pairs, is called the domain. The set of numbers {2, 4, 5}, which includes the second numbers in the ordered pairs, is called the range.

References: Cartesian Coordinates, Domain, Image, Inverse Relations, Ordered Pairs, Range.

ARROWHEAD

Reference: Kite.

ASSOCIATIVE LAW

References: Addend, Commutative Law, Distributive Law.

ASYMMETRY

Reference: Symmetry.

ASYMPTOTE

A straight line is an asymptote of a curve if the line and the curve get closer and closer together, but never actually meet. In figure a, the straight line is an asymptote of the curve. In this example as x gets larger and larger, the curve and the asymptote get closer and closer together. The distance between them is forever diminishing.

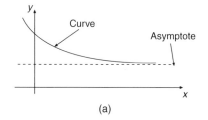

(a)

Example. Bill plans to build a chicken run in the shape of a rectangle that will have an area of 96 square meters. Not being sure about its dimensions, he calls the length of the run x meters and the width y meters. He writes down a few possible values for x and y. For example, when the length is $x = 12$ meters the width is $y = 8$ meters, so that the area is 96 square meters. Being a well-organized person, he enters the various values of x and y into a table in order to analyze them.

x (length) in meters	3	4	6	8	12	16	24	32
y (width) in meters	32	24	16	12	8	6	4	3

Figure b is the graph of the values from the table for the length and width in meters of the chicken run. This graph is called a rectangular hyperbola.

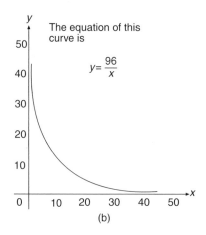

The equation of this curve is

$$y = \frac{96}{x}$$

(b)

Bill soon realizes that as x gets bigger, y gets smaller, and the bigger x gets, the smaller y gets. In fact, as x gets bigger and bigger, y gets closer and closer to zero. For example, when $x = 1000$, $y = 0.096$, but of course in practical terms Bill would not use these values as dimensions of his chicken run. The curve and the x-axis get closer and closer together as x gets larger. This means that the x-axis is the horizontal asymptote to the curve. Similarly, the y-axis is the vertical asymptote to the curve.

Reference: Exponential Curve, Hyperbola, Logarithmic Curve, Rectangular Hyperbola, Table of Values, Tangent.

AVERAGE

This is a quantity that refers to the three different statistical terms mode, median, and mean. The mean is also known as the arithmetic mean, which is sometimes incorrectly called the average.

Reference: Mean, Median of a Set of Data, Mode.

AVERAGE SPEED

The average speed for a journey is calculated by dividing the total distance traveled to complete the journey by the total time taken for the whole journey. If there are any stops during the journey, they are usually included in the total time taken for the journey. The formula for average speed is

$$\text{Average speed} = \frac{\text{total distance traveled}}{\text{total time taken}}$$

Care must be taken to ensure that the units of distance and time are compatible. For an average speed that is measured in kilometers per hour (km/h), the distance must be in kilometers and the time in hours. Some other units of speed in general use are as follows:

- Centimeters per second, cm s^{-1} or cm/s
- Meters per second, m s^{-1} or m/s
- Miles per hour, mi h^{-1} or mi/h

Example.   John drives his family out into the countryside for afternoon tea at Madge's Tearoom. The outward journey from home of 75 km takes them $1\frac{1}{4}$ h. Tea with Madge takes $\frac{1}{2}$ h, and the family return home by the same route. This return journey takes $1\frac{1}{2}$ h. Find his average speed for the whole trip.

Solution. Write

$$\text{The total distance traveled} = 75 + 75$$

$$= 150 \, \text{km}$$

$$\text{The total time taken} = 1\frac{1}{4} + \frac{1}{2} + 1\frac{1}{2}$$

$$= 3\frac{1}{4} \, \text{h}$$

$$\text{Average speed} = \frac{150}{3\frac{1}{4}} \qquad \text{Using the formula for average speed}$$

$$= 46.2 \quad \text{(to 1 dp)}$$

The average speed for the whole trip is 46.2 km/h.

Note. Suppose a journey was in two parts of equal distances. The speed for the first part was 30 km/h and the speed for the second part was 40 km/h. If we calculated the mean of these two speeds it would yield a different answer from the average speed. The mean speed is $(30 + 40) \div 2 = 35$ km/h. Check in the entry Algebraic Fractions for the correct answer to the average speed for this problem.

References: Average, Gradient, Mean.

AXES

Reference: Cartesian Coordinates.

AXIOM

Reference: Theorem.

AXIS OF SYMMETRY

An axis of symmetry is a straight line that divides a shape into two identical halves that are mirror images of each other. An axis of symmetry is often labeled "m" on a figure (see figure a).

A point P on one side of the axis of symmetry has an image point P′ on the other side, and P and P′ are equidistant from the axis of symmetry. This is true for other points and their images like Q and Q′. The straight line joining P to P′ is at right angles

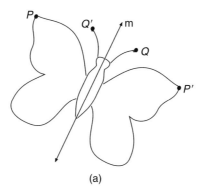

(a)

to the axis of symmetry. An axis of symmetry is sometimes called a mirror line or line of symmetry.

Some shapes have more than one axis of symmetry, like the square (figure b). Other shapes have no axes of symmetry, like the parallelogram (figure c).

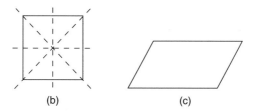

(b)   (c)

To find more information on the symmetry properties of various shapes, search for them under the respective entries.

References: Image, Mediator, Mirror Line, Object, Rotational Symmetry.

AXIOM

Reference: Theorem.

B

BALANCING AN EQUATION

This is a method of solving equations in which the equals sign is the point of balance of the equation. To begin with, the expression on the left-hand side of the equals sign is equal to the expression on the right-hand side. The balance of the two sides of the equation is maintained by performing the same operation on each side. The process of applying these operations leads to solving the equation. The operations are adding, subtracting, multiplying, dividing, squaring, and taking the square root.

Example. Solve the equation $2x - 3 = -8 + x$.

Solution. The left-hand side of the equation is $2x - 3$, which is balanced by the right hand side of the equation, which is $-8 + x$. To solve this equation, we need to finish up with x on left-hand side of the equation and a number called the solution on the right-hand side. Whatever operation is done on one side of the equation must, at the same time, be done on the other side of the equation, thus "balancing the equation." This technique of solving by balancing the equation is demonstrated in this example. Write

$$2x - 3 = -8 + x$$
$$2x - 3 - x = -8 + x - x \qquad \text{Subtracting } x \text{ from both sides}$$
$$x - 3 = -8 \qquad \text{Simplifying, because } 2x - x = x \text{ and } +x - x = 0$$
$$x - 3 + 3 = -8 + 3 \qquad \text{Adding 3 to both sides}$$
$$x = -5 \qquad \text{Simplifying because } -3 + 3 = 0 \text{ and } -8 + 3 = -5$$

The solution of the equation is $x = -5$.

For setting out purposes, we ensure that there is only one equals sign per line of working, and the equals signs are kept in a vertical line directly underneath each other.

References: Equations, Inverse Operation, Linear Equation, Quadratic Equations, Solving an Equation.

BAR CHART

Reference: Bar graph.

BAR GRAPH

The bar graph, also called a bar chart or column graph, is used in statistics to display data. A bar graph consists of a number of vertical bars, or rectangles, of equal widths whose heights are proportional to the frequencies of certain quantities. If the data are discrete, or in separate categories, a small space is left between each bar. For continuous data a histogram is used. Data which may be contained in a frequency table are displayed much more effectively in a bar graph so that comparisons of quantities can be made more easily and with a greater impact. Whenever a bar graph is drawn, make sure it has a title, the axes are numbered, or named, and labeled, and the heights of the columns represent the frequencies.

Example. The frequency table shows how the 20 people in Peter's city office travel to work. Draw a bar graph to display these data.

Method of travel	Bus	Car	Cycle	Train	Walk
Frequency (f)	6	2	7	1	4

Solution. Columns in the bar graph represent each of the methods of travel to work, and the heights of the columns represent their frequencies (see figure). The graph is given a suitable title. The frequency axis is numbered, and the methods of transport are labeled on the horizontal axis. The heights of the columns represent the frequencies.

If desired, the bars can be drawn horizontally instead of vertically.

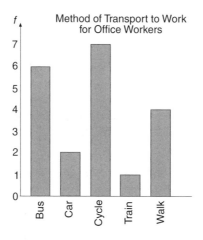

References: Bar Chart, Column Graph, Continuous Data, Data, Discrete Data, Distribution, Frequency, Frequency Table.

BASE (GEOMETRY)

When a polygon is drawn, the base is the side at the bottom of the shape. The polygon can be rotated so that another of its sides is the base. For example, suppose the polygon is a triangle. Each side of a triangle in turn can be the base, depending on how the triangle is drawn. The same triangle is drawn in figure a in three different positions showing how each side in turn is the base of the triangle. The altitude of a triangle is its perpendicular height, which is the shortest distance from its highest point to its base.

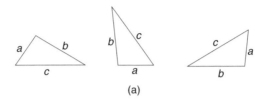

(a)

In the first position of the triangle in figure a the side c is the base, in the second position side a is the base, and in the third position side b is the base. When calculating the area of a triangle it may be advantageous to carefully select one side as the base in preference to the others. This point is illustrated in the following example.

Example. Calculate the area of the right-angled triangle ABC in figure b. The measurements are in meters.

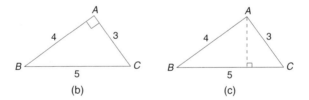

(b)   (c)

Solution. In order to find the area of the right-angled triangle ABC, we will use AC as the base of the triangle and AB as the height.

$$\text{Area of triangle } ABC = \tfrac{1}{2} \times 3 \times 4 \qquad \text{Area of triangle} = \tfrac{1}{2} \times \text{ base } \times \text{ height}$$
$$= 6 \text{ m}^2$$

If, instead, BC is used as the base, then the dashed line in figure c is the height. Since the measurement for the height is not known, the area of the triangle is not readily calculated.

For a solid shape the base is the face of the solid at the bottom. In each of the solids in figures d–f the base is shaded.

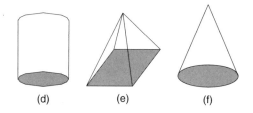

(d)   (e)   (f)

References: Cone, Cylinder, Polygon, Pyramid.

BASE (NUMBERS)

References: Binary Numbers, Exponent, Number Bases.

BEARINGS

A bearing is a direction in which to travel. It is an angle measured in a clockwise direction from due north. A bearing is expressed in degrees using three figures, and is used in navigation. Bearings can also be written as compass points.

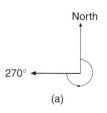

(a)

The arrow in figure a shows a bearing of 270°, which is an angle measured in degrees clockwise from due north. We say that a bearing of 270° is the direction of the arrow in the figure. For bearings which only require two figures, like 45°, a zero is included at the beginning to make it up to the required three figures. For example, a bearing of 45° is written as 045°.

Figure b shows some of the main points of the compass with the equivalent three-figure bearings alongside them. Due north can be expressed as 000°, which is a zero turn, or as 360°, which is a full turn.

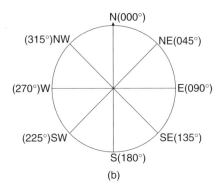

(b)

Example. There is a lighthouse at the point L in figure c and on a bearing of $150°$ is a buoy at the point B. The lighthouse keeper visits the buoy, and then sets a bearing from B to travel back to the lighthouse. On what bearing does he travel to get from B to L?

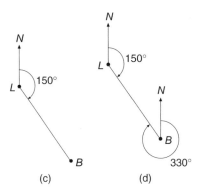

(c)   (d)

Solution. To solve this problem, we draw a vertical line at the point B to represent due north, because we are taking the bearing from B (see figure d). The next step is to start at B and find the angle turned through from north clockwise in order to travel on the route BL. The fact that the two north lines are parallel enables a geometry theorem to be used:

Angle $NBL = 30°$ Sum of cointerior angles $= 180°$

The reflex angle $NBL = 330°$ Sum of angles at a point $= 360°$

The bearing of L from B is $330°$.

References: Angles at a Point, Cointerior Angles, Compass Points, Geometry Theorems.

BEDMAS

This is a mnemonic for remembering the order of operations when doing calculations with numbers; the order is as follows:

- B, Brackets
- E, Exponents
- D, Division
- M, Multiplication
- A, Addition
- S, Subtraction

BEDMAS is an extremely useful aid in working out problems with complex operations. The instruction to do brackets first means to work out the inside of the brackets before performing the next operation. It is sometimes helpful to insert brackets of your own to further clarify the order of operations. If you enter the numbers and the operations into a scientific calculator in the exact order in which they are written, from left to right, the calculator will automatically perform the operations in the correct order. If there is more than one set of brackets, work from the inside outward. The mnemonic lists division before multiplication, but they have equal ranking. Similarly, addition and subtraction have equal ranking. The following examples show how to obtain answers without using a calculator.

Example 1. Work out $(-4 - 2) \times -7$.

Solution. Write

$$(-4 - 2) \times -7 = (-6) \times -7 \qquad \text{Inside of brackets is done first}$$
$$-6 \times -7 = 42$$

Example 2. Work out $12 - 7 \times 3 + 8$.

Solution. Write

$$12 - 7 \times 3 + 8 = 12 - (7 \times 3) + 8 \qquad \text{Inserting brackets to clarify that}$$
$$\text{multiplication is done next}$$
$$= 12 - 21 + 8$$
$$= (12 - 21) + 8 \qquad \text{Inserting brackets to clarify the next step}$$
$$= -9 + 8$$
$$= -1$$

Example 3. Work out $2 - 4 \times (-2 - 3)^2$.

Solution. Write

$$
\begin{aligned}
2 - 4 \times (-2 - 3)^2 &= 2 - 4 \times (-5)^2 & \text{Working inside the brackets first} \\
&= 2 - 4 \times 25 & \text{Exponents next} \\
&= 2 - (4 \times 25) & \text{Inserting brackets to clarify the next step} \\
&= 2 - 100 \\
&= -98
\end{aligned}
$$

Example 4. Work out

$$
\frac{4 + 6}{2 - 12}
$$

Solution. Division cannot be done first, because the expression is not the same as $4 + 6 \div 2 - 12$, which gives an answer of -5. The expression is the same as $(4 + 6) \div (2 - 12)$. Write

$$
\begin{aligned}
\frac{4 + 6}{2 - 12} &= \frac{(4 + 6)}{(2 - 12)} & \text{Inserting brackets to clarify that division is done last} \\
&= \frac{10}{-10} & \text{Working inside the brackets first} \\
&= -1
\end{aligned}
$$

References: Brackets, Operations.

BIAS

This is a word used in statistics to describe an unfair sample that has been chosen from a population. A sample is biased if the sample does not accurately represent the population from which it has been chosen. A biased sample cannot be used to make reliable predictions about the population from which it has been drawn. It is necessary, so that the errors in sampling are not repeated, to recognize the reason(s) why a sample is biased and know how to choose another sample that has no bias.

There is a famous story about a poll that was taken by an American magazine just prior to the presidential election in 1936. A huge, 2 million sample was selected from readers of the magazine, who were then contacted by telephone. The results of the questionnaire predicted a huge defeat for Franklin D. Roosevelt. In reality the opposite happened and Roosevelt was elected by a landslide. The sampling was biased because the sample was restricted only to readers of that particular magazine, and then only to telephone subscribers. A sample can be biased for a variety of reasons. Some reasons are given in the following example.

Example. A sample of 100 people is to be chosen from the population of a town to investigate the attitude of the adult townsfolk toward smoking. Which of the following sampling methods are biased?

1. Every tenth person leaving a supermarket in the center of the town is surveyed. More women than men use a supermarket, so this survey is biased toward women.

2. People are asked to respond to a newspaper advertisement requesting views on smoking, and the first 100 people who respond are surveyed.

The people who respond will be readers of that particular newspaper and therefore not a random selection. This also is biased.

3. Every 50th adult person on the Electoral Roll is chosen until 100 names are obtained. Each one is visited and surveyed.

There is no bias in this method of sampling.

4. The largest factory in the town is chosen and 100 workers are randomly selected from the work force.

This selection may not accurately represent the population of the town for a variety of reasons: The work force of this factory may be predominantly one sex, for example, and so the selection is biased toward that sex. Another reason is that the largest factory could manufacture cigarettes! This sampling method is biased.

References: Population, Random Sample, Sample, Statistics, Survey.

BILLION

Common usage, especially in the United States, puts this number at 1000 million, which is 1,000,000,000, or 10^9. In some countries, such as the United Kingdom and Germany, it is regarded as 10^{12}, which is one million million.

Reference: Index.

BIMODAL

When a set of data, or a frequency distribution, has two separate modes, then the data are said to be bimodal, and the frequency curve of the distribution has two "humps."

Example. Forty students took a short mathematics test that was graded out of 10. The frequency distribution of the results was recorded in a table. Draw a suitable bar graph of the results and identify the mode.

Mark out of 10 (x)	0	1	2	3	4	5	6	7	8	9	10
Frequency (f)	0	1	3	5	7	6	4	7	5	1	1

Solution. The mode is the mark with the highest frequency. The highest frequency is 7, and occurs twice. The modes of this distribution are $x = 4$ and $x = 7$. Since there are two modes in this distribution, we say the distribution is bimodal. A suitable graph is a bar graph (see figure), because the data are discrete.

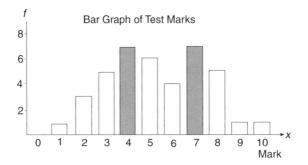

References: Bar Graph, Data, Discrete, Frequency, Frequency Distribution, Mode.

BINARY DIGIT

A binary digit is either a 0 or a 1. Often abbreviated "bit."

Reference: Binary numbers.

BINARY NUMBERS

These are numbers which are written in base 2, whereas the numbers we usually calculate with are in base 10, which are called denary numbers. Suppose you have a set of weights in grams (abbreviated g) and use them to weigh out certain quantities. The weights are 1, 2, 4, 8, and 16 g. The table shows the weights needed to weigh certain quantities.

Number of Each Weight Needed					Quantity
16 g	8 g	4 g	2 g	1 g	to be Weighed
				1	1 g
			1	0	2 g
			1	1	3 g
	1	0	0	0	8 g
	1	1	0	0	12 g
	1	1	1	1	15 g
1	0	0	0	1	17 g
1	0	0	1	1	19 g

The numbers 1, 10, 11, 1000, 1100, 1111, 10001, and 10011, stating the numbers of weights needed to weigh certain quantities, are examples of binary numbers. Binary numbers can be added together as demonstrated in the following example.

Example. Find the weights needed to weigh 27 g.

Solution. The answer can be found by adding together the weights that are needed to weigh 12 g and 15 g, which together make 27 g. The weights required for 12 g are 1100, and the weights required for 15 g are 1111, which can be added using columns, as shown in the table.

16 g	8 g	4 g	2 g	1 g
	1	1	0	0
1	1$_1$	1	1	1
1	**1**	**0**	**1**	**1**

The process known as "carrying" to the next column is used. Since we are working in base 2, we carry over to the next column when the total in a column is 2 or greater. From the columns we can see that $1100 + 1111 = 11011$.

Example. Luke counted 22 sheep as they went into the paddock. How would you write 22 as a base 2 number?

Solution. The column headings needed are 1, 2, 4, 8, and 16. It is unnecessary to go to the next column of 32, because there are only 22 sheep. Starting with the greatest column heading of 16, we subtract 16 from 22, leaving 6. The next column heading is 8 and no subtraction is necessary, because 8 is greater than 6. The next column heading is 4, which we subtract from 6, leaving 2. The next column heading is 2, which we subtract from 2, leaving 0. There is zero left over for the last column of units. We express this in numbers as follows:

$$22 = (16 \times 1) + (8 \times 0) + (4 \times 1) + (2 \times 1) + (1 \times 0)$$

This is abbreviated to 10110_2 with the subscript 2, indicating that we are working in base 2. Using columns, we write

16	8	4	2	Unit
1	0	1	1	0

Base 2 numbers can be added, subtracted, and multiplied using methods similar to those for base 10 numbers. It is important to remember the following rules for binary numbers.

1. The only digits that can be used are 0 and 1.
2. $1 + 1 = 10, 1 + 1 + 1 = 11, 1 + 1 + 1 + 1 = 100$, etc.
3. $10 - 1 = 1, 11 - 1 = 10, 111 - 1 = 110, 110 - 1 = 101, 100 - 1 = 11$.
4. Borrowing is sometimes necessary when numbers are subtracted. When we borrow from the next column we borrow a 2, or a 4, or an 8, or a 16,..., depending on which column we borrow from.
5. $1 \times 1 = 1, 1 \times 0 = 0$. Also, when multiplying a number by 10, we simply add a zero to the number, and when multiplying by 100, we add two zeroes, and so on. For example, $101 \times 10 = 1010$.

The first 20 counting numbers in binary form are as follows:

1	**1**	6	**110**	11	**1011**	16	**10000**
2	**10**	7	**111**	12	**1100**	17	**10001**
3	**11**	8	**1000**	13	**1101**	18	**10010**
4	**100**	9	**1001**	14	**1110**	19	**10011**
5	**101**	10	**1010**	15	**1111**	20	**10100**

Binary numbers are the basis of digital computing. Electronic computers depend on the flow of electric current. Since every binary number can be written using only two digits, 0 and 1, they can be represented by off and on in an electrical circuit. A row of light bulbs can be set up to represent binary numbers. The number illustrated in the figure is 10010.

Reference: Decimal System, Number Bases.

BINOMIAL

A binomial is an algebraic expression that contains two terms. Some binomials can be factored using common factors or the difference of two squares. Examples of binomials are $x^2 - 3x$, $2xy + 4$, $-x^3y^4 + 7$, and $4xy^5 - 7a$.

References: Coefficient, Common Factor, Difference of Two Squares, Powers.

BISECT

To bisect a geometrical figure is to divide it into two congruent or equal parts. The geometrical constructions, using ruler and compasses, for perpendicular bisection of a line segment is described under the entry Perpendicular Bisector, and the bisection of an angle is described under the entry Angle Bisector. In both cases the resulting construction line is a mirror line.

References: Angle Bisector, Congruent Figures, Mediator, Mirror Line, Perpendicular Bisector.

BOX AND WHISKER GRAPH

This graph is sometimes called a box plot. It is used in statistics and displays the spread of data about the median. The items involved in box and whisker graphs are listed as follows:

- Greatest value
- Upper quartile, UQ
- Median, M
- Lower quartile, LQ
- Least value
- Range
- Interquartile range

The appearance of the graph is illustrated in figure a.

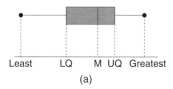

```
Least    LQ    M UQ  Greatest
              (a)
```

The shaded region is the box, and the solid lines to each side of the box are the whiskers. It is a good idea to draw small dots at the ends of the whiskers to clearly indicate where they start and end. The dashed lines are not part of the graph and the width of the box is not important, but the length of the box is a measure of the interquartile range. The total length from end to end of the whiskers is the range of the data. This graph is very informative and is a method of comparing one set of data with another. A drawback to this presentation of data is that it gives no information about frequencies and mode, which the column graph provides. The graph can also be drawn in a vertical position.

Example. The marks out of 10 in a mathematics test taken by 30 students are recorded in the frequency table.

(a) Calculate the median and upper and lower quartiles.

(b) Draw the box and whisker graph for this frequency distribution.

Mark out of 10 (x)	0	1	2	3	4	5	6	7	8	9	10
Frequency (f)	0	0	2	4	3	6	8	5	1	1	0

Solution. To understand how to calculate the median, upper quartile, and lower quartile, search for the methods that are explained under their particular entries.

(a) The number of marks is 30 and there are two middle ones, the 15th, which is 5, and the 16th, which is 6. The median is the mean value of 5 and 6:

$$\text{Median} = \frac{5+6}{2}$$
$$= 5.5$$

The lower quartile is the eighth mark counted from the bottom of the list:

$$\text{LQ} = 4$$

The upper quartile is the eighth mark counted from the top of the list:

$$\text{UQ} = 6$$

(b) The top mark is 9 and the bottom mark is 2. The box and whisker graph can now be drawn as in figure b.

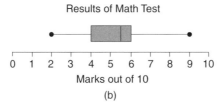

Results of Math Test

0 1 2 3 4 5 6 7 8 9 10

Marks out of 10

(b)

References: Box Plots, Column Graph, Lower Quartile, Median of a Set of Data, Range, Spread, Statistics, Upper Quartile.

BOX PLOT

This another name for a box and whisker graph.

BRACES

These are a type of brackets; the symbol for braces is { }. They are used to indicate a set, and the elements of the set are written inside the braces. For example, the set of winter sports listed at Luke's Community Center is {baseball, basketball, hockey, soccer}.

Reference: Brackets.

BRACKETS

Brackets, which are used to enclose something, is a general name for any of the following:

1. Braces { }, which are used in set theory to enclose a set. For an example see the entry Braces.

2. Curved brackets ( ), sometimes called parentheses, which are used in algebra to enclose terms that represent a single quantity or expression.

Example. Darren has a square sandpit and wishes to extend it in one direction by 2 meters. Write down an expression for the total area of the new sandpit.

Solution. Since the length of the original sandpit is not known, let its length be x meters (see figure). The total length of the new sandpit is $x + 2$ meters and its width is x meters. Since the expression $x + 2$ is a single quantity, which is the length of the sandpit, it is enclosed in brackets and written as $(x + 2)$. Write

$$\text{Area of new sand pit} = \text{width} \times \text{length}$$

$$= x(x + 2)$$

This expression can be expanded to be $x^2 + 2x$ if it is desired to write the expression without brackets.

3. Square brackets [ ], which are used in conjunction with curved brackets in expressions where one set of curved brackets is insufficient to make clear what is intended. When simplifying expressions that contain a mixture of brackets, the rule is to work from the inside out.

Example. Simplify $4[x + (x - 2)^2]$.

Solution. Write

$$4[x + (x - 2)^2] = 4[x + x^2 - 4x + 4] \qquad \text{Expanding inside brackets first}$$
$$= 4[x^2 - 3x + 4] \qquad \text{Collecting terms, } x - 4x = -3x$$
$$= 4x^2 - 12x + 16 \qquad \text{Expanding square bracket last}$$

4. Angle brackets $\langle \ \rangle$. The numbers, or terms, between these brackets are a sequence. For example, the sequence of triangle numbers $= \langle 1, 3, 6, 10, 15, \ldots \rangle$.

References: Braces, Collecting Terms, Expanding Brackets, Parentheses, Sequence, Triangle numbers.

BREADTH

Another, more common name for breadth is width. The breadth of a geometrical shape is the distance measured across the shape at its widest place in a direction at right angles to the shape's length. The length of a shape is its longest measurement. The breadth and length of the ellipse are shown in figure a.

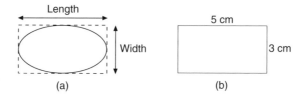

(a)   (b)

The units for measuring breadth are millimeters, centimeters, meters, and kilometers, and are the same as those for measuring length. The word breadth may be used in place of the word width in the formula for finding the area of a rectangle (see figure b).

References: Area, Ellipse, Length, Rectangle.

$$C$$

CALCULATE

To calculate means to use numbers to find the answer to a question. When the question is about an algebraic expression we use simplify, instead of calculate, as the instruction.

References: Algebra, Circle, Indices.

CALCULATOR

A direct algebraic logic, abbreviated DAL, scientific calculator is recommended to aid your study of the mathematics in this book. DAL means that you enter information into the calculator in the same order that you normally write it down. Different calculators have slightly different instructions, so it has been decided not to include in this book the calculator processes. Instead you are referred to the calculator manual.

CANCELING

When fractions are canceled down they are rewritten as equivalent fractions in their simplest form.

Example. Cancel down the fraction 15/72.

Solution. Find the greatest common factor of 15 and 72 that is a positive integer greater than 1. This factor is 3. Write each of the numbers 15 and 72 as a product of two factors using 3 as one of the factors. Then cancel out the threes, since $3 \div 3 = 1$:

$$\frac{15}{72} = \frac{5 \times 3}{24 \times 3}$$
$$= \frac{5}{24} \times \frac{3}{3}$$
$$= \frac{5}{24}$$

Example. Cancel down the fraction 1080/1701.

Solution. By inspection it is found that 3 and 9 are both factors of 1080 and of 1701:

$$\frac{1080}{1701} = \frac{40}{63} \times \frac{9}{9} \times \frac{3}{3}$$

$$= \frac{40}{63} \qquad \text{Canceling the 9's and 3's}$$

Algebraic fractions can also be canceled down by writing the numerator and denominator as products of factors.

Example. Simplify $2a^2b/6ab^3$.

Solution. Identify the common factor of $2a^2b$ and $6ab^3$, which is $2ab$. Write each term as the product of two factors, where one factor is $2ab$. The steps of working are as follows:

$$\frac{2a^2b}{6ab^3} = \frac{a \times 2ab}{3b^2 \times 2ab}$$

$$= \frac{a}{3b^2} \times \frac{2ab}{2ab}$$

$$= \frac{a}{3b^2}$$

Example. Simplify

$$\frac{x^2 + x - 12}{x^2 - 16}$$

Solution. Canceling cannot take place at present because the numerator and denominator are not yet written as the product of factors. The numerator is a quadratic expression and the denominator is the difference of two squares. Write

$$\frac{x^2 + x - 12}{x^2 - 16} = \frac{(x-3)(x+4)}{(x-4)(x+4)} \qquad \text{Factoring}$$

$$= \frac{(x-3)}{(x-4)} \times \frac{(x+4)}{(x+4)}$$

$$= \frac{(x-3)}{(x-4)} \qquad \text{Canceling the common factor } (x+4)$$

References: Algebraic Fractions, Difference of Two Squares, Fractions, Equivalent Fractions, Factor, Numerator, Product of Prime Factors, Quadratic Equations, Rational Expression.

CAPACITY

Capacity refers to the volume of liquid that a container holds. The units of capacity are as follows:

- Milliliter (ml). One milliliter is equivalent to a volume measure of 1 cubic centimeter (cm^3, or cc). It is roughly the capacity of a teaspoon.
- Liter (l). One liter is equal to 1000 milliliters. It is roughly the capacity of a large jug, or about $4\frac{1}{2}$ cups.
- Kiloliter (kl). One kiloliter is equal to 1000 liters. It is the capacity of a cube that has a volume of 1 cubic meter ($1 \, m^3$).

References: Cube (geometry), Metric Units, Volume.

CAPITAL

This is a sum of money that a person has to start a company, or the money accumulated by a person that may be invested at a profit.

Reference: Principal.

CARRY

Three of the column headings for base 10 numbers are shown in figure a; the greatest digit that can be written in any column is 9. When adding numbers together one column at a time, it is possible to obtain a sum greater than 9, which is too large to be written in that column. When this happens, 10, or multiples of 10, are subtracted from the sum and carried to the next column of higher value, and the remnant is written in the column in which you are adding. This process is demonstrated in the following example.

100	10	Units

(a)

Example. Add together 57, 29, and 47 using hundreds, tens, and units columns.

Solution. Arrange the numbers 57, 29, and 47 in columns. We add each column in turn, starting with the units column (see figure b):

100	10	Units
	5	7+
	2	9
1	4 2	7
1	**3**	**3**
	13	23

(b)

- Units column: $7 + 9 + 7 = 23$, which is two 10's and three units. The two 10's are carried to the column of 10's as a 2, and the 3 is written in the units column.
- Tens column: $5 + 2 + 4 + 2 = 13$, which is really 130. The 100 is carried to the column of 100's as a 1, and the 30 is written as a 3 in the 10's column.
- Hundreds column: There is just a 1 in this column, which is the sum for the column.

Reference: Digit.

CARTESIAN COORDINATES

This is a system for fixing the position of a point in a plane according to its distances from two perpendicular number lines, called axes. An example is a good way to explain how Cartesian coordinates work.

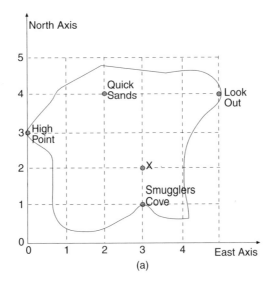

(a)

A map is drawn of Treasure Island with an East axis and a North axis (see figure a). These axes are numbered at equal intervals. The position of any point on the map is known by stating both its number on the East axis and its number on the North axis. For Quick Sands the East number is 2 and the North number is 4. This reference for Quick Sands is abbreviated to (2, 4), with the East number written before the North number. We say the coordinates of the point Quick Sands is (2, 4) with respect to these axes.

The point where the two axes meet is called the origin, and instead of writing two zeros, we write one zero for the point of intersection.

René Descartes (1596–1650) invented the grid system made up of two axes drawn on a grid of squares, and it bears his name: "Cartesian." He called the East axis the x-axis and the North axis the y-axis and included negative numbers on these axes. The two numbers in a bracket that identify a point, like Quick Sand (2, 4), are called ordered pairs, as well as the coordinates of the point. The first number in the bracket, 2, is called the x coordinate, and the second number in the bracket, 4, is called the y coordinate. Plotting a point means marking the position of the point on a set of axes.

Example. A, B, and C are three of the points of a square, which are drawn on the grid shown in figure b. Write down the coordinates of A, B, and C, and find the coordinates of D, the fourth vertex of the square.

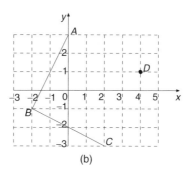

(b)

Solution. The coordinates are $A(0, 3)$, $B(-2, -1)$, $C(2, -3)$. The coordinates of D, the fourth vertex of the square, can be found by drawing a straight line through the vertex C parallel and equal in length to the line BA. This is done by counting two squares along and four squares up from C to reach the point D. The coordinates of D are (4, 1). Cartesian coordinates can be used to solve practical problems like the one below.

Example. Amanda operates a ferry service to the mainland. If a passenger has 80 kilograms of luggage, she charges him $16, for a passenger with 40 kilograms of luggage the charge is $12, and for a passenger with no luggage the charge is $8. What does she charge a passenger with 60 kilograms of luggage?

Solution. The problem can be solved using Cartesian coordinates by first writing the information regarding weight of luggage and charges in dollars as ordered pairs. The ordered pairs are (80, 16), (40, 12), and (0, 8). Axes are drawn on a grid (see figure c), with the horizontal axis labeled w for weight and the vertical axis labeled c for charges. The three points are plotted on the axes and joined up with a line, which happens to be straight.

The point on the straight-line graph is found where $w = 60$ kg, and the value of c at this point is 14. The cost for a passenger with 60 kg of luggage is $14.

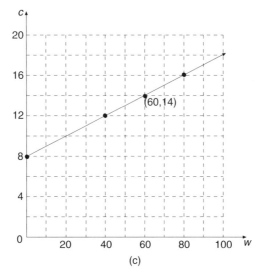

(c)

Alternatively, the equation of the straight-line graph is found to be $c = 0.1w + 8$, using the gradient-intercept method. The value of c is found by substituting $w = 60$ into this equation.

Reference: Gradient-Intercept Method.

CELSIUS

This unit is used as a measure of temperature. It was formerly called centigrade, which means "100 steps."

Reference: Temperature.

CENSUS

This is an official process of counting all the people in a population, and at the same time gathering other information about them such as their religion, age, gender, income, possessions, etc. Countries conduct a census of their population every few years. When an investigation is done into a sample of a population it is called a survey. A survey may take the form of administering a questionnaire, measuring some quantity, or counting some items. A survey is usually carried out on a sample of a population and the information gathered is used to make predictions and draw conclusions about the whole population. For example, a city council may survey a sample of its residents about where to build a new library, and use the information from the sample to gauge the view of all the city residents.

References: Bias, Population, Questionnaire, Random Sample, Sample, Statistics.

CENTER OF A CIRCLE

Two methods of finding the center of a circle are as follows:

1. Paper folding. This is best done if the circle is cut out, but it is not essential. Fold the circle exactly in half and crease the paper (see figure a). Open the circle out and repeat the process, but producing a different fold. The point where the two folds intersect is the center of the circle.

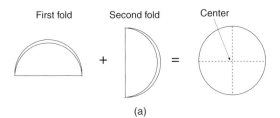

(a)

2. Construction using compasses. This method is appropriate if the circle cannot be folded. For a large circle, such as a flowerbed or a lawn, a piece of rope with a stick fastened at each end will serve as an improvised compass.

Choose any two points on the circumference of the circle and call them A and B (see figure b). Open the compasses to any distance, but greater than half the length from A to B. Draw a semicircle with A as the center. Using the same radius, draw a semicircle with B as the center. Draw in the straight line which passes through the two points where the semicircles intersect. This line is called the mediator of the points A and B, and is also called the perpendicular bisector of the line segment AB. Repeat the steps to construct the mediator for another pair of points C and D that lie on the circumference of the circle. The point where the two mediators intersect is the center of the circle.

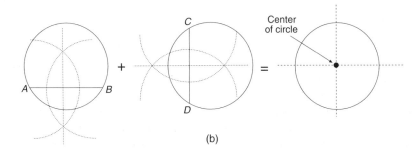

(b)

References: Circle, Circumference, Mediator, Perpendicular Bisector, Radius, Semicircle.

CENTER OF ENLARGEMENT

Reference: Enlargement.

CENTER OF ROTATION

Reference: Rotation.

CENTI

A prefix which means one-hundredth part of some quantity. One hundredth written as a fraction is 1/100. For example, one-hundredth part of a meter is called a centimeter, which is abbreviated cm, where c stands for centi and m stands for meter. Metric units are mainly based on dividing (or multiplying) some quantity by 1000, so the prefix centi does not have much use except in centimeter and in cent, which is 1/100 of a dollar.

References: Meter, Metric Units.

CENTIGRADE

Reference: Temperature.

CENTIMETER

References: Centi, Metric Units.

CENTRAL TENDENCY

For a normal distribution that has been ranked in order of size, the three averages mean, mode, and median tend to be near the center of the ranking. We say that they are measures of central tendency for the distribution. The larger the quantity of data, the more likely it is that the three averages will be closer and closer to each other in value and closer to the center. For a small quantity of data that is not normally distributed the three averages are not expected to be near the center of the ranked data.

The study of central tendency offers an opportune time to discuss the three averages mean, mode, and median as representative values. They can be used to represent a quantity of data in order to compare that quantity of data with another. Suppose we are comparing and contrasting the goal-scoring capabilities of two soccer teams, United and Rovers, who are of similar ability. The goals each team has scored in their last 17 games are given in the following table:

United	2	3	9	2	2	9	3	1	4	4	3	0	1	1	3	10	0
Rovers	3	3	4	1	4	3	0	5	2	2	4	3	4	4	2	2	4

In order to analyze the two sets of data, it is useful to rank the scores for each team:

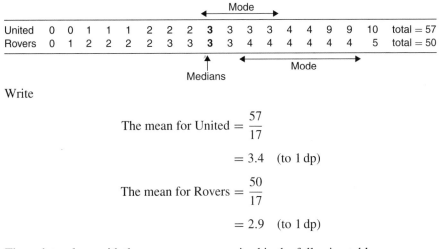

Write

$$\text{The mean for United} = \frac{57}{17}$$

$$= 3.4 \quad \text{(to 1 dp)}$$

$$\text{The mean for Rovers} = \frac{50}{17}$$

$$= 2.9 \quad \text{(to 1 dp)}$$

These data, along with the range, are summarized in the following table:

	Mean	Mode	Median	Range
United	3.4	3	3	10
Rovers	2.9	4	3	5

The representative values (mean, mode, and median) are useful for comparing two sets of data, provided they truly represent their set. This means that the representative value will stand in place of the whole set of data and accurately reflect the properties of the set. We will compare each of the values mean, mode, and median for United's and Rovers' data and see if they represent each team's goal-scoring abilities.

None of the three values represents United, because they are so inconsistent and the range of 10 goals is so large. On the other hand, the data for Rovers is more closely clustered about the mean and median, which is a sign of their consistency. For Rovers their mean and median would be good representative values, but not the mode.

Conclusion: If the two teams played each other, I would expect Rovers to win, because they consistently, game by game, score more goals than United. In two games United scored high, but they lack consistency. This conclusion does not take into consideration the number of goals scored against either team, since we have no knowledge of this information.

References: Average, Data, Mean, Median of a Set of Data, Mode, Normal Distribution, Range.

CGS SYSTEM OF UNITS

This system of measuring quantities was based upon the metric system and used centimeters (cm) to measure length, grams (g) to measure mass, and seconds (s) to measure time. This system of units was replaced in 1960 by the Système International d'Unités, or SI units. The SI units use meters (m), kilograms (kg), and seconds (s) as the basic units of length, mass, and time, respectively. Other units in this system are multiples, and fractions, of 1000 of these basic units. Exceptions to this rule of multiples (and fractions) of 1000 are 10 millimeters = 1 centimeter, 100 centimeters = 1 meter, 10000 square meters = 1 hectare, and 100 hectares = 1 square kilometer.

References: Centimeter, Gram, Hectare, Kilogram, Meter, Metric Units, Millimeter, Second, SI Units.

CHANCE

Chance is another name for probability.

Reference: Probability.

CHANGING THE SUBJECT OF A FORMULA

The formula to find the area of a circle if the radius is known is $A = \pi R^2$. The subject of this formula is A, the term on the left-hand side of the equals sign. Suppose you knew the value of the area of the circle and wanted to calculate its radius. In other words, you need a formula where R is the subject. The process of rearranging a formula to make another variable the subject is called changing the subject of a formula.

The rules for changing the subject of a formula are the same as those for solving an equation. The first step of changing the subject of a formula is to rewrite the formula with the term that will be the subject of the formula on the left of the equals sign, as illustrated in the following examples.

Example. The formula for the area of a circle is $A = \pi R^2$. Find the radius of the circle if the area is 5.57 cm^2.

Solution. The first step is to make R the subject of the formula:

$$A = \pi R^2$$

$$\pi R^2 = A \qquad \text{Rewriting with the term in } R \text{ on the left of the equals sign}$$

$$R^2 = \frac{A}{\pi} \qquad \text{Dividing both sides by } \pi \text{ to isolate the term } R^2$$

$$R = \pm\sqrt{\frac{A}{\pi}} \qquad \text{Taking the square root of both sides of the equation}$$

$$R = \sqrt{\frac{A}{\pi}}$$ Discarding the $\pm$ sign, because radius cannot be negative

$$R = \sqrt{\frac{5.57}{\pi}}$$ Substituting $A = 5.57$

$R = 1.33$ cm   (to 3 sf)   Using the calculator value for π

The answer is rounded to three significant figures, since $A = 5.57$ cm^2 was given to 3 sf in the question.

Example. When she was young, Isabel's grandmother had a formula she quoted to her at bedtime. The number of hours sleep young children need is given by the rule

$$H = 7 + \frac{19 - y}{2}$$

where H is the number of hours of sleep and y is the age of the child. Rearrange the formula to make y the subject.

Solution. Write

$$7 + \frac{19 - y}{2} = H$$ Rewriting with the term in y on the left of equals sign

$$7 + \frac{19 - y}{2} - 7 = H - 7$$ Subtracting 7 from both sides of the equation

$$\frac{19 - y}{2} = H - 7$$ Simplifying $7 - 7 = 0$

$$19 - y = 2(H - 7)$$ Multiplying both sides by 2, and using brackets

$$-y = 2(H - 7) - 19$$ Subtracting 19 from both sides

$$-y = 2H - 14 - 19$$ Expanding the brackets

$$-y = 2H - 33$$ $-14 - 19 = -33$

$$y = -2H + 33$$ Multiplying both sides by -1, which changes the signs

Note. It is obvious that grandmother's formula only works for children. For example, a person aged 50 would need -8.5 hours of sleep!

References: Balancing an Equation, Formula, Perimeter, Rounding, Solving an Equation, Square Root.

CHORD

A chord is a straight line segment joining two points on a curve. The curve we will study is the circle. In figure a, *AB* is a chord of the circle. The perpendicular bisector, or mediator, of the chord of a circle always passes through the center of the circle.

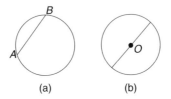

(a)   (b)

The longest possible chord, which passes through the center *O* of the circle, is called the diameter (see figure b).

When one or both ends of a chord are extended the resulting line is called the secant of the circle. Two examples are shown in figure c.

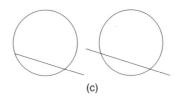

(c)

References: Center of a Circle, Circle, Mediator, Perpendicular Bisector.

CIRCLE

This is the set of points in a plane that are equidistant from a fixed point *O* (see figure a). The fixed point is called the center of the circle. The concept of a circle is explained here.

(a)

Amanda is schooling up her horse George for a show. She stands in one place, at point *O*, holding the end of a rope. The other end of the rope is tied to George, who is at *G*. As George trots around her, Amanda turns so that she is always facing him, but keeps the rope tight and the same length. George is always the same distance from Amanda, all the positions of George as Amanda does a full turn form a circle. The different parts of the circle are illustrated here in figure b, but more information is available about them under their own separate headings.

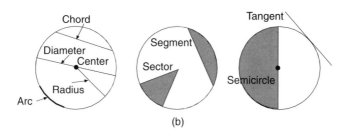

(b)

If the radius of a circle is R, then the area of the circle is given by the formula $A = \pi R^2$. The circumference of the circle is given by the formula $C = 2\pi R$, or $C = \pi D$, where D is the diameter of the circle. The circumference of a circle is its perimeter.

Example. Luke's father is going to build him a fishpond in the garden. Luke insists on having a circular pond. To hold 10 fish, the pond must have a diameter of 3 meters. Calculate the circumference and the area of the pond.

Solution. Write

$$C = \pi \times 3 \qquad \text{Substituting } D = 3 \text{ in the formula } C = \pi D$$

$$C = 9.42 \quad \text{(to 2 dp)} \qquad \text{Using } \pi \text{ in the calculator}$$

The circumference of the pond is 9.42 meters.
    Now write

$$A = \pi \times 1.5^2 \qquad \text{Substituting } R = 1.5 \text{ into the formula } A = \pi R^2,$$
$$\text{where } D \div 2 = R$$

$$A = 7.07 \quad \text{(to 2 dp)} \qquad \text{Using } \pi \text{ in the calculator}$$

The area of the pond is 7.07 square meters.

References: Area, Center of a Circle, Changing the Subject of a Formula, Chord, Diameter, Graphs, Perimeter, Radius, Secant, Sector of a Circle, Segment of a Circle, Semicircle, Tangent.

CIRCLE GRAPH

Reference: Graphs

CIRCLE GEOMETRY THEOREMS

A list of the theorems is given here. They are explained in detail under their separate entries.

- The angle at the center of a circle is twice the angle at the circumference. See Angles at the Center and Circumference of a Circle.
- The angle in a semicircle is a right angle. See Angle in a Semicircle.
- Angle in the Alternate Segment.
- Angles on the same arc are equal. See Angles on the Same Arc.
- Converse of angle in a semicircle theorem. See Angle in a Semicircle.
- Converse of angles on the same arc are equal. See Cyclic Quadrilateral.
- Converse of opposite angles of a cyclic quadrilateral $= 180°$. See Cyclic Quadrilateral.
- The exterior angle of a cyclic quadrilateral is equal to the interior opposite angle. See Cyclic Quadrilateral.
- Radius is perpendicular to the tangent. See Tangent and Radius Theorem.
- The sum of opposite angles of a cyclic quadrilateral $= 180°$. See Cyclic Quadrilateral.
- Tangents from a common point are equal. See Tangents from a Common Point.
- Intersecting chords theorems. See Intersecting Chords.

Reference: Geometry Theorems.

CIRCULAR FUNCTIONS

This is another name for trigonometric functions. Three circular functions with their abbreviations in brackets are sine (sin), cosine (cos), and tangent (tan). They are defined in two ways:

1. An elementary definition, which involves a right-angled triangle, which is given in the entry Trigonometry.

2. A more comprehensive definition, given here, which involves a circle, hence the name circular functions. The definitions of sin, cos, and tan relate to an angle of θ degrees, which must be carefully explained.

The circle in figure a has a radius of 1 unit, and is called the unit circle. The origin of the axes is O, the center of the circle. $P(x, y)$ is any point on the circle, and the length OP is 1 unit. Angle θ is measured counterclockwise from the positive direction of the x-axis. As the radius OP rotates about the point O, the angle θ increases in size.

Figure b shows that θ increases from $0°$ to $90°$ to $180°$ to $270°$. After a full turn the angle $\theta = 360°$, which is at the same place as $\theta = 0°$. For convenience the four sectors of the circle are named the first, second, third, and fourth quadrants, respectively, as shown in figure b.

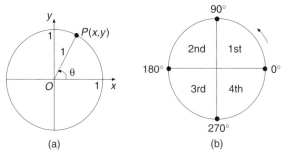

(a)   (b)

Using the unit circle, we can now state the definitions of the three circular functions in terms of the coordinates of the point P:

$$\cos \theta = x, \qquad \sin \theta = y, \qquad \tan \theta = \frac{y}{x}, \quad \text{provided } x \text{ is not equal to zero}$$

The circular functions $\sin \theta$, $\cos \theta$, and $\tan \theta$ can take positive and negative values, depending on the values of x and y in each of the quadrants of the circle.

- First quadrant: When θ is between $0°$ and $90°$, both the x and y coordinates are positive, so both $\cos \theta$ and $\sin \theta$ are positive, and so is $\tan \theta$.
- Second quadrant: When θ is between $90°$ and $180°$, x is negative and y is positive, so $\cos \theta$ is negative and $\sin \theta$ is positive, and $\tan \theta$ is negative.
- Third quadrant: When θ is between $180°$ and $270°$, x is negative and y is negative, so $\cos \theta$ is negative and $\sin \theta$ is negative, and $\tan \theta$ is positive.
- Fourth quadrant: When θ is between $270°$ and $360°$, x is positive and y is negative, so $\cos \theta$ is positive and $\sin \theta$ is negative, and $\tan \theta$ is negative.

The same cycle is repeated for angles from $360°$ to $720°$.

The graphs of the three circular functions are sketched in figure c, but only for values of θ from $0°$ to $360°$, they can be extended horizontally for all real values of θ. The tan graph has asymptotes at $\theta = 90°$, $270°$, $450°$, ..., because the tan of each of these angles is undefined since $x = 0$ at these angles. These asymptotes are repeated every $180°$.

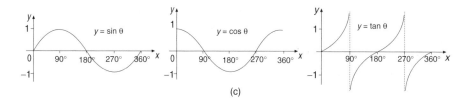

(c)

References: Asymptote, Trigonometry.

CIRCUMCENTER

This is the center of the circle that passes through the three vertices of a triangle, or vertices of any polygon. This circle is called the circumcircle of the triangle ABC (see figure a). This circumcenter can be outside the triangle if it has an obtuse angle. A circumcircle can always be drawn through the vertices of any triangle, but this is not necessarily true for other polygons.

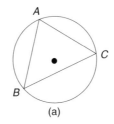

(a)

Example. Figure b is a map of a park that has an arts center (A), a museum (M), and the park gates (G). The town council has decided to build a toilet block (T) (see figure c), which is to be equal distances from G, A, and M. Find where to build the toilet block.

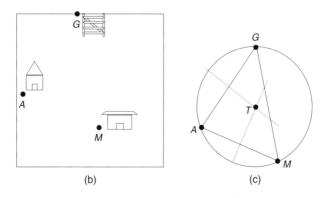

(b)   (c)

Solution. The method will be a geometrical construction on the map. If the toilet block (T) is to be equidistant from each of the three points A, M, and G, then its position will be at the circumcenter of the circle that passes through the triangle AMG. The steps of the construction refer to figure c. Use a ruler to join the three points G, A, M. With ruler and compasses, construct the perpendicular bisectors of two sides of the triangle. The point where they intersect is the center of the circumcircle of triangle GAM. The method is explained in the entry Center of a Circle, method 2. The center of the circle T is the place where the toilet block should be erected, because T is equidistant from the points G, A, and M.

Reference: Center of a Circle.

CIRCUMCIRCLE

Reference: Circumcenter.

CIRCUMFERENCE

The circumference of a circle refers to either the curved boundary line of the circle or the length of the boundary of the circle. Circumference is a word that also applies to other closed geometric figures. The circumference (C) of a circle can be calculated using the formula $C = \pi D$, where D is the diameter of the circle. An example of this calculation is given in the entry Circle.

References: Circle, Revolution.

CIRCUMSCRIBE

This means to draw a circle, or any closed curve, around the outside of a polygon so that the circle passes through all the vertices of the polygon.

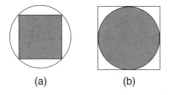

(a)   (b)

Figure a shows how a square peg fits snugly into a round hole. The round hole circumscribes the square peg. The sides of the square are chords of the circle. Inscribe means to draw a circle inside a polygon so that all the sides of the polygon just touch the circle. Figure b shows the view from the top of a round cake in a square box. The circle is inscribed in the square. The sides of the square are tangents to the circle.

References: Chord, Polygon, Tangent.

CLASS INTERVAL

The following grouped frequency table shows the marks out of 100 obtained by 100 students who took a mathematics examination:

Mark	0–9	10–19	20–29	30–39	40–49	50–59	60–69	70–79	80–89	90–100
Frequency	1	3	12	15	20	23	12	8	4	2

The marks from 0 to 100 are divided up into the intervals 0–9, 10–19, 20–29, etc., in order to make the data more manageable. These subdivisions are called class intervals. In the table above there are 10 class intervals. Class limits are the greatest and least values in an interval. In the table, the class limits are 0 and 9, 10 and 19, 20 and 29, etc.

Reference: Arithmetic Mean.

CLASS LIMITS

Reference: Class Interval.

COEFFICIENT

The coefficient of a variable in an algebraic term is the number that multiplies the variable. The coefficient of the variable x in the term $3x$ is 3, since 3 multiplies the variable x. Additional examples are

- The coefficient of $x^2 y$ in the term $-5x^2 y$ is -5.
- The coefficient of xy in the term $3x + 5xy$ is 5, and the coefficient of x is 3.
- The coefficient of xy in the term $2 + xy$ is 1, because $1xy$ is written as xy.

Binomial coefficients were used by Blaise Pascal (1623–1662) to expand brackets of the type $(a + b)^n$, where n is a whole number. At the time he was trying to solve problems in probability. Pascal expanded these brackets by multiplying the brackets together for $n = 1, 2, 3, 4, \ldots$, and noticed a pattern in the coefficients (see table). Using that pattern, he was able to expand brackets of the type $(a + b)^n$, where n is a whole number.

Brackets	Expansion	Coefficients				
$(a+b)^0$	1	1				
$(a+b)^1$	$a + b$	1	1			
$(a+b)^2$	$a^2 + 2ab + b^2$	1	2	1		
$(a+b)^3$	$a^3 + 3a^2 b + 3ab^2 + b^3$	1	3	3	1	
$(a+b)^4$	$a^4 + 4a^3 b + 6a^2 b^2 + 4ab^3 + b^4$	1	4	6	4	1

Arranging the coefficients in the shape of an isosceles triangle, Pascal noticed a pattern, which is explained below (see figure). In the expansion of $(a + b)^n$, the algebraic terms in a and b can be easily worked out. The power of a reduces by one each time, from n to zero. The power of b increases by one each time, from zero to n.

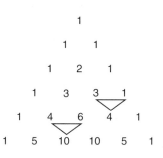

The coefficients can be worked out by continuing Pascal's triangle, which is the name given to this table of coefficients. Each number, other than 1, is the sum of the two numbers standing above it to the left and to the right. For example, $10 = 4 + 6$, and $4 = 3 + 1$. The table can be continued further as needed.

Example. Expand the bracket $(a + b)^5$.

Solution. The line of coefficients, from Pascal's triangle, that is to be used is 1, 5, 10, 10, 5, 1. In the expansion of $(a + b)^5$ the terms in a start at a^5, and the next terms are a^4, a^3, a^2, a^1, and a^0, which is 1. The terms in b start at b^0, which is 1, and the next terms are b^1, b^2, b^3, b^4, and b^5. Combining the coefficients, the terms in a and the terms in b, and remembering that $a^1 = a$, we obtain the expansion

$$(a + b)^5 = a^5 + 5a^4b + 10a^3b^2 + 10a^2b^3 + 5ab^4 + b^5$$

References: Brackets, Expanding Brackets, Pascal's Triangle, Powers, Variable, Whole Numbers.

COINTERIOR ANGLES

Reference: Alternate Angles.

COLLECTING DATA

Collecting data is the gathering together and recording of information, which in many cases is numerical, in order to analyze it, draw conclusions, and use it to help make decisions or predictions. Data are collected in response to a problem, and the method of collecting the data depends on the type of problem.

Example. The highway going past Washington High School is busy all day with cars and trucks. At the end of the school day, students leaving the campus on motorcycles, in cars, or as pedestrians have to join the traffic or cross the road. There is a possibility that a student may be injured, despite being constantly reminded by teachers to be vigilant. The school's Parent Teachers Association has decided that something must be done. But what?

Solution. First, how bad is the situation? The traffic numbers should be counted, at the appropriate times of day, to establish just how great the volume of traffic is. The problem may be no worse than in other areas of the city, except that this is outside a school.

Second, the opinion of parents and members of the public should be sought to find their reactions to the following suggestions:

- Pedestrian crossing
- Foot bridge over the road
- A tunnel under the road
- Speed restriction for vehicles
- Speed bumps at various intervals along the road to slow the traffic
- An entrance lane joining the school gate to the highway
- Reroute traffic away from the school
- Traffic lights at the end of the school drive

Third, ask the public and/or students for alternative views and suggestions. These data are collected in the form of a survey. This may take the form of

- An observation: traffic count at the side of the road outside the school gates
- A questionnaire: the public's reaction to various options
- An interview: to gain alternative views

References: Data, Questionnaire, Survey.

COLLECTING TERMS

Reference: Algebra.

COLLINEAR

A set of points is collinear if all the points lie on the same straight line. A straight line can always be drawn through two points, but for three or more points this is not necessarily so.

Example 1. Prove that the three points $A(1, -2)$, $B(4, 4)$, and $C(6, 8)$ are collinear.

Solution. The points A, B, and C are plotted on axes, and squares are counted to find gradients (see figure a). Write

$$\text{Gradient of } BC = \frac{\text{rise}}{\text{run}}$$

$$= \frac{4}{2}$$

$$= 2$$

$$\text{Gradient of } AC = \frac{\text{rise}}{\text{run}}$$

$$= \frac{10}{5}$$

$$= 2$$

The line segments BC and AC are parallel, because they have equal gradients. Also, they have a common point C, which makes the points A, B, and C collinear.

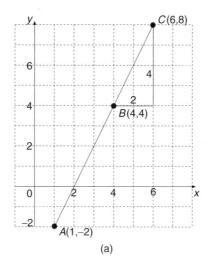

(a)

Word similar to collinear is concurrent, which means that two or more straight lines or curves all pass through the same point. In other words, all the lines intersect at a point. Collinear and concurrent are two of the four options for a set of straight lines:

1. They are parallel.
2. They are concurrent.
3. They are skew, which means not parallel and not concurrent.
4. They are a combination of some of these three.

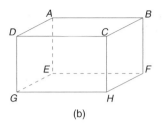

(b)

Example 2. Figure b is a sketch of Jacob's room. Find one example for each of the following types of lines:

1. Three parallel lines
2. Three concurrent lines
3. Two skew lines

Solution.

1. *AB*, *DC*, and *GH* are parallel lines.
2. *DC*, *HC*, and *BC* are concurrent lines, and their point of concurrence is point *C*.
3. *CD* and *FH* are skew lines, because they are neither parallel nor concurrent.

References: Concurrent, Gradient, Parallel, Simultaneous Equations.

COLORING MAPS

Reference: Four-color problem.

COLUMN

A column describes a vertical or upright line of numbers, or terms, in a table or vector. A row is at right angles to a column.

References: Table of Values, Vector.

COLUMN GRAPH

This is the same as a bar graph, except that the bars are drawn vertically as columns, whereas the bars on a bar graph can be drawn either vertically or horizontally.

Reference: Bar graph.

COMBINATIONS

Another name for combinations is selections, which is really a more appropriate word to describe the process. A selection is the number of ways a set of r elements can be chosen from a set of n elements when the order of the choice of the elements is not taken into account. This difficult definition is best explained with an example.

Example 1. Ann asked for volunteers for her debating team at Plato High School. There are 3 students needed to form a debating team, and she had 5 volunteers. In how many different ways can the team of 3 be selected from 5 students?

Solution. Suppose the 5 volunteers were A, B, C, D, and E. The selections, or combinations, of 3 students from the 5 volunteers can be written down in the following orderly way:

$$ABC \quad ABD \quad ABE \quad ACD \quad ACE \quad ADE$$
$$BCD \quad BCE \quad BDE$$
$$CDE$$

This makes 10 combinations in total, so there are 10 ways of selecting 3 people from 5. This combination of selecting 3 people from 5 people is written as $^5c_3 = 10$. This value of $^5c_3$ can be worked out to give the answer 10, without writing down all the combinations, using a scientific calculator. An alternative way of writing the combination $^5c_3$ is $\binom{5}{3}$, both notations are in common use. Another term that is sometimes confused with combinations is permutations, which are often called arrangements. An arrangement is the number of ways a set of r elements can be chosen from a set of n elements when the order of the choice of the elements is taken into account. Suppose we extend the example of the debating team further with the following example.

Example 2. In how many different ways can Ann choose a team of 3 debaters from 5 volunteers if the order of the selection is taken into account? Ann may decide to make the first person, in the order of selection, the first speaker, and the second person in the order of selection the second speaker, and so on.

Solution. We proceed as before to obtain the 10 selections or combinations:

$$ABC \quad ABD \quad ABE \quad ACD \quad ACE \quad ADE$$
$$BCD \quad BCE \quad BDE$$
$$CDE$$

Now, however, each selection can be arranged six times; for example, one can rearrange CDE as

$$CDE \quad CED \quad DCE \quad DEC \quad ECD \quad EDC$$

The are 10 selections, and each selection can be arranged 6 times, so altogether there are $10 \times 6 = 60$ arrangements. There are 60 ways of choosing 3 people from 5, if the order does count. We write this permutation, or arrangement, of selecting 3 people from 5 people as $^5\mathbf{p}_3$, which is equal to 60.

This permutation $^5\mathbf{p}_3$ can be worked out using a scientific calculator to give the answer 60, without writing down all the possible arrangements.

Reference: Factorial.

COMMON FACTOR

Reference: Factor.

COMMON MULTIPLE

Reference: Multiple.

COMMUTATIVE LAW

Under this entry the meaning of three laws will be explained: the commutative law, the associative law, and the distributive law.

The *commutative law* involves two elements of a set, say a and b, and an operation $*$ which can be performed on any pair of elements of the set. The commutative law states

$$a * b = b * a$$

This law emphasizes that the order of the elements can be reversed. This formal definition can be understood more easily using an example in which the set containing a and b is the set of real numbers and the operation $*$ is multiplication.

Example 1. Is multiplication of real numbers commutative? Give two examples that verify your answer.

Solution. In these examples the commutative law involves two numbers, and the operation that combines them is multiplication. Suppose the two numbers are 5 and 7. Does $5 \times 7 = 7 \times 5$? Yes, because $5 \times 7 = 35$, and $7 \times 5 = 35$.

Suppose the two numbers are 1.3 and 5.4. Does $1.3 \times 5.4 = 5.4 \times 1.3$? Yes, because $1.3 \times 5.4 = 7.02$ and $5.4 \times 1.3 = 7.02$.

The addition of real numbers is also commutative: for example, $3 + 6 = 6 + 3$. The division of real numbers is not commutative: for example, $8 \div 2 \neq 2 \div 8$, where $\neq$ means "is not equal to." Subtraction is also not commutative for the set of real numbers.

For an operation to be commutative for the elements of a set it must be true for all possible pairings of elements of the set. If there is one pair of elements that fails the test, then we say the set is not commutative for that operation.

The *associative law* involves three elements of a set, say a, b, and c, and an operation $*$ which can be performed on any pair of elements of the set at a time. The associative law states

$$(a * b) * c = a * (b * c)$$

It must be remembered that the law of BEDMAS is to be applied, and the inside of the brackets is worked out first. Applying the associative law to an example will aid understanding of the concept. Suppose the set containing a, b, and c is the set of real numbers and the operation is addition. The following working suggests that the associative law is true for addition of numbers:

$$(3 + 5) + 2 = 3 + (5 + 2)$$

$$(3 + 5) + 2 = 8 + 2 \qquad \text{and} \qquad 3 + (5 + 2) = 3 + 7$$

$$= 10 \qquad\qquad\qquad\qquad = 10$$

The associative law is true for multiplication of numbers, but is not true for subtraction and division of numbers. The following example verifies that the division of numbers is not associative.

Example 2. Is this statement true?

$$(18 \div 6) \div 2 = 18 \div (6 \div 2)$$

$$(18 \div 6) \div 2 = 3 \div 2 \qquad \text{and} \qquad 18 \div (6 \div 2) = 18 \div 3$$

$$= 1.5 \qquad\qquad\qquad\qquad = 6$$

The *distributive law* involves three elements of a set, say a, b, and c, and two operations $*$ and #. Each operation can be performed on any pair of elements of the set at a time. The distributive law states that $*$ is distributive over # and is expressed as

$$a * (b \# c) = a * b \# a * c$$

An important practical use of the distributive law is in algebra and the two operations are multiplication and addition. When we expand brackets we are using the fact that multiplication is distributive over addition. Factoring is a process that uses the distributive law in reverse.

Example 3. Expand $3(x + 4)$.

Solution. Write

$$3(x + 4) = 3 \times (x + 4) \qquad \text{Inserting the} \times \text{sign}$$
$$= 3 \times x + 3 \times 4 \qquad \text{Multiplication is distributive over addition.}$$
$$= 3x + 12 \qquad \text{Writing the product } 3 \times x \text{ without the} \times \text{sign.}$$

It is interesting to note that addition is not distributive over multiplication. This fact is demonstrated in the following example: $a + (b \times c) \neq a + b \times a + c$.

References: Addend, BEDMAS, Expanding Brackets, Operations.

COMPACT FORM OF DECIMALS

When a decimal is written in the form 0.358 it is in compact form, and when it is written as

$$3 \times \tfrac{1}{10} + 5 \times \tfrac{1}{100} + 8 \times \tfrac{1}{1000}$$

it is written in expanded form. The expanded form gives meaning to the column headings in which the figures of the compact decimal are written. In a similar way, when a number is written in the form 287 it is in compact form. When it is written as $2 \times 100 + 8 \times 10 + 7 \times 1$ the number is in expanded form.

Reference: Binary Numbers.

COMPACT FORM OF NUMBERS

Reference: Compact Form of Decimals.

COMPARATIVE COSTS

In a supermarket the same product may be marketed differently and shoppers are keen to get the best buy, or in other words, the best value for their money. In order to find the best buy, the costs are compared as in the following example.

Example. A 10-kg bag of Gala apples costs $15.70 and a 3-kg pack of the same apples is on special at $4.65. Which is the better value for the money?

Solution. We compare the two prices by calculating the cost per kilogram in each case.

- 10-kg bag:

$$\text{Cost per kg} = \$15.70 \div 10$$
$$= \$1.57 \text{ per kg}$$

- 3-kg pack:

$$\text{Cost per kg} = \$4.65 \div 3$$
$$= \$1.55 \text{ per kg}$$

The 3-kg pack of apples is the better buy, because it costs less per kilogram.

Reference: Rate.

COMPASS POINTS

The main 16 points of the compass are shown in figure a. There is a 22.5° angle between each of the 16 points. The bearings are listed here:

N is due north	NNE is north of northeast
NE is northeast	ENE is east of northeast
E is due east	ESE is east of southeast
SE is southeast	SSE is south of southeast

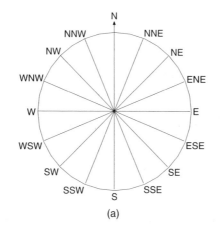

(a)

S is due south	SSW is south of southwest
SW is southwest	WSW is west of southwest
W is due west	WNW is west of northwest
NW is northwest	NNW is north of northwest

Compass bearings can also be written as a mixture of compass points and angles, by expressing the direction as three instructions, in the following way. The compass bearing of W 30°S is shown in figure b, and its direction is explained by following three instructions:

1. W: Face west.
2. 30°: Turn through an angle of 30°.
3. S: Turn away from west toward the south.

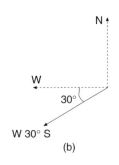

(b)

Reference: Bearings.

COMPLEMENT OF AN ANGLE

Reference: Complementary Angles.

COMPLEMENTARY ADDITION

This is a method of using addition to subtract two numbers. Before electronic cash registers were installed in shops, the shopkeeper would frequently use a form of complementary addition when giving change to a customer. Suppose John bought a pen for $17 and offered the shopkeeper a $50 bill in payment. The shopkeeper starts counting out the change, and his starting point is the cost of the pen, which is $17. He gives John $3, which brings the amount up to $20. Then he gives him $30, to bring the amount up to $50.

Complementary addition is used to subtract two numbers in the following example.

COMPLEMENTARY ANGLES

Two angles are complementary if they add together to make a 90° angle. Figure a shows a drop-leaf table with the leaf initially in a horizontal position. Then it swings about the edge of the table through angle a, and then through angle b until the leaf is vertical. The two angles a and b are complementary angles, because they add together to make 90°. This means $a + b = 90°$. We say that angle a is the complement of angle b, and vice versa.

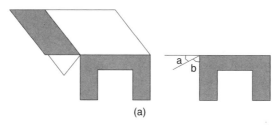

(a)

Similarly, two angles are supplementary angles if their sum is 180°. Suppose there is a trapdoor in the floor of a barn, and the hinged lid is free to open (figure b). It swings about the hinge through the angles a and b until the door is fully open. The two angles a and b are supplementary angles, because they add together to make 180°. This means that $a + b = 180°$. We say that angle a is the supplement of angle b and vice versa.

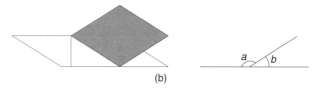

(b)

Two angles are conjugate angles if their sum is 360°. The weighing scale in the drug store has a circular dial, and weighs up to 150 kg (see figure c). When Nathan gets on the scales the pointer turns round the scale through an angle a. If it turns through a further angle b, the pointer will be back at the top of the scale. The two angles a and b are conjugate angles, because they add together to make 360°. This means that $a + b = 360°$. We say that angle a is the conjugate of angle b, and vice versa.

(c)

References: Adjacent Angles, Angles at a Point, Cointerior Angles.

COMPLEMENTARY EVENTS

Complementary events occur in considerations of probability and may be confused with two other similar terms used in probability, called mutually exclusive events and independent events. These three terms are defined in this entry, using examples to make clear their differences. The methods of calculating the probabilities of complementary events, mutually exclusive events, and independent events are explained under the entry Probability of an Event.

Amanda is playing a game which involves rolling a die, which has 6 faces numbered 1–6. Suppose there are three events in which we are interested. One event is rolling a 6, and let this be called S. Another is rolling an even number, and let this be called E. Finally, let O be the event of rolling an odd number.

Two events that cannot happen at the same time are called *mutually exclusive events*. In our example, with one roll of the die you cannot roll an even number and an odd number at the same time, so E and O are mutually exclusive events. Also, S and O are mutually exclusive, because one die cannot show a 6 and an odd number at the same time. But, if you roll a 6, you have achieved event S and event E at the same time, so S and E are not mutually exclusive.

Two events are *complementary* if the two following conditions are both true:

1. The two events are mutually exclusive, which means both cannot happen at the same time.
2. Together the two events make up the whole sample space for the experiment of rolling a die. This means that they are the only two possible events that can take place.

In our example, rolling an odd number and rolling an even number are complementary events, because they are mutually exclusive and they make up the whole sample space, since every number rolled must be odd or even. This means that E and O are complementary events. However, S and O are not complementary events. They meet the first requirement to be mutually exclusive, but they do not make up the whole sample space, because they do not include the numbers 2 and 4.

Two events are *independent* when the outcome of one event has no effect on the outcome of the other event. Independent events usually take place when one event follows another, as in the following example. Liz has a bag that holds three red, one green, and two blue marbles. She draws one marble from the bag, has a look at its color, and replaces it back in the bag. After shaking the bag to ensure the marbles are thoroughly mixed, she then draws another marble from the bag. There are two events here:

- Drawing the first marble, which we call event F.
- Drawing another marble, which we call event A.

These two events are independent, because the outcome of F does not affect the outcome of A, and vice versa. For example, whether the first marble is red, green, or blue has no effect on the outcome of the next draw, because it is replaced in the bag before the next draw takes place. On the other hand, if the first marble is not replaced, then its outcome does affect the second draw, because there are less of that color to select on the next draw, and then F and A are not independent events.

References: Dice, Event, Probability of an Event.

COMPLETING THE SQUARE

Completing the square is something that is done to quadratic expressions to make them perfect squares. Quadratic expressions can be graphed more easily when they are written as perfect squares. Also, completing the square is a method of solving quadratic equations, especially those which do not have factors.

A quadratic expression, like $x^2 - 8x + 16$, is a perfect square when it is written as $(x - 4)^2$. Other examples of perfect squares are

- $(x + 1)^2$, which is equal to $x^2 + 2x + 1$
- $(x - 3)^2$, which is equal to $x^2 - 6x + 9$
- $(x + 6)^2$, which is equal to $x^2 + 12x + 36$

The quick rule for recognizing perfect squares is

$$x^2 + bx + c \quad \text{is a perfect square if} \quad c = \left(\frac{b}{2}\right)^2$$

Sometimes a number needs to be added to a quadratic expression to make it into a perfect square.

Example 1. What must be added to $x^2 + 10x + 20$ to make it a perfect square, and what is the perfect square?

Solution. Write

$$b = 10 \qquad \text{Matching } x^2 + 10x + 20 \text{ with the formula } x^2 + bx + c$$

$$c = \left(\frac{10}{2}\right)^2 \qquad \text{Substituting } b = 10 \text{ into the formula } c = \left(\frac{b}{2}\right)^2$$

$$c = 25 \qquad 5^2 = 25$$

$x^2 + 10x + 25$ is a perfect square, so 5 needs to be added to $x^2 + 10x + 20$ to make it a perfect square. The perfect square is $(x + 5)^2$. By adding 5 to $x^2 + 10x + 20$, we have completed the square on $x^2 + 10x + 20$.

Example 2. Write $x^2 - 6x + 2$ in the form of $(x - p)^2 + q$.

Solution. Write

$$b = -6$$
Matching $x^2 - 6x + 2$ with the formula
$$x^2 + bx + c$$

$$c = \left(\frac{-6}{2}\right)^2$$
Substituting $b = -6$ into the formula $c = \left(\frac{b}{2}\right)^2$

$$c = 9$$
$$(-3)^2 = 9$$

$$x^2 - 6x + 2 = x^2 - 6x + 9 - 7$$
Since $x^2 - 6x + 9$ is a perfect square

$$= (x - 3)^2 - 7$$
Replacing $x^2 - 6x + 9$ by $(x - 3)^2$

Completing the square is also used to solve quadratic equations, as shown in the following example.

Example 3. Solve the quadratic equation $x^2 - 6x + 3 = 0$ by completing the square, leaving the answers to 2 dp.

Solution. Write

$$x^2 - 6x + 3 = 0$$

$$x^2 - 6x = -3$$
Subtracting 3 from both sides of the equation

$$x^2 - 6x + 9 = -3 + 9$$
Adding 9 to both sides of equation, which completes the square on the left side

$$x^2 - 6x + 9 = 6$$

$$(x - 3)^2 = 6$$
Replacing, $x^2 - 6x + 9$ by $(x - 3)^2$

$$x - 3 = \pm\sqrt{6}$$
Taking the square root, and remembering the $\pm$

$$x - 3 = \sqrt{6} \quad \text{or} \quad x - 3 = -\sqrt{6} \qquad \text{Writing as two answers}$$

$$x = \sqrt{6} + 3 \quad \text{or} \quad x = -\sqrt{6} + 3 \qquad \text{Adding 3 to both sides of the}$$
equations

$$x = 5.45 \quad \text{or} \quad x = 0.55 \quad \text{to 2 dp} \qquad \text{Using a calculator}$$

References: Balancing an Equation, Perfect Square, Quadratic Equations, Quadratic Graphs.

COMPONENTS OF A VECTOR

The concept of components of a vector in two dimensions is introduced through an example. The notation used is that if R is a vector it is written as **R**.

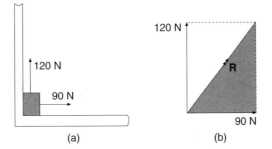

(a)   (b)

Example. Two brothers, Nathan and Jacob, are playing in their playroom, and get into an argument over a heavy toy in the corner of the room. Each fastens a rope to it and they begin a tug-of-war, with Nathan pulling along one wall with a force of 120 newtons, and Jacob pulling along the other wall with a force of 90 newtons. A plan view of the room is shown in figure a. Find the resultant pull in newtons on the toy, and the direction in which the toy moves. The symbol for newtons is N.

Solution. If a scale drawing of the two forces is made as in figure b, the diagonal of the rectangle will be the resultant force **R** acting on the toy, which has the same effect on the toy as the two forces of 120 and 90 N. This means that a single force **R** can be used to find out what is happening to the movement of the toy. The two forces 120 and 90 N are said to be the components of the resultant force **R**. Using Pythagoras' theorem in the shaded right-angled triangle, we can find the magnitude of **R**. The magnitude is the size. Write

$$R^2 = 90^2 + 120^2$$

$$R^2 = 8100 + 14,400$$

$$R^2 = 22,500$$

$$R = 150 \qquad \text{Square root of } 22,500$$

The resultant pull on the toy is 150 newtons.

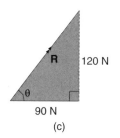

(c)

Finding θ, using trigonometry in the same right-angled triangle, will enable us to state the direction of the resultant of 150 N (see figure c):

$$\tan \theta = \frac{120}{90} \qquad\qquad \tan = \frac{\text{opposite side}}{\text{adjacent side}}$$

$$\theta = \tan^{-1}\left(\frac{120}{90}\right) \qquad \text{If } \tan a = b, \text{ then } a = \tan^{-1} b$$

$$\theta = 53.1° \quad \text{(1 dp)} \qquad \text{Using inverse tan on the calculator}$$

The resultant makes an angle of 53.1° with Jacob's force of 90 N.

In general terms, a set of two components of a vector in two dimensions can now be defined, using the concepts developed in the example. Suppose the vector **R** has vector components **H** horizontally and **V** vertically, and the angle between these components is θ degrees (see figure d).

The horizontal component of **R** is $\mathbf{H} = \mathbf{R} \cos \theta$. The vertical component of **R** is $\mathbf{V} = \mathbf{R} \sin \theta$.

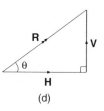

(d)

Example 1. A small aircraft is taking off and at a certain instant its velocity is 150 km h^{-1}, and it is flying at an angle of 10° with respect to the horizontal runway, (see figure e). Find its horizontal and vertical components of velocity.

Vector Diagram

(e)

Solution. Write

$$\mathbf{H} = 150 \times \cos 10° \qquad \text{Horizontal component}$$
$$\mathbf{V} = 150 \times \sin 10° \qquad \text{Vertical component}$$
$$\mathbf{H} = 147.72 \quad \text{(to 2 dp)} \qquad \text{and} \qquad \mathbf{V} = 26.05 \quad \text{(to 2 dp)}$$

The horizontal and vertical components of velocity are 147.27 and 26.05 km h^{-1}, respectively. This means that the aircraft is gaining altitude at a velocity of 26.05 km h^{-1}.

The components of a vector do not have to be horizontal and vertical, as shown in the following example.

Example 2. Tom is pulling his toboggan with a force of 90 newtons up a slope. The rope fastened to the toboggan makes an angle of 30° with the slope (figure f). Find the components of his pulling force that are parallel to the slope and at a right angle to the slope.

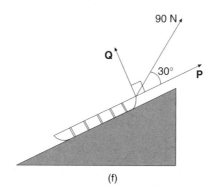

(f)

Solution. Let the components of his pulling force be **P** and **Q** as shown in figure f. Write

$$\mathbf{P} = 90 \times \cos 30° \qquad \text{and} \qquad \mathbf{Q} = 90 \times \sin 30°$$

The components of Tom's pulling force are 77.94 newtons (to 2 dp) parallel to the slope and 45 newtons at a right angle to the slope.

References: Pythagoras' Theorem, Square Root, Trigonometry, Vector.

COMPOSITE BAR GRAPHS

When two or more bar graphs showing similar information are drawn together on the same axes the resulting graph is called a composite bar graph. In this example Ann,

Helen, and Liz are the three strikers for the Ramblers soccer team, and the composite bar graph in the figure shows their goal-scoring feats over three seasons. Composite bar graphs are useful for showing trends or changes that have taken place over a period of time. For example, it is clear from the graph that Liz is the only striker who is scoring more goals in successive years.

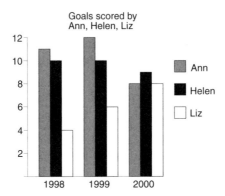

Reference: Bar Graph.

COMPOSITE SHAPES

These are shapes that are made up of two or more other shapes. The processes of finding the area and the perimeter of a composite are explained in the example.

Example. The composite shape of a running track is made up of a rectangle, which is shown shaded in the figure, and two semicircles. All the dimensions are in meters. Find the area and perimeter of the composite shape.

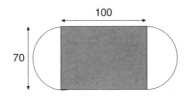

Solution. The perimeter of this composite shape is made up of two straight lines, each of length 100 meters, and the circumference C of a whole circle of diameter 70 meters. Write

$C = \pi D$	Formula for circumference of a circle
$C = \pi \times 70$	Substituting $D = 70$
$C = 219.9$ (to 1 dp)	Using π in a calculator

Therefore

$$\text{Total perimeter} = 100 + 100 + 219.9$$
$$= 419.9\,\text{m}$$

The perimeter of the composite shape is 419.9 meters.

The area of this composite shape is made up of the area of a whole circle and the area of a rectangle. Write, for the circle,

$\text{Area} = \pi R^2$	Formula for area of a circle
$\text{Area} = \pi \times 35^2$	$R = 70 \div 2$
$\text{Area} = 3848.5\,\text{m}^2$ (to 1 dp)	

and for the rectangle,

$\text{Area} = \text{length} \times \text{width}$	Formula for area of a rectangle
$\text{Area} = 100 \times 70$	$\text{Length} = 100$ and $\text{width} = 70$
$\text{Area} = 7000\,\text{m}^2$	

Therefore

$$\text{Area of composite shape} = 3848.5 + 7000$$
$$= 10{,}848.5\,\text{m}^2$$

References: Area, Circumference, Circle, Perimeter, Rectangle, Semicircle, Square.

COMPOSITE TRANSFORMATIONS

These are sometimes called combined transformations. When a final image is formed after two or more transformations such that the first image becomes the object for the second transformation, then a composite transformation has taken place. Often capital letters are used to represent transformations. For example, R is often used to represent a rotation and M a reflection, since m is the first letter of the word mirror. The transformations that may be combined with each other are reflection, rotation, translation, and enlargement. The composite transformations that are explained here are reflection in two parallel mirrors and reflection in two perpendicular mirrors.

1. *Reflection in two parallel mirrors.* Suppose M is the transformation reflection in the mirror m and N is the reflection in the mirror n, where m and n are parallel mirrors placed 5 units apart. The object is a flag F that is placed 3 units in front of the first mirror m. The first transformation of the flag F is reflection in the mirror m to give the image F'. This is written as $M(F) = F'$, where F' is the image of F. Then

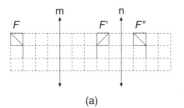

(a)

F' becomes the object for the second transformation N, which is a reflection in the mirror n, and is written as $N(F') = F''$.

The composite transformation M followed by N on the flag F to give the final image F'' is written as $NM(F) = F''$. Note that in the composite transformation $NM(F)$, the reflection M is done first, followed by N. It can be seen that the composite transformation $NM(F)$ is equivalent to a translation of $\binom{10}{0}$ (see the entry Combinations for this notation), where 10 units is twice the distance between the mirrors.

If F is reflected in the mirror n first and then in m, the composite transformation $MN(F)$ is equivalent to a translation of $\binom{-10}{0}$.

2. *Reflection in two perpendicular mirrors.* As before, M is a reflection in the mirror m and N is a reflection in the mirror n, but on this occasion the mirrors are at right angles instead of parallel (see figure b). The composite transformation $NM(F)$ is equivalent to a rotation of $180°$ about the point of intersection of the two mirrors.

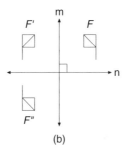

(b)

If the angle between the mirrors was $\theta°$ instead of $90°$, the composite transformation would be a rotation of $2\theta°$ about the point of intersection of the mirrors.

References: Enlargement, Parallel, Perpendicular Lines, Reflection, Rotation, Transformation Geometry, Translation.

COMPOUND INTEREST

Compound interest is an application of percentages to the investment of money in the business world. The terminology used is defined as follows:

- The money that a customer invests in the bank is called the principal, denoted by P.

- The principal earns interest, and this interest is money the bank gives to the customer as a reward for using their money, and is denoted by I.
- The rate of interest is the percentage of their principal given to the customer, denoted by $R\%$.
- Per annum means "for each year."
- The amount is the principal plus the interest, and is denoted by A.

A straightforward example of calculating interest is worked through here before compound interest is explained.

Example 1. Jo invests $200 in the bank at 8% per annum. What interest will this investment earn after 1 year?

Solution. Write

$$I = 8\% \times 200 \qquad I = R \times P \text{ for 1 year}$$
$$I = 0.08 \times 200 \qquad 8\% = 0.08$$
$$I = 16$$

The interest after 1 year is $16.

For a *simple interest* investment the interest on $200 at 8% per annum is $16 every year. If the principal of $200 is invested at a *compound interest* rate of 8% per annum, the interest grows bigger each year. The idea behind compound interest is that the interest is added to the principal and thereafter also earns interest. If the interest of $16 at the end of the first year is added to the principal, then Jo has a new principal of $216 to invest for the second year. The interest grows each successive year, and we say the interest is compounded. A useful formula for calculating compound interest is

$$A = P \left(1 + \frac{R}{100} \right)^n$$

In this formula A is the amount in the bank (principal + interest) after an investment period of n years when P is invested at compound interest of $R\%$ per annum.

Example 2. Jo invests $200 in her bank at an interest rate of 8% per annum for 3 years, and the interest is compounded annually. What will her investment amount to after 3 years?

Solution. The following formula can be applied to Jo's investment:

$$A = 200 \times \left(1 + \frac{8}{100}\right)^3 \qquad \text{Substituting } P = 200, \ R = 8, \text{ and } n = 3$$

$$A = 200 \times (1 + 0.08)^3 \qquad 8 \div 100 = 0.08$$

$$A = 200 \times 1.08^3$$

$$A = \$251.94 \quad \text{(to 2 dp)} \qquad \text{Using a calculator}$$

In the examples we have studied so far the unit of time has been 1 year. But the interest rate can be compounded for a different unit of time, such as every 6 months as in the next example.

Example 3. Mr. and Mrs. Millar deposited $6000 with a savings firm. The money is to accrue interest at the rate of 6% per annum, compounded 6-monthly. Find the maturity value at the end of 7 years if no further deposits or withdrawals are made.

Solution. Write

$$A = P\left(1 + \frac{R}{100}\right)^n \qquad \text{Formula for compound interest}$$

$$A = 6000 \times \left(1 + \frac{3}{100}\right)^{14} \qquad \begin{array}{l}\text{Substitute } P = 6000, \ R = 3 \text{ (divide annual} \\ \text{rate by two for 6-monthly rate),} \\ n = 14 \text{ (the time period is 6 months and} \\ \text{there are 14 time periods in 7 years)}\end{array}$$

$$A = \$9075.54 \quad \text{(to 2 dp)} \qquad \text{Using a calculator}$$

The maturity value after 7 years = $9075.54.

If the same principal of $6000 is invested at the same interest rate of 6%, but is compounded annually instead of 6-monthly, a smaller answer is obtained for the amount:

$$A = 6000 \times \left(1 + \frac{6}{100}\right)^7$$

$$A = \$9021.78 \quad \text{(to 2 dp)}$$

References: Percentage, Simple Interest.

CONCAVE

Concave is a word that is applied to polygons and curves. A *concave polygon* is a polygon that has at least one interior angle greater than $180°$. In other words, a concave polygon has at least one interior angle that is a reflex angle. A concave polygon is sometimes described as a reentrant polygon.

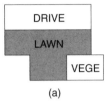

(a)

Figure a is a plan of Amanda's garden. The lawn is a concave octagon, and has two angles greater than 180°. All triangles are convex, because it is impossible to have a concave triangle, since none of its angles can be greater than 180°.

A *convex polygon* has none of its interior angles greater than 180°.

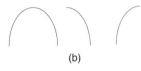

(b)

Curved graphs have concavity which is described in the following way. Graphs which are the shapes shown in figure b are said to be *concave down*. Graphs which are the shapes shown in figure c are said to be *concave up*.

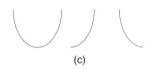

(c)

In figure d the parabola $y = -x^2$ is concave down and the parabola $y = x^2$ is concave up.

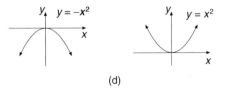

(d)

Points on a curve where a curve changes concavity are called *points of inflection*. In figure e the cubic curve $y = x^3$ is concave down for $x < 0$ and concave up for $x > 0$. The origin, where this curve changes concavity, is a point of inflection.

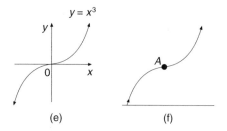

(e)   (f)

A rocket is fired into the air and its second stage is fired just as the first stage reaches its maximum height. In figure f the point A where the second stage is fired is a point of inflection.

References: Angle, Cubic Graphs, Graphs, Octagon, Parabola, Polygon, Reflex Angle, Triangle.

CONCAVE DOWN / UP

Reference: Concave.

CONCENTRIC CIRCLES

Circles are concentric when they have the same center (see figure). When a small stone is dropped into the center of a bowl of still water the waves travel outward from the center of the bowl in the form of concentric circles.

The target that is used in the game of darts is made up of concentric circles made of wire and fastened to a circular board.

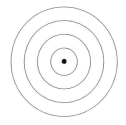

References: Annulus, Circle.

CONCURRENT

Lines are concurrent if they all pass through the same point. An example is given in the entry Orthocenter.

References: Altitude, Collinear, Orthocenter, Simultaneous Equations.

CONCYCLIC POINTS

Reference: Cyclic Quadrilateral.

CONE

A cone is a three-dimensional shape similar to a pyramid, except that its base is a closed curve. The cone we shall deal with has a circle for its base and it is a right cone, which means its axis of symmetry is at right angles to its base. The parts of the cone are labeled in figure a. A generator is the shortest straight line drawn on the curved surface of the cone from the vertex to a point on the circumference of the base.

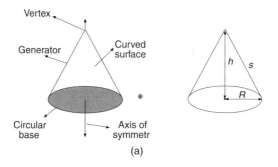

(a)

The curved surface of the cone is its whole surface, but does not include the circular base. The slant height s is the length of a generator. The perpendicular height h, which is also referred to as the altitude of the cone, is the distance from the vertex to the center of the base. The radius of the circular base is R. The volume of the cone is given by

$V = \frac{1}{3} \times$ area of base $\times$ perpendicular height   Area of circular base $= \pi R^2$

$V = \frac{1}{3}\pi R^2 h$

The area of the curved surface, not including the base, is

$$A = \pi R s$$

The area of the circular base is

$$A = \pi R^2$$

There is a relationship between R, h, and s, using Pythagoras' theorem:

$$s^2 = R^2 + h^2$$

The net of a cone is a sector of a circle. The center O of the circle becomes the vertex of the cone, and the radius R of the sector of the circle becomes the slant height s of the cone. The area of the sector of the circle becomes the curved surface area of the cone. The net of the cone is folded so that the two radii R and R come together as one generator of the cone.

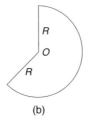

(b)

Example. Madge has bought her children an artificial Christmas tree (see figure c). They put it together and discover it is in the shape of a cone. Julie measures the slant height s of the tree to be 45 centimeters and Jane measures the radius of the circular base to be 18 centimeters. Calculate the volume and the area of the curved surface of the conical tree.

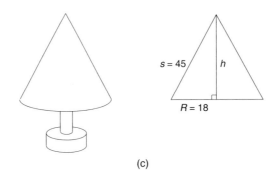

(c)

Solution. Write, to find the volume,

$$45^2 = h^2 + 18^2 \qquad \text{Using Pythagoras' theorem}$$

$$h^2 = 45^2 - 18^2 \qquad \text{Rewriting with } h \text{ on the left of}$$
$$\text{the } = \text{sign}$$

$$h = \sqrt{1701} \qquad \text{Taking the square root to obtain } h$$

$$V = \tfrac{1}{3}\pi R^2 h \qquad \text{Formula for volume of a cone}$$

$$V = \tfrac{1}{3} \times \pi \times 18^2 \times \sqrt{1701} \qquad \text{Substituting } R = 18, h = \sqrt{1701}$$

$$\text{Volume of tree} = 13{,}990 \text{ cm}^3 \quad \text{(to 4 sf)} \qquad \text{Using value of } \pi \text{ in calculator}$$

For the area

$$A = \pi R s$$ Formula for area of a curved surface

$$A = \pi \times 18 \times 45$$ Substituting $R = 18$ and $s = 45$

Area of curved surface $= 2545 \, \text{cm}^2$ (to 4 sf) Using value of π in calculator

References: Altitude, Cylinder, Frustum, Net, Pyramid, Pythagoras' Theorem, Radius, Sector of a Circle.

CONGRUENT FIGURES

Two figures are congruent if they are identical in shape and size so that one figure can be laid exactly on top of the other figure. If one figure needs turning over in order to lie exactly on top of the other, then the two figures are *oppositely congruent*. If one of the figures does not need turning over, the two figures are *directly congruent*. The two pentominoes in figure a are oppositely congruent, since one of them needs turning over for it to lie exactly on top of the other. The two pentominoes in figure b are directly congruent, because one will lie exactly on top of the other without one of them being turned over, but one of them will need turning around.

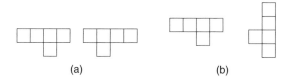

(a) (b)

For the transformations of rotation and translation the object and the image are directly congruent, so they are called *direct transformations*. For reflection the object and the image are oppositely congruent, so reflection is an opposite, *indirect*, transformation. Reflection may be called an *indirect transformation*. When the object and the image are congruent shapes the transformation is called an *isometric*. This means that rotation, translation, and reflection are isometrics. Enlargement is not an isometric, since the object and the image are not congruent, because they are not the same size.

References: Circle Geometry Theorems, Geometry Theorems, Direct Transformation, Enlargement, Image, Indirect Transformation, Isometry, Object, Pentominoes, Reflection, Rotation, Translation.

CONIC SECTIONS

Conic sections are a set of curves {circle, parabola, ellipse, hyperbola} that are formed when a plane surface cuts a cone. Each of these curves is discussed in turn.

Circle. This conic section is formed when the plane surface is parallel to the base of the cone (see figure a).

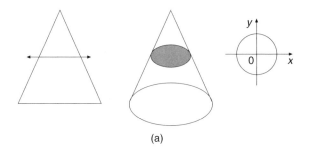

(a)

Parabola. This conic section is formed when the plane surface is parallel to one of the generators of the cone (see figure b).

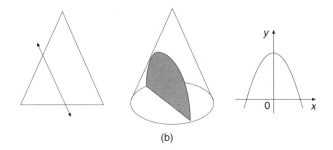

(b)

Ellipse. This conic section is formed when the plane surface is not parallel to one of the generators or the base of the cone (see figure c).

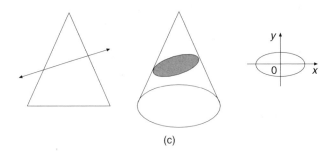

(c)

Hyperbola. This conic section is formed when the plane surface is perpendicular to the bases of two cones (see figure d). Two cones are needed, because the hyperbola has two branches.

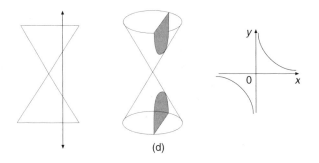

(d)

References: Circle, Cone, Cross Section, Ellipse, Graphs, Hyperbola, Parabola, Plane, Quadratic Graphs.

CONJUGATE ANGLES

Angles are conjugate if their sum is 360°.

References: Angles at a Point, Complementary Angles.

CONSECUTIVE NUMBERS OR TERMS

These are numbers or terms that are next to each other in a sequence. For example, the consecutive counting numbers are $\langle 1, 2, 3, 4, 5, \ldots \rangle$. The consecutive odd numbers, used, for example, for numbering houses, on the same side of the street, are $\langle 1, 3, 5, 7, 9, \ldots \rangle$. The formula for these odd numbers is $\langle 2n - 1 \rangle$, where n may be replaced one after the other by successive counting numbers. The consecutive even numbers, for houses on the other side of the street, are $\langle 2, 4, 6, 8, 10, \ldots \rangle$, and the formula for these even numbers is $\langle 2n \rangle$, where n may be replaced one after the other by successive counting numbers.

References: Formula, Sequence.

CONSTANT OF PROPORTIONALITY

Reference: Proportion.

CONSTANTS

The symbols we use to represent numbers whose values are fixed are called constants. These symbols may be the actual numbers, or they may be letters of the alphabet which represent numbers. Constants occur in formulas, and they also can be part of algebraic terms and expressions. A constant is used when the value of a number has not been given, but may be given at a later time. It is a value that is fixed and is unchanging for a particular calculation.

In algebra, the convention is that lower case letters near the end of the alphabet, like x, y, z, and also t and u, are used for variables. Other letters of the alphabet, like a, b, c, and k, are used for constants. Upper case letters, in particular, like C and K, are often used for constants. On the other hand, a variable is a quantity which can be assigned any of a set of values. A variable is used to represent any number, and its value(s) may be found by solving an equation. Furthermore, a variable may be replaced by a range of values if we wish to draw a graph of a relation, or a variable may be replaced by one value at a time in a formula.

In trigonometry we often use letters of the Greek alphabet, especially α, β, and θ, to stand for angles which are variables. The Greek letter π is used as a constant for the ratio of a circle's circumference to its diameter. A constant can stand for a fixed physical quantity, such as c for the speed of light or g for the acceleration due to gravity.

An example that illustrates constants and variables is the formula $y = mx + c$ for the equation of a straight-line graph. For a particular straight line, say $y = 2x + 3$, the gradient $m = 2$ and the y intercept $c = 3$ are constants, but for these values for the constants, the variables x and y can take an infinite number of different values. For another straight line the constants m and c can take different values, which in turn generate another set of values for the variables x and y.

Another example is the formula $A = \pi R^2$ for finding the area of a circle, in which π is a constant in all the calculations. The variable R, for the radius of the circle, can take an infinite number of different values and A will take the same number of corresponding values. In the algebraic term $3xy$, 3 is the constant and x and y are variables.

References: Algebra, Area, Circle, Formula.

CONSTRUCTIONS

The constructions, done using ruler and compasses only, explained under this entry are as follows:

- Angles of 60°, 30°, 90°, 45°
- Regular hexagon
- Rectangle
- One line parallel to another

1. *Construction of an angle of 60°.* Draw a straight line with a point A on it (see figure a). Open the compasses to any radius, and with the center at A draw a large arc, and where this arc cuts the straight line call it point B. Using the same radius, with the center at B, draw an arc to cut the first arc at point C. Join points A and C with a straight line. Angle BAC is 60°.

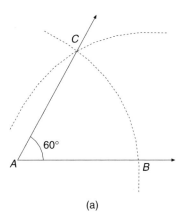

(a)

2. *Construct an angle of 30°.* First construct an angle of 60°. Then bisect the 60° angle to make an angle of 30°. See the entry Angle Bisector for this method.

3. *Construct an angle of 90°.* Draw a straight line with a point A on the line (see figure b). Open compasses to any radius, and with the center at A draw a semicircle to cut the line at the points C and D. Open the compasses out a little further, and with the center at point C draw an arc. Using the same radius, and with the center at D, draw another arc to cut at the point E that arc you have just drawn. Draw the line AE. Angle DAE is an angle of 90°, and AE is the perpendicular bisector of the line segment CD.

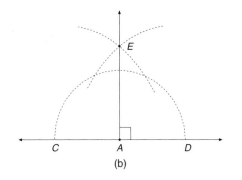

(b)

4. *Construct an angle of 45°.* First construct an angle of 90°. Then bisect the 90° angle to make an angle of 45°. See the entry Angle Bisector for this method.

5. *Construct a regular hexagon,* which has six sides each equal to a length of 5 cm. First mark a point O where you want the center of the regular hexagon to be. (See figure c). With the center at O, draw a circle of radius 5 cm. Mark a point A at the top of the circle, or at any point you wish, on the perimeter of the circle. Using the same radius of 5 cm, starting at the point A, step off the other vertices

B, *C*, *D*, *E*, and *F*. Join up the points *A*, *B*, *C*, *D*, *E*, and *A* to make the regular hexagon.

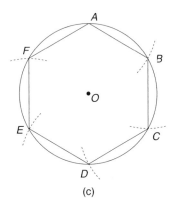

(c)

6. *Construct a rectangle* with sides of lengths 6 and 4 centimeters. Draw a line with a point *A* on it (see figure *d*). From *A* measure a distance of 6 centimeters to another point *B*. At point *A* and at point *B* construct two lines at 90° to the line *AB*, using the method described earlier. Along the first line measure a distance of 4 centimeters and mark the point *D*, and on the second line do likewise and mark the point *C*. Join the points *C* and *D* with a straight line. The required rectangle is *ABCD*.

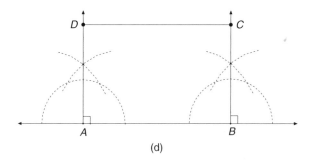

(d)

7. *Construct one line parallel to another*. In figure e, *AB* is a straight line and *P* is a point that is not on the line. Construct a straight line that is parallel to the line *AB*, and also passes through the point *P*. Center the compasses on the point *A* and with a radius of length *AP* draw an arc to intersect the line *AB* at a point called *C*. Using the same radius, with the center at the point *C* draw an arc that will intersect at *Q* a similar arc drawn from the point *P*. Draw the line *PQ*. The line *PQ* is parallel to the line *AB*.

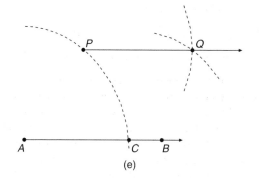

(e)

References: Angle, Arc, Angle Bisector, Bisect, Hexagon, Perpendicular Bisector, Parallel, Radius, Rectangle.

CONTINUOUS CURVE

Reference: Discontinuous.

CONTINUOUS DATA

Reference: Data.

CONTINUOUS DISTRIBUTION

Reference: Frequency Distribution.

CONVERSE OF A THEOREM

This applies to theorems in geometry, but also it can apply to other mathematical statements or assertions. Refer to the entry Cyclic Quadrilateral for an example. The converse of a theorem is best explained by quoting a well-known theorem and then its converse. The example quoted here is Pythagoras' theorem, which can be written as an "If . . . then . . . " statement (see figure a):

(a)

If a triangle *ABC* is right-angled at *B*, then $b^2 = a^2 + c^2$.

The converse of Pythagoras' theorem reverses the "If ... then ..." statement:

If $b^2 = a^2 + c^2$ is true for a triangle, then the triangle is right-angled at *B*.

In Pythagoras' theorem, the theorem and its converse are both true, but this is not the case with every theorem. When the ancient Egyptians built the square-based pyramids, they were able to construct very accurate right angles using the converse of Pythagoras' theorem. The Great Pyramid of Gizeh was erected about 2900 BC and it has been calculated that the right angles of its base are accurate to within 1 part in 27,000. It is not known for certain how the builders achieved that degree of accuracy, but it is thought that the following method might have been used.

A rope in the shape of a loop was divided up into 12 equal parts by 12 knots (see figure b). The rope was arranged into the shape of a triangle with sides of length 3, 4, and 5 units. Three stakes were fixed into the ground, one at each vertex of the triangle, to ensure the rope was taut. The angle in the triangle that was opposite to the longest side was the right angle.

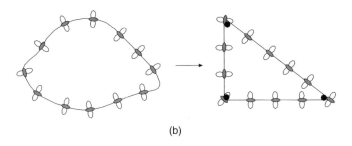

(b)

Greater accuracy is obtained using larger triangles. Corners of a soccer pitch can be marked out using three people and three measuring tapes of lengths 300, 400, and 500 meters.

In addition to geometry theorems that have converses, there are statements and assertions which have converses. Two examples are explained here in the form of pairs of statements.

Example 1. (a) If a number is even, (b) then it is exactly divisible by 2, without a remainder.

The converse assertion reverses these statements: (a) If a number is exactly divisible by 2 without a remainder, (b) then the number is even.

Both the statement and its converse are true.

Example 2. The converse of the following is not necessarily true: (a) If *A* is the father of *B*, (b) then *B* is the child of *A*.

The converse states that (a) If *B* is the child of *A*, (b) then *A* is the father of *B*.

The converse of this statement is not necessarily true, because A could be the mother of B and not the father.

References: Circle Geometry Theorems, Cyclic Quadrilateral, Geometry Theorems, Pythagoras' Theorem.

CONVERSION

Conversions are used to change the units of a quantity. For example, the units for measuring temperature are degrees Celsius (°C) and degrees Fahrenheit (°F). It may be necessary to change the units from one to the other. In order to change the units, a formula is needed, as illustrated in the following example.

Example. The temperature of a warm, sunny day is 24°C. What temperature is this on the Fahrenheit scale?

Solution. Write

$$F = \tfrac{9}{5} C + 32$$ This is the formula for converting from °C to °F

$$F = \tfrac{9}{5} \times 24 + 32$$ Substituting $C = 24$

$$F = 75.2°F$$ Using the calculator

References: Currency Conversions, Temperature.

CONVERSION FACTOR

The idea of a conversion factor is explained using the following example.
    Madge owns a ladies dress shop and is having a winter sale. She decides to reduce all items of clothing by 10%. This may be referred to as giving a discount of 10%. A dress that usually sells for $124 will be reduced in price by 10%. Write

$$\text{Discount} = 10\% \text{ of } \$124$$

$$= \frac{10}{100} \times \$124 \qquad 10\% = \frac{10}{100}$$

$$= \$12.40 \qquad \text{Using the calculator}$$

The discount is the price reduction of the dress, and this reduction is $12.40. The sales price is $124 − $12.40 = $111.60. This sales price can be obtained in one step by multiplying the normal price of $124 by the figure 0.9:

$$\$124 \times 0.9 = \$111.60$$

This multiplying figure of 0.9 is called the conversion factor for converting the normal price of goods to a sale price.

References: Conversion, Discount, Percentage.

CONVEX

Reference: Concave.

COORDINATES

Reference: Cartesian Coordinates.

COPLANAR

Points, lines, and polygons are coplanar if they lie in the same plane. Lying in the same plane means lying in the same flat surface. The problem about whether or not points, lines, and polygons are coplanar only arises in three-dimensional figures, where more than one plane exists. The figure of a three-dimensional house is used to explain the term coplanar.

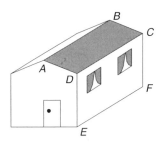

Coplanar points: The points *A*, *B*, *C*, and *D* in the figure are coplanar, because they all lie in the same plane, which is the sloping roof. The points *A*, *B*, *C*, and *F* are not coplanar, because the point *F* does not lie in the same plane as the other points *A*, *B*, and *C*.

Coplanar lines: The line segments *AB* and *CD* are coplanar, because they lie in the same plane, which is the sloping roof of the house. *DC* and *EF* are also coplanar, and their common plane is the wall. In fact the line segments *AB* and *EF* are also coplanar, and their common plane is *ABFE*, which is an inclined plane and is not drawn in the figure. The three line segments *AB*, *CD*, and *EF* are not all coplanar.

Coplanar polygons: The two square windows are coplanar polygons, and their common plane is the wall of the house, but they are not coplanar with the rectangular door.

Skew lines: When lines that are not parallel and not coplanar they are called skew lines. The lines *AD* and *CF* are skew lines. Skew lines do not intersect and are not parallel.

References: Line, Line Segment, Parallel, Plane, Point, Polygon.

COROLLARY

This is an additional theorem that follows from a main theorem that has already been proved. As an example, consider the angle at the center theorem, which is a circle geometry theorem. The main theorem is (see figure a): The angle subtended at the center *O* of a circle by the arc *AB* is equal to twice the angle that is subtended by the same arc at any point *C* on the circumference of the same circle. In symbols what this means is that

$$\text{Angle } AOB = 2 \times \text{angle } ACB$$

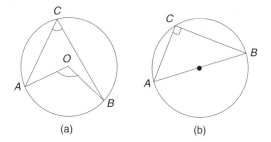

(a)   (b)

The corollary of this main theorem is as follows: Suppose the minor arc *AB* is made longer so that it becomes a semicircle (see figure b). Then the angle *AOB* at the center is now 180°, and the angle *ACB* at the circumference will be 90°. The corollary of the angle at the center theorem is now: The angle in a semicircle is a right angle.

Reference: Geometry Theorems.

CORRESPONDENCE

There are a number of terms, listed below, which are related and can best be explained together under this entry:

- Relation
- Correspondence
- Function
- Mapping

A *relation,* which is also known as a relationship, is simply defined to be a set of ordered pairs (x, y). A relation defines a connection between the first member x of each ordered pair and its corresponding second member y. The set of first numbers x is called the domain of the relation, and the set of second numbers y is called the range of the relation. The relation between x and y can be expressed in four ways:

- As a set of ordered pairs
- As an arrow graph
- As an equation
- As a Cartesian graph

1. *A relation as a set of ordered pairs.* If Luke has $2 more than Liz, and Liz has $2 more than Harry, then a relation between the members of the set {Luke, Liz, and Harry} could be "has $2 more than." The set of ordered pairs for this relation is

$$R = \{(\text{Luke, Liz}) (\text{Liz, Harry})\}$$

where R stands for the relation "has $2 more than."

Another example of a relation expressed as a set of ordered pairs is the following: Suppose the relation "is two more than" exists between the set of counting numbers less than 10. This relation can be expressed as the set of ordered pairs

$$R = \{(3, 1) (4, 2) (5, 3) (6, 4) (7, 5) (8, 6) (9, 7)\}$$

The domain of $R = \{3, 4, 5, 6, 7, 8, 9\}$ and the range of $R = \{1, 2, 3, 4, 5, 6, 7\}$.

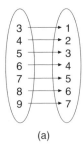

(a)

2. *A relation as an arrow diagram.* The relation "is two more than" can be expressed as an arrow diagram, using the ordered pairs (see figure a).

3. *A relation as an equation.* The equation for the relation "is two more than" is

$$y = x - 2$$

where x is the set of counting numbers $\{3, 4, 5, 6, 7, 8, 9\}$. When writing the relation as an equation it is usual to list the domain at the same time so that we know the numbers for which the equation is defined.

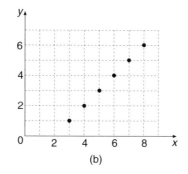

(b)

4. *A relation as a Cartesian graph.* The ordered pairs of the relation can be plotted as points on Cartesian axes, and it can be seen in figure b that they lie on a straight-line graph. The points cannot be joined up, because the numbers for x were stated as counting numbers. If the numbers for x and y were real numbers, the points could be joined up to make a continuous straight line.

Having explained the four different ways of expressing a relation, we now discuss the meaning of the word *correspondence*. There are four basic types of relations and each one is called a correspondence. These four types of relations are:

- One-to-one correspondence
- Many-to-one correspondence
- One-to-many correspondence
- Many-to-many correspondence.

Examples of each of these four correspondences are given below as arrow diagrams and ordered pairs.

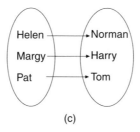

(c)

1. *One-to-one.* "Is the wife of" is an example of a one–one relation (see figure c). Helen is the wife of Norman, Margy is the wife of Harry, and Pat is the wife of Tom. The ordered pairs are {(Helen, Norman) (Margy, Harry) (Pat, Tom)}. In this correspondence, one member of the domain maps onto only one member of the range, and each member of the domain and range occurs only once. Therefore, it is called a one-to-one correspondence.

2. *Many-to-one.* "Is the son of" is an example of a many–one relation (see figure d). Paul is the son of Anne, Mark is the son of Anne, and John is the son of Ken. The ordered pairs are {(Paul, Anne) (Mark, Anne) (John, Ken)}. In this correspondence none of the members of the domain are repeated, but at least one member of the range is repeated, which is Anne. Therefore, it is called a many-to-one correspondence.

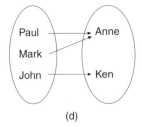

(d)

3. *One-to-many.* "Is the parent of" is an example of a one–many relation (see figure e). Anne is the parent of Paul, Anne is the parent of Mark, and Ken is the parent of John. The ordered pairs are {(Anne, Paul) (Anne, Mark) (Ken, John)}. In this correspondence at least one member of the domain is repeated, which is Anne, but none of the members of the range is repeated. Therefore, it is called a one-to-many correspondence.

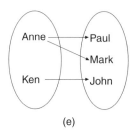

(e)

4. *Many-to-many.* "Is the brother of" is an example of a many–many relation (see figure f). John is the brother of Harry, John is the brother of Helen, Bill is the brother of Harry, and Bill is the brother of Helen. The ordered pairs are {(John, Harry) (John, Helen) (Bill, Harry) (Bill, Helen)}. In this correspondence at least one of the members of the domain is repeated, and at least one of the members of the range is repeated. Both John and Bill of the domain are repeated, and both Harry and Helen in the range are repeated. Therefore, it is called a many-to-many correspondence.

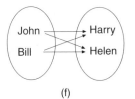

(f)

The next term to explain is a *function*. When a correspondence is one-to-one or a many-to-one, the relation is called a function. A function is sometimes called a mapping. If a relation is written as a set of ordered pairs, it is easy to recognize a function, because no members of the domain are repeated.

Example 1. Is the relation $R = \{(1, 2)(2, 3)(3, 2)(5, 1)\}$ a function?

Solution. Yes, because each member of the domain $\{1, 2, 3, 5\}$ is not repeated in the domain of the ordered pairs.

Example 2. Is the relation $P = \{(2, 1)(3, 2)(2, 3)(1, 5)\}$ a function?

Solution. No, because one member of the domain of $\{\mathbf{2}, 3, \mathbf{2}, 1\}$ is repeated in the domain, which is 2.

If a relation is given in the form of a graph, we can apply the "vertical line test" to see if it is a function. In figure g, if a vertical line, which is shown dashed, is drawn anywhere on the graph and it intersects the graph in at most one point, then the graph is of a function. On the other hand, if the straight line intersects the graph in more than one point, then the graph is not of a function. In the examples in figure g, graph A is of a function and graph B is not of a function.

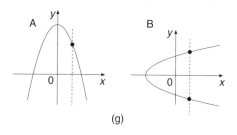

(g)

References: Arrow Graph, Cartesian Coordinates, Domain, Equations, Graphs, Ordered pairs, Range.

CORRESPONDING ANGLES

Reference: Alternate Angles.

CORRESPONDING SIDES

Suppose a triangle ABC is enlarged to obtain its image $A'B'C'$. The image of AB is $A'B'$ and they are a pair of corresponding sides. The corresponding sides AB and $A'B'$ are opposite equal angles A and A'. The other two pairs of corresponding sides are AC and $A'C'$, and BC and $B'C'$.

Corresponding sides occur whenever one polygon is an enlargement of another, and we say the polygons are similar.

References: Congruent Figures, Enlargement, Similar Figures.

COSINE

Cosine is usually abbreviated cos.

Reference: Trigonometry.

COSINE RULE

The cosine rule is a set of trigonometric formulas connecting the three sides of a triangle with the cosine of one of the angles of the triangle. It is mainly used in triangles that are not right-angled, because simpler methods are used for right-angled triangles, as explained under the entry Trigonometry. The notation used for the cosine rule, and also for the sine rule, is that the sizes of the three angles of the triangle are referred to using capital letters, say A, B, and C, and the lengths of the sides are lower case letters, say a, b, and c. The convention is that side a is opposite to angle A, and similarly for sides b and c.

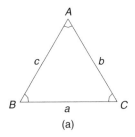

(a)

The cosine rule states that in any triangle

$$a^2 = b^2 + c^2 - 2bc \cos A$$

and this rule is used for finding the length of side a.
There are two more variations of the same rule:

$$b^2 = a^2 + c^2 - 2ac \cos B \qquad \text{This is used for finding the length of side } b$$
$$c^2 = a^2 + b^2 - 2ab \cos C \qquad \text{This is used for finding the length of side } c$$

Each of these three forms of the cosine rule can be rearranged in order to use them to calculate the angles of a triangle. The rearranged forms are

$$\cos A = \frac{b^2 + c^2 - a^2}{2bc} \qquad \text{For finding angle } A$$

$$\cos B = \frac{a^2 + c^2 - b^2}{2ac} \qquad \text{For finding angle } B$$

$$\cos C = \frac{a^2 + b^2 - c^2}{2ab} \qquad \text{For finding angle } C$$

It is important at this stage to recognize which type of triangle can be solved using the cosine rule and which type of triangle requires the sine rule. The rules explaining which to use are listed here:

1. There is no need to use these two rules for right-angled triangles, because it is easier to use sine, cosine, or tangent.

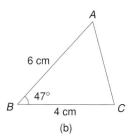

(b)

2. If the lengths of two sides and the angle between the two sides are given in the problem, then use the cosine rule to find the third side b. In the example in figure a, the angle of 47° lies between the two sides of 6 cm and 4 cm, and therefore the cosine rule is used to find the length of side b.
3. If the lengths of three sides of the triangle are given, use the rearranged form of the cosine rule to find one of the angles.
4. For all other triangles use the sine rule.

Example 1. On her bedroom wall Amanda has hung a small picture, but it does not hang straight (see figure b). The lengths of the strings are $AC = 8.4$ cm $AB = 7.3$ cm, and the angle $BAC = 84°$. Use the cosine rule to find the length of BC.

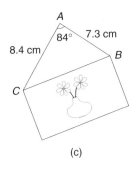

(c)

Solution. Write down what we know about the triangle ABC:

$$b = 8.4 \text{ cm}, \quad c = 7.3 \text{ cm}, \quad A = 84°$$

We are given two sides and the angle between them, and are to find the length of the third side, which is a. Since we are finding the length of the side of a triangle, there are three versions of the cosine rule from which to choose. We use the following to find a: $a^2 = b^2 + c^2 - 2bc \cos A$. Write

$a^2 = 8.4^2 + 7.3^2 - 2 \times 8.4 \times 7.3 \times \cos 84°$	Substituting $b = 8.4, c = 7.3$, and $A = 84°$
$a^2 = 70.56 + 53.29 - 12.82$	Multiplying terms, and rounding to 2 dp
$a^2 = 111.03$	Adding and subtracting the terms
$a = \sqrt{111.03}$	Taking the square root
$a = 10.54 \quad$ (to 2 dp)	

The length of BC is 10.54 cm.

Take care with your calculator when the angle you are working with is obtuse. Say the angle A is obtuse; then the expression $2bc \cos A$ is negative, because the cosine of an obtuse angle is negative. The next example demonstrates how to find the size of an angle using the rearranged form of the cosine rule.

Example 2. The line QR in figure c represents a grassy bank 3 meters long; a ladder QP, which is 4.1 meters long, rests against a brick wall and reaches 2 meters up the wall. Find the size of angle PRQ. There is a peg at the point Q to stop the ladder slipping down the grassy bank.

Solution. Draw the triangle PQR, but rename it ABC, because the cosine rule refers to a triangle ABC. Write down what we know about triangle ABC: $a = 3$, $b = 2$,

$c = 4.1$. We are given the lengths of three sides, therefore we use the cosine rule to find angle C. Write

$$\cos C = \frac{a^2 + b^2 - c^2}{2ab} \qquad \text{Cosine rule for finding angle } C$$

$$\cos C = \frac{3^2 + 2^2 - 4.1^2}{2 \times 3 \times 2} \qquad \text{Substituting } a = 3, b = 2, \text{ and } c = 4.1$$

$$\cos C = -0.3175 \qquad \text{Squaring terms and simplifying}$$

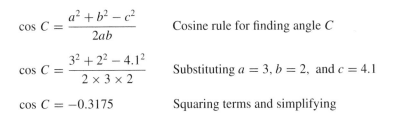

(d)

The negative sign indicates that angle C is obtuse:

$$C = \cos^{-1}(-0.3175) \qquad \text{Use inv cos on the calculator}$$

$$= 108.5° \quad \text{(to 1 dp)}$$

Angle $PRQ = 108.5°$ \qquad Replacing C by PRQ

A trigonometric formula for the area of a triangle is introduced here. When two sides and the angle between the two sides of a triangle are known, the area of the triangle can be found using a trigonometric formula. Using the triangle notation where the three angles are A, B, and C and the lengths of the sides are a, b, and c, we have three versions of this area formula for the triangle, and we choose the appropriate one according to the information given about the triangle.

The three versions of the formula for the area of any triangle are

$$\text{Area} = \tfrac{1}{2}bc \sin A, \qquad \text{Area} = \tfrac{1}{2}ab \sin C, \qquad \text{Area} = \tfrac{1}{2}ac \sin B$$

Example 3. John makes gourmet cheeses and he has designed eight triangles of cheese to fit into an octagonal box. The longest diagonal of the box is 10 cm. Find the area of the lid of the box.

Solution. The lid is made up of eight congruent triangles (see figure d) and one of them is triangle ABC. The area of the lid is eight times the area of triangle ABC.

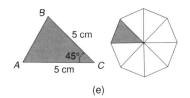

(e)

In triangle *ABC*,

$BC = 5$ cm	Half the longest diagonal of the octagon.
$a = 5$ cm	Using the triangle notation.
$b = 5$ cm	$a = b$
Angle $ACB = 45°$	$360°$ in a full turn $\div$ 8 angles $= 45°$
$C = 45°$	Angle ACB is angle C

Area of triangle $ABC = \frac{1}{2}ab \sin C$	Choosing the appropriate formula
$= \frac{1}{2} \times 5 \times 5 \times \sin 45°$	Substituting $a = 5, b = 5, C = 45°$
$= 8.839$ (to 3 dp)	Using the calculator
Area of lid $= 70.7 \text{ cm}^2$	8 sectors $\times 8.839 = 70.7$ to 1 dp

References: Cosine, Obtuse Angle, Octagon, Sine, Sine Rule, Tangent, Trigonometry.

COST PRICE

Amanda runs a sports shop. When she buys a pair of running shoes from a manufacturer to sell in the shop the price she pays for them is called the cost price (CP). The price at which she sells them to a customer is called the selling price (SP). The difference between these two prices is called her profit or markup. This can be written as a formula:

$$\text{Profit} = \text{SP} - \text{CP}$$

The formula for percentage profit is

$$\% \text{ profit} = \frac{\text{profit}}{\text{cost price}} \times 100$$

Example 1. Amanda buys in a pair of golf shoes for $189 and sells them to a customer for $289.50. Find her percentage profit on the golf shoes.

Solution. Write

Profit = $289.50 − $189.00   Using the formula Profit = SP − CP

Profit = $100.50

$\%\text{ profit} = \dfrac{100.50}{189} \times 100$ Using the formula $\%\text{ profit} = \dfrac{\text{profit}}{\text{cost price}} \times 100$

% profit = 53.2%   (to 1 dp)   Using the calculator

The problem may be restated, as in the following example, to find the selling price.

Example 2. Amanda buys in a netball from the supplier for $38. For how much should she sell it to make a profit of 64%?

Solution. We can regard the cost price as 100% and the selling price as $100 + 64 = 164\%$. Write

Selling price = 164% × cost price

$SP = \dfrac{164}{100} \times 38$ $164\% = \dfrac{164}{100}$ and CP = $38

$SP = 63.32$ Using the calculator

The selling price is $63.32.

The problem may be restated again to find the cost price, as in the next example.

Example 3. Amanda puts a markup of 60% on a baseball bat, which she then sells for $152. What is the cost price?

Solution. The cost price is 100% and the selling price is $(100 + 60) = 160\%$. Write

Cost price $= \dfrac{100}{160} \times$ selling price

$CP = 0.625 \times 152$ $SP = \$152$ and $\dfrac{100}{160} = 0.625$

$CP = 95$ Using the calculator.

The cost price of the baseball bat is $95.

Reference: Percentage.

COUNTEREXAMPLE

This is a particular example which, if it can be found, is used to disprove a formula or theory which is generally believed to be true. A general theory cannot be true if there is at least one example where the theory fails to be true.

Example. It has long been the dream of mathematicians to get a formula for prime numbers. One such formula put forward was $P = n^2 - n + 41$, where n is any counting number and P is a prime number. Find a counterexample that disproves this formula.

Solution. To find a counter example we can substitute any of the counting numbers $n = 1, 2, 3, \ldots$ into the formula until a value for P turns up which is not a prime number:

$$n = 1 \qquad P = 1^2 - 1 + 41$$
$$P = 41 \qquad \text{Which is a prime number}$$
$$n = 7 \qquad P = 7^2 - 7 + 41$$
$$P = 83 \qquad \text{Which is a prime number}$$
$$n = 13 \qquad P = 13^2 - 13 + 41$$
$$P = 197 \qquad \text{Which is a prime number}$$
$$n = 41 \qquad P = 41^2 - 41 + 41$$
$$P = 1681 \qquad \text{Which is not a prime number}$$

1681 is not a prime number, because it has three factors, which are $1, 41$, and 1681. A counterexample is when $n = 41$, which disproves the formula.

References: Formula, Prime Numbers, Substitution.

COUNTING NUMBERS

Reference: Integers.

CRITICAL POINT

Reference: Stationary Point.

CROSS MULTIPLY

Reference: Linear Equation.

CROSS SECTION

This is the flat surface obtained when a plane cuts straight through a solid shape. A good analogy is that of a saw cutting through a block of wood. The "saw cut" can be made at any angle. This is illustrated in the figure, as the saw cuts through a solid block of wood.

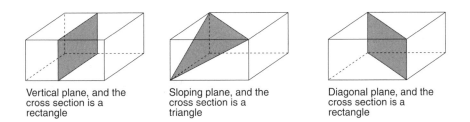

Vertical plane, and the cross section is a rectangle

Sloping plane, and the cross section is a triangle

Diagonal plane, and the cross section is a rectangle

References: Conic Sections, Cuboid, Cylinder, Parallel, Prism, Pyramid, Solids.

CUBE (ALGEBRA)

We cube a number, a quantity, or terms in algebra by multiplying the number by itself three times. For example, we cube the number 4 by working out $4 \times 4 \times 4 = 64$. We say that 4 cubed $= 64$, or the cube of 4 is 64. Using the exponent notation, the abbreviation for $4 \times 4 \times 4$ is 4^3, and we can write $4^3 = 64$.

Example 1. Find the volume of a cube that has sides of length 4 cm.

Solution. Write

$$\text{Volume} = 4 \times 4 \times 4 \qquad \text{Volume} = \text{length} \times \text{width} \times \text{height}$$

$$= 4^3 \qquad \text{Using the index notation}$$

The volume of the cube is 64 cubic centimeters.

Numbers can be cubed using a scientific calculator and you are referred to the calculator handbook. Terms in algebra can also be cubed, as explained in the following examples.

Example 2. Find the cube of each of these terms: (a) $3w^2$, (b) $-2x/y^2$.

Solution. In algebra it is common to use a dot to stand for $\times$ when multiplying terms. For term (a),

$(3w^2)^3 = 3w^2 \cdot 3w^2 \cdot 3w^2$ Multiplying the term by itself three times

$\qquad = 3 \cdot 3 \cdot 3 \cdot w^2 \cdot w^2 \cdot w^2$ Arranging the numbers together, and also the terms

$\qquad = 27w^6$ $3 \cdot 3 \cdot 3 = 27$, and adding the exponents when multiplying

For term (b),

$\left(\dfrac{-2x}{y^2}\right)^3 = \dfrac{-2x}{y^2} \cdot \dfrac{-2x}{y^2} \cdot \dfrac{-2x}{y^2}$ Multiplying the term by itself three times

$\qquad = \dfrac{-8x^3}{y^6}$ $-2 \times -2 \times -2 = -8$ and $y^2 \cdot y^2 \cdot y^2 = y^6$

The inverse of cubing is cube rooting, or taking the cube root. The cube root of a number n is the number which, when cubed, gives n. This statement is also true for a quantity or an expression in algebra. For example, the cube root of 64 is 4, because $4^3 = 64$. This is abbreviated using the cube root notation: $\sqrt[3]{64} = 4$. Cube roots can also be written using the fraction 1/3 as the exponent. The cube root of 64 can be written as $64^{1/3} = 4$. The cube root of a negative number is also negative. For example, the cube root of -64 is -4. Numbers can be cube rooted using a calculator, and once again you are referred to the calculator handbook.

Terms in algebra can also be cube rooted, as shown in the following examples.

Example 3. Find the cube root of each of the following: (a) x^6, (b) $27/y^3$, (c) $-8w^6$.

Solutions. For (a), write

The cube root of $x^6 = \left(x^6\right)^{1/3}$ A cube root can be written as a power of 1/3.

$\qquad = x^2$ $6 \times \dfrac{1}{3} = 2$, and using a law of indices

Note. When finding a cube root, the exponent is divided by 3.
For (b),

The cube root of $\dfrac{27}{y^3} = \dfrac{3}{y}$ Taking cube roots of the top line and the bottom line

For (c),

The cube root of $-8w^6 = -2w^2$ The cube root of -8 is -2, and the cube root of w^6 is w^2

Example 4. If the volume of a cube is 216 cubic meters, find the length of one side of the cube.

Solution. The length of one side of the cube is the cube root of 216. Write

$$\text{The length of one side} \ = \ \sqrt[3]{216}$$

$$= 6 \qquad \text{Using a calculator}$$

The length of one side of the cube is 6 meters.

References: Algebra, Cube, Exponent, Hexomino, Inverse.

CUBE (GEOMETRY)

A cube is one of the five Platonic solids. It is also called a regular hexahedron, which means a solid with six identical plane faces. Each of its faces is a square; they are the same size, and meet at right angles. If the length of each side of the cube is x units, then the volume of the cube $= x^3$ cubic units (see figure a).

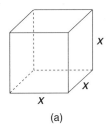

(a)

The surface area of the cube $= 6x^2$, and is made up of the areas of six squares added together, where each square is of area x^2.

The net of a cube, which is made up of six squares, can take many forms, and two of them are drawn in figure b.

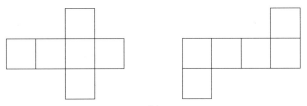

(b)

References: Congruent, Hexomino, Net, Platonic Solids, Right Angle, Square, Surface Area, Volume.

CUBE NUMBERS

These numbers are the cubes of the counting numbers:

Counting number	1	2	3	4	...	N
Cube number	$1^3 = 1$	$2^3 = 8$	$3^3 = 27$	$4^3 = 64$	...	N^3

The first four cube numbers are $\{1, 8, 27, 64\}$.

Reference: Cube (algebra).

CUBE ROOT

Reference: Cube (algebra).

CUBIC EQUATIONS

A cubic equation is an equation in which 3 is the highest power to which the variable is raised. The following examples demonstrate how to solve a variety of cubic equations.

Example 1. Solve $x^3 = -8$.

Solution. Write

$$x = \sqrt[3]{-8} \qquad \text{Cube root of } -8 = -2$$

$$x = -2 \qquad \text{By inspection, or calculator}$$

Example 2. Solve $5x^3 = 4$.

Solution. Write

$$x^3 = 0.8 \qquad \text{Divide both sides of equation by 5}$$

$$x = 0.93 \qquad \text{Cube root of 0.8 using calculator, to 2 dp}$$

Example 3. Solve $5(x + 3)^3 = 84$.

Solution. Write

$$(x + 3)^3 = 16.8 \qquad \text{Divide both sides of equation by 5}$$

$$x + 3 = 2.56 \qquad \text{Cube root of 16.8 is 2.56 to 2 dp}$$

$$x = -0.44 \qquad \text{Subtract 3 from both sides of equation}$$

Example 4. Solve $(x + 3)(x - 5)(5 - 2x) = 0$.

Solution. If three brackets multiply together to make zero, then the equation is solved by equating each bracket to zero and obtaining three answers. Write

$$x + 3 = 0$$

$$x = -3 \qquad \text{Subtracting 3 from both sides of the equation}$$

$$x - 5 = 0$$

$$x = 5 \qquad \text{Adding 5 to both sides of the equation}$$

$$5 - 2x = 0$$

$$-2x = -5 \qquad \text{Subtracting 5 from both sides of the equation}$$

$$x = 2.5 \qquad \text{Dividing both sides of the equation by } -2.$$

The three solutions are $x = -3, 5, 2.5$

Example 5. Solve $4(x + 3)(x - 5)(5 - 2x) = 0$.

Solution. Write

$$(x + 3)(x - 5)(5 - 2x) = 0 \qquad \text{Dividing both sides of the equation by 4}$$

Then proceed as in the previous example to obtain the same three solutions as in Example 4.

Some cubic equations may have a squared bracket, as in the following example, in which case there are three solutions, but one is a repeated solution.

Example 6. Solve $(x + 3)(x - 5)^2 = 0$.

Solution. Write

$$(x + 3)(x - 5)(x - 5) = 0 \qquad \text{Using an index law that } b^2 = b \times b$$

The solutions are $x = -3, 5$ (twice).
We say that $x = 5$ is a repeated solution, or a repeated root.

Example 7. David loves the outdoors and he made a sketch of a hillside with a small lake at the foot of the hill. He modeled the scene with an equation and drew its graph,

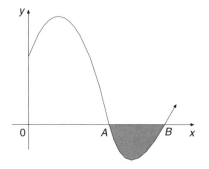

which is shown in the figure. All lengths are in meters. How long is the small lake if the equation of the graph is

$$y = \tfrac{1}{100}(x + 20)(x - 50)(x - 80)?$$

Solution. The curve crosses the x-axis at $x = A$ and $x = B$, and the length of the lake is $B - A$. Write

$\tfrac{1}{100}(x + 20)(x - 50)(x - 80) = 0$	The curve crosses the x-axis when $y = 0$
$(x + 20)(x - 50)(x - 80) = 0$	Dividing both sides of the equation by $\tfrac{1}{100}$
$x + 20 = 0, \quad x - 50 = 0, \quad x - 80 = 0$	Equating each bracket to zero
$x = -20, \quad x = 50, \quad x = 80$	These are the three solutions
$A = 50, \quad B = 80$	The curve crosses the x-axis at $x = A$ and $x = B$
Length of lake $= 30$ meters	Length of lake $= B - A$

References: Cube Root, Cubic Equations, Cubic Graphs, Linear Equation.

CUBIC GRAPHS

Reference: Quadratic Graphs.

CUBOID

This is a hexahedron, which is a solid shape that has six rectangular faces that meet at right angles. The hexahedron is not regular; otherwise its faces would be squares and it would be a cube. The cuboid is also known as a rectangular block. The dimensions of the cuboid are length, width, and height, denoted by L, W, and H, respectively (see figure a).

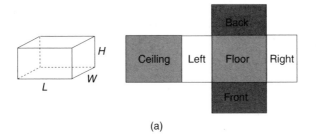

(a)

The net of the cuboid, which is made up of six rectangles, can take many arrangements; only one of them is drawn in figure a. These six rectangles are made up of three pairs of congruent rectangles. They can be thought of as floor and ceiling, front and back, and right and left.

The volume of the cuboid is given by

$V = $ Length $\times$ width $\times$ height,   cubic units

$V = LWH$ Expressed in shortened form

The units of volume may be cubic centimeters (cm^3), cubic meters (m^3), and so on.

The surface area of the cuboid is the surface area of the net, which is made up of three pairs of rectangles:

$A = 2 \times L \times W + 2 \times L \times H + 2 \times W \times H$ square units

$A = 2(LW + LH + WH)$ In factored form

The units of area may be square centimeters (cm^2), square meters (m^2), and so on.

Example. A shoebox with a lid is drawn in figure b, with the measurements in centimeters. The drawing is not to scale. Calculate the volume and surface area, and draw the net of the open box and the lid.

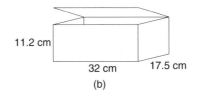

(b)

Solution. Write

         Volume $= L \times W \times H$ Formula for the volume of a cuboid

         Volume $= 32 \times 17.5 \times 11.2$ Substituting $L = 32$, $W = 17.5$,
                                                 $H = 11.2$

Volume of shoebox $= 6272$ cm^3 Using a calculator

We will take the surface area of the shoebox to mean the surface area of the cardboard needed to make the box, not including the tabs used to glue the box together. Write

Area of box $= 2(LW + LH + WH)$ Formula for the area of a cuboid

$= 2(32 \times 17.5 + 32 \times 11.2 + 17.5 \times 11.2)$ Substituting $L = 32$, $W = 17.5$, $H = 11.2$

$= 2(560 + 358.4 + 196)$

$= 2228.8 \text{ cm}^2$

The area of the surface of the shoebox is 2228.8 cm^2.

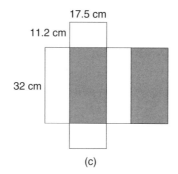

(c)

The net of the shoebox is made up of the open box and the lid. Figure c is not drawn to scale; the key dimensions are marked on the figure.

References: Cube, Net.

CUMULATIVE FREQUENCY GRAPH

The cumulative frequency graph is explained using an example. It is especially useful for finding the median, upper quartile, lower quartile, and interquartile range of a set of continuous data.

David's son William attends a small high school and all the students in his school are doing a statistics project about their heights. There are 100 students in the school and their heights are measured and recorded in a frequency table. Their heights in the table are entered in class intervals of 5 cm. The class interval "140–" means that the heights are from 140, to 145, including 140, but not including 145. Note that the final cumulative frequency of 100 is always the total frequency.

Height in cm	Frequency	Cumulative Frequency (Number of Students less than Greatest Height in the Class Interval)
140–	4	4 who are less than 145
145–	7	11
150–	17	28
155–	22	50
160–	24	74
165–	15	89
170–	9	98
175–180	2	100

The word cumulative sounds like the word accumulate, which means gathering together. In fact, we gather together the frequencies as we move down the table. In the first row of the cumulative frequency table there are 4 students with heights less than 145 cm. Therefore, the coordinates (145, 4) are associated with this row of data. In the second row there are 11 students (made up of $4 + 7$) with heights less than 150 cm. Therefore, the coordinates (150, 11) are associated with the second row, and so on. By plotting, on a set of axes, the coordinates (145, 4), (150, 11), (155, 28), (160, 50), (165, 74), (170, 89), (175, 98), (180, 100), and, of course, (140, 0), we obtain a cumulative frequency graph of the heights of students at Luke's school. The coordinates are joined with a series of straight-line segments. The cumulative frequency curve should start on the horizontal axis, with a cumulative frequency of zero, which is the point (140, 0). Provided the frequency is large enough, we can expect the cumulative frequency graph to be approximately S-shaped, and resemble the graph sketched in figure a. The cumulative frequency curve is sometimes called an ogive.

(a)

The broken horizontal axis in figure b indicates that there are some missing numbers. These numbers are omitted to save space, otherwise a large part of the Cartesian plane would have no graph drawn on it. The median is the middle of 100 students, which is the 50th person. The upper quartile (UQ) is the middle of the top half, which is the 75th person. The lower quartile (LQ) is the middle of the bottom half, which is the 25th person. The heights representing the median and the upper and the lower quartiles can be read from the graph, as accurately as possible:

Median = 160 cm	The height of the 50th student
Upper quartile = 166 cm	The height of the 75th student
Lower quartile = 154 cm	The height of the 25th student
Interquartile range = 12 cm	The upper quartile minus the lower quartile = $166 - 154$

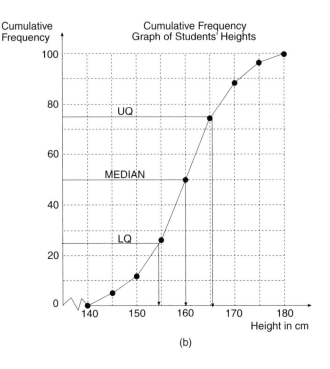

(b)

CURRENCY CONVERSIONS

Different countries of the world have different currencies and when visiting another country it is necessary to exchange some of your money for some of theirs in order to be able to buy goods in that country. The rate of exchange of one currency for another varies from day to day, and your bank will have the latest rate.

Example 1. Ray and Jacky, who live in England, are taking an overseas trip to the United States to see some friends. They need to change their currency from pounds ($£$) into dollars ($\$$). They visit their bank and are told that one pound converts to 2.8563 dollars, that is, $£1 = \$2.8563$. They decide to convert $£1500$ to dollars. How many dollars will they get?

Solution. To change pounds into dollars they must multiply the pounds by 2.8563, which is the conversion factor. This figure is often given to four or more decimal places. This kind of accuracy will be needed if a large amount of money is being exchanged. Write

$$£1500 \times 2.8563 = \$4284.45 \qquad \text{Multiply the number of pounds by the conversion factor}$$

They will get $\$4284.45$ in exchange for $£1500$.

Example 2. At the end of their time in the United States they have $157.00 that is unspent, and at the airport convert this back into pounds ready for use at home. How many pounds will they get if the conversion rate has not changed?

Solution. The conversion factor to change from dollars to pounds is 1/2·8563, which is equal to 0.3501 to 4 dp. Write

$$\$157.00 \times 0.3501 = \pounds54.9657 \qquad \text{Multiply the number of dollars by the}$$
conversion factor.

They will get £54.97, to the nearest penny, in exchange for $157.

CYCLE

A cycle is one repetition of a periodic graph, which is a graph that repeats at regular intervals. An example of a periodic graph is a sine curve, whose equation is $y = \sin x$. The cycle of this sine curve is highlighted in the figure and is from $x = 0°$ to $x = 360°$. This portion of the curve is repeated every 360°, therefore we say the period of this cyclic curve is 360°. The period of a curve is defined using x values.

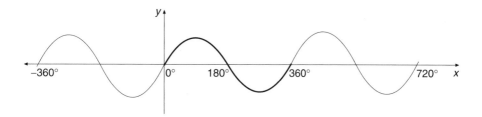

References: Amplitude, Circular Functions, Graphs.

CYCLIC

References: Cycle, Cyclic Quadrilateral.

CYCLIC QUADRILATERAL

A quadrilateral is cyclic if its four vertices all lie on a circle. Since the four vertices all lie on the same circle, we say that the four points are concyclic points. In figure a,

ABCD is a cyclic quadrilateral, and the four points *A*, *B*, *C*, and *D* are concyclic points. The following geometry theorems relate to the cyclic quadrilateral:

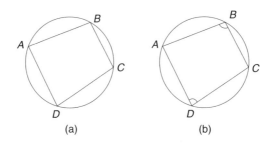

(a)   (b)

Theorem 1. The opposite angles of a cyclic quadrilateral are supplementary, which means they add up to 180°. In figure b,

$$\text{Angle } B + \text{angle } D = 180°$$

Angles *A* and *C* are also supplementary.

Theorem 2. The exterior angle of a cyclic quadrilateral is equal to the interior opposite angle (see figure c). An exterior angle is obtained by extending one of the sides, say *BA*, to the point *E*. The theorem states that

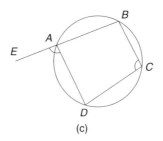

(c)

$$\text{Angle } EAD = \text{angle } BCD$$

This theorem applies at each vertex of the cyclic quadrilateral when the side is produced to form an exterior angle.

Theorem 3. This theorem is the converse of Theorem 1. It states as follows:
  If one pair of opposite angles of a quadrilateral are supplementary, then the quadrilateral is cyclic.

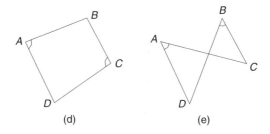

(d)   (e)

This means that if, in figure d, angle DAB + angle $BCD = 180°$, then $ABCD$ is a cyclic quadrilateral. It follows that a circle can be drawn through the points A, B, C, and D. For an example of this converse theorem, refer to the entry Angle in a Semicircle.

Theorem 4. Another converse of a theorem can be used to prove that a quadrilateral is cyclic, and is stated as follows (see figure e):

If angle DAC = angle CBD, then the quadrilateral $ABCD$ is cyclic.

This statement is the converse of the theorem stating that the angles on the same arc are equal.

Example. In figure f, O is the center of the circle and angle $CAO = 35°$. Calculate the sizes of angles ADC and ABC.

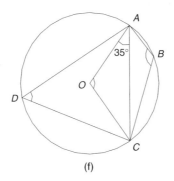

(f)

Solution. Write

Angle $ACO = 35°$ Triangle AOC is isosceles, because AO and OC are radii

Angle $AOC = 110°$ Sum of angles of triangle $= 180°$

Angle $ADC = 55°$ Angle at the center is twice the angle at the circumference

Angle $ABC = 125°$ Sum of opposite angles of a cyclic quadrilateral $= 180°$

References: Angle at the Center and Circumference of a Circle, Angle Sum of a Triangle, Geometry Theorems, Isosceles Triangle, Quadrilateral, Supplementary Angles.

CYCLOID

This is the locus traced by a point on the circumference of a circle when the circle rolls, without slipping, along a line. In the figure the many positions of the point on the rolling circle are joined with a dashed line to show what the cycloid looks like. Imagine the point as being on the tread of a tire as the tire rolls along a flat road.

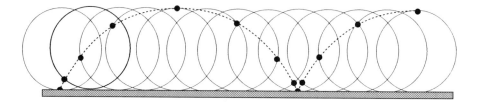

References: Circle, Locus.

CYLINDER

A cylinder is a solid formed by three surfaces. The cylinder described here is made up of two circles in parallel planes joined to a tube formed by rolling up a rectangle (see figure a). The cylinder we shall study is an upright one, so that its axis of symmetry is at right angles to its base; this is usually called a right cylinder. A cylinder is a prism of circular cross section.

The net of the cylinder is made up of a rectangle that rolls up to make a tube, and two circles. Each circle is joined to the rectangle at a point. Thus

$$\text{Length of the rectangle} = \text{circumference of the circle}$$

$$\text{Length} = 2\pi R$$

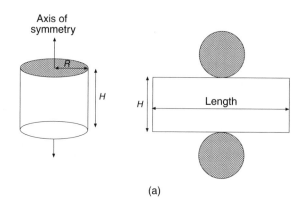

(a)

The formulas for the volume and surface area of the cylinder are stated here. The height of the cylinder is H units and the radius of each circular end is R units:

$$\text{Volume of cylinder} = \pi R^2 H \quad \text{cubic units}$$

$$\text{Surface area of cylinder} = 2\pi R H + 2\pi R^2 \quad \text{square units}$$

Example. Jack has bought a can of beans from the shop. He measures the dimensions of the can and records them in a sketch as shown in figure b. Find the volume of the can and the area of tin used in its manufacture. Draw a net of the can showing the main dimensions.

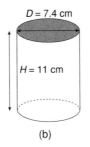

$D = 7.4$ cm

$H = 11$ cm

(b)

Solution. Write

$$R = 3.7\,\text{cm} \qquad \text{Radius = one-half of the diameter}$$

$$\text{Volume} = \pi R^2 H \qquad \text{Formula for the volume of a cylinder.}$$

$$= \pi \times 3.7^2 \times 11 \qquad \text{Substituting } R = 3.7, H = 11$$

$$= 473 \quad \text{(to 3 sf)} \qquad \text{Using } \pi \text{ in the calculator}$$

The volume of the can is 473 cubic centimeters.
Next, write

$$\text{Surface area} = 2\pi R H + 2\pi R^2 \qquad \text{Formula for the surface area of a cylinder}$$

$$= 2 \times \pi \times 3.7 \times 11 + 2 \times \pi \times 3.7^2 \qquad \text{Substituting } R = 3.7, H = 11$$

$$= 342 \quad \text{(to 3 sf)} \qquad \text{Using } \pi \text{ in the calculator}$$

The surface area of the can is 342 square centimeters.

The net of the can is shown in figure c.

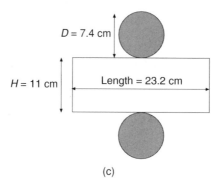

(c)

The dimensions L, H, and D of the net that are shown in figure c are calculated in the following way:

$C = 2\pi R$	Formula for the circumference of a circle
$C = 2 \times \pi \times 3.7$	Substituting $R = 3.7$, which is half of the diameter
$C = 23.2$ (to 3 sf)	Using π in the calculator

The length of the rectangle is equal to the circumference of the circular ends: L is 23.2 cm.

The width of the rectangle is equal to the height of the can: $H = 11$ cm.

The diameter of the circular ends is given as 7.4 cm.

References: Area, Circle, Circumference, Diameter, Prism, Rectangle, Right Angle, Volume.

D

DATA

The set of quantities or information gathered from a sample of a population when a survey or an experiment is done is called the data. Statisticians are people who collect, display, and analyze data in order to draw conclusions and make predictions.

Example. The heights of the students at Washington College is a set of data. The nationalities of the teachers at Washington College is another set of data. Data can be classified according to the data tree shown in the figure:

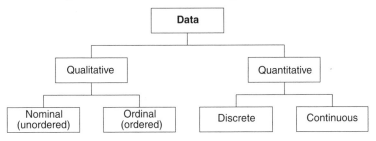

1. Qualitative data are descriptive data, for example, eye color, nationality, gender, types of jobs, attitudes toward politicians, and so on.

2. Quantitative data are numerical data, which are about numbers, for example, the number of workers in a factory, the number of children in a family, the temperature of a liquid, and so on. These data can be ranked in order of size.

3. Nominal data are qualitative data that cannot be ranked in order. For example, the color of cars cannot be ranked, since one color cannot be regarded as higher or lower than another color.

4. Ordinal data are qualitative data that can be ranked in some sort of order. For example, a person's attitude toward television can be ranked as three levels: likes it, finds it OK, dislikes it.

5. Discrete data are quantitative data obtained by counting, and such that the items of data cannot be subdivided. For example the number of people in a family is a counting number and cannot be subdivided into half a person.

6. Continuous data are quantitative data obtained by measuring, with each item of measurement having an infinite number of possible values, restricted only by the limitations of the measuring device, for example, the speed of cars passing a certain mark on a highway.

References: Experiment, Population, Sample, Survey.

DECAGON

A decagon is a polygon with 10 sides. Figure a shows some decagons; the last one is a regular decagon. The regular decagon has 10 sides that are of equal length, and its 10 angles are each equal to 144°. It has 10 axes of symmetry, and the order of rotational symmetry is 10. The regular decagon is made up of 10 congruent isosceles triangles whose angles are 36°, 72°, and 72° (see figure b). The regular decagon will not tessellate.

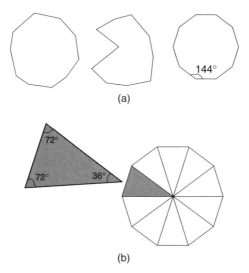

(a)

(b)

References: Axis of symmetry, Congruent Figures, Isosceles Triangle, Order of Rotational Symmetry, Polygon, Regular Polygon, Tessellations.

DECAY

Reference: Exponential Decay.

DECILE

Reference: Percentiles.

DECIMAL

A decimal is also known as a decimal fraction. A decimal is equivalent to a proper fraction that has a denominator of either 10, 100, 1000, or further powers of 10, but is written without a denominator and with a decimal point. A decimal can be negative.

Example 1. Write the fraction $\frac{3}{10}$ in the form of a decimal.

Solution. This fraction has a denominator of 10, so it can readily be expressed in decimal form. As a decimal it is written without a denominator, but it does have a decimal point: $\frac{3}{10} = 0.3$. The zero indicates there is no whole number, the 3 represents $\frac{3}{10}$, and the decimal point separates them. Each figure that is written after the decimal point represents a proper fraction. The first figure after the decimal point represents tenths. For example, $0.8 = \frac{8}{10}$ and $0.9 = \frac{9}{10}$. If there are two figures after the decimal point, they are equivalent to hundredths. For example, $0.75 = 75/100$, which cancels to $\frac{3}{4}$. Also, $0.08 = 08/100$, which is written as 8/100, and cancels to 2/25.

If there are three figures after the decimal point they are equivalent to thousandths, and so on. For example, $0.257 = 257/1000$.

Decimals involving two or more figures after the decimal point can be expressed as the sum of tenths, hundredths, thousandths, and so on. For example,

$$0.257 = \tfrac{2}{10} + \tfrac{5}{100} + \tfrac{7}{1000}$$

If a fraction does not have a denominator of 10 or of powers of 10, it can still be converted into decimal form. The method is explained in the following example.

Example 2. Convert these fractions into decimal form: (a) $\frac{3}{5}$, (b) $\frac{2}{25}$, (c) $\frac{5}{7}$.

Solution. For (a), write

$$\tfrac{3}{5} = \tfrac{3}{5} \times \tfrac{2}{2} \qquad \text{Multiplying by } \tfrac{2}{2} \text{ ensures that the denominator is 10 and the fraction has the same value}$$

$$= \tfrac{6}{10}$$

$$= 0.6 \qquad \text{Which is in decimal form}$$

For (b),

$$\tfrac{2}{25} = \tfrac{2}{25} \times \tfrac{4}{4} \qquad \text{Multiplying by } \tfrac{4}{4} \text{ ensures that the denominator is 100}$$

$$= \tfrac{8}{100}$$

$$= 0.08 \qquad \text{Which is in decimal form}$$

For (c),

$$\tfrac{5}{7} = 5 \div 7$$ 7 cannot be multiplied by a whole number to obtain a power of 10, so the fraction is turned into a division

$$= 0.714285714\ldots$$ Using the calculator for division

$$= 0.714 \quad \text{(to 3 dp)}$$ See later in this entry for recurring decimals

Example (c) indicates that fractions can be changed into decimals using a calculator for dividing the numerator by the denominator. This means that the methods used in (a) and (b) need only be employed if no calculator is available!

Decimals can be converted into fractions using the methods outlined in the following examples.

Example 3. Convert the following decimals into fractions, and leave your answers in their simplest form: (a) 0.25, (b) 0.80, (c) 2.857.

Solution. For (a), write

$$0.25 = \tfrac{25}{100}$$ Two figures after the decimal point indicate that the denominator of the fraction is 100

$$= \tfrac{1}{4} \times \tfrac{25}{25}$$ 25 is a common factor of 25 and 100

$$= \tfrac{1}{4}$$ Canceling the 25's

For (b),

$$0.80 = 0.8$$ The zero after the 8 indicates that there are no hundredths

$$= \tfrac{8}{10}$$ One figure after the point indicates tenths

$$= \tfrac{4}{5} \times \tfrac{2}{2}$$ 2 is a common factor of 8 and 10

$$= \tfrac{4}{5}$$ Canceling the 2's

For (c),

$$2.857 = 2 + \tfrac{857}{1000}$$ The 2 is a whole number, because it is before the decimal point

$$= 2\tfrac{857}{1000}$$ The fraction will not cancel

All fractions, when converted into decimals, are either terminating or recurring:

Terminating decimals have a finite number of decimal places. Refer to the earlier examples of $\tfrac{3}{5}$ and $\tfrac{2}{25}$ in this entry.

Recurring decimals have an infinite number of decimal places, but repeat a certain pattern of numbers over and over again. Recurring decimals should not be confused with irrational numbers, which also have an infinite number of decimal places, but do not have a repeating pattern of numbers. Irrational numbers cannot be converted into fractions. For example, the irrational number π may be expressed as a decimal to an ever-increasing number of decimal places, but cannot be expressed as a fraction, because there is no repeating pattern to the numbers. All recurring decimals can be converted into fractions. Recurring decimals are often called repeating decimals.

Example 4. Convert these fractions into recurring decimals: (a) $\frac{5}{9}$, (b) $\frac{4}{7}$

Solution. For (a), write

$$\frac{5}{9} = 5 \div 9$$

$$= 0.55555555\ldots \qquad \text{Using calculator to give repeating 5's}$$

$$= 0.\dot{5} \qquad \text{The dot above the 5 is shorthand for "0.5 recurring"}$$

For (b), write

$$\frac{4}{7} = 4 \div 7$$

$$= 0.571428571\ldots \qquad \text{Using the calculator for division; there is an infinite number of decimal places}$$

If we look closely at the pattern, we can see that it repeats itself: $0.\underline{571428}\,\underline{571428}$ and so on. Therefore we write

$$\frac{4}{7} = 0.\dot{5}7142\dot{8} \qquad \text{With a dot above the first and last numbers that repeat}$$

$$\frac{4}{7} = 0.\overline{571428} \qquad \text{An alternative notation for a recurring group of numbers}$$

Every fraction can be converted into a decimal or into a recurring decimal. For some fractions the decimals do no recur immediately. For example $119/990 = 0.12\dot{0}\dot{2}$. Decimals may be added, subtracted, multiplied, and divided, using a calculator.

References: Canceling, Decimal System, Denominator, Finite Decimals, Fractions, Infinite, Multiple, Numerator, Percentage.

DECIMAL PLACES

Reference: Accuracy.

DECIMAL POINT

References: Decimal, Decimal System.

DECIMAL SYSTEM

The decimal system is a number system based on 10 and powers of 10, and is sometimes called the denary system. It originated in India and became established in Europe by the Arabs. The decimal system is used for counting, for money, and for weights and measures. The decimal number system uses the digits 0, 1, 2, 3, 4, 5, 6, 7, 8, and 9 to write numbers. The value of each of these digits 0–9 depends on its place in that decimal number. For example, in the decimal number 7325, the 7 represents a value of $7 \times 10^3 = 7000$, the 3 represents a value of $3 \times 10^2 = 300$, the 2 represents a value of $2 \times 10^1 = 20$, and the 5 represents a value of $5 \times 10^0 = 5$. In the following table column headings are the powers of 10 and tell us the value of each position that the digits 0–9 can occupy. Here the columns are only shown as far as 10^4:

$10^4 = 10,000$	$10^3 = 1000$	$10^2 = 100$	$10^1 = 10$	$10^0 = 1$

When negative powers are used the column headings are for numbers between 0 and 1. These numbers are called decimals. The whole number and the decimal are separated by a decimal point.

The column headings for decimals as far as 10^{-4} are shown in the following table:

$10^{-1} = 0.1$	$10^{-2} = 0.01$	$10^{-3} = 0.001$	$10^{-4} = 0.0001$

The need for counting numbers larger than 4 or 5 probably arose for a variety of reasons. Shepherds needed to count large numbers of their sheep and cows, once they had herds, to see if any were missing. Farmers needed to know about the lengths of seasons and when the rains were expected so they could sow seeds, and when to harvest crops. They needed to count so that some sort of calendar could be set up. About 4000 years before the birth of Christ the Egyptians knew there were 365 days in a year.

It is thought that early peoples may have learned to count on their fingers very much as young children first learn to count today. Once they reached a count of 10 fingers they would need another pair of hands to which to transfer their "10" while they counted up to 10 again. That is when an assistant would have come in handy. Once they had counted up to 10, 10 times, and each 10 transferred to the assistant, they would have needed a second assistant with another pair of hands, and so on.

References: Decimal, Powers.

DECOMPOSITION

The word decomposition means to break down a number, or quantity, into smaller parts. The number 456 represents $400 + 50 + 6$, which can be rearranged into $400 + 40 + 16$. This latter form of the number is of the same value, but has undergone decomposition. When subtracting numbers it is sometimes helpful to use decomposition of one of the numbers. This process, which is often called "borrowing," is explained in the following example.

Example. Work out $456 - 38$.

Solution. The number 38 is arranged under the 456 using the columns 100's, 10's, and units (see table). The first stage of the process is to subtract the 8 from the 6 in the units column, which cannot be done using positive numbers. Now we use the decomposition of 56 which is replaces $50 + 6$ by $40 + 16$, and then subtract the 8 from the 16. The next stage is to subtract 3 from 4 in the tens column, and finally subtract 0 from 4 in the hundreds column.

Reference: Carry.

DECREASING FUNCTION

The concept of a strictly decreasing function is explained using a specific function $y = x^2$, which is defined for all real numbers. Suppose A in figure a is the point with coordinates $(-3, 9)$ and B is the point $(-2, 4)$. Over the interval from A to B, the x values increase in size from -3 to -2, but the y values decrease in size from 9 to 4. As x increases, y decreases, and we say that the function $y = x^2$ is strictly decreasing over the interval from A to B. The gradient of the line segment AB is negative, and this also is an indication that the function is strictly decreasing over this interval. In

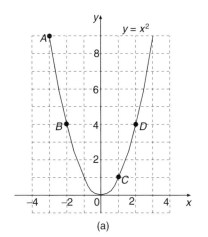

(a)

fact the function $y = x^2$ is strictly decreasing for all values of x that are less than zero.

Consider another two points on the curve, $C(1, 1)$ and $D(2, 4)$. Over this interval from C to D, the x values increase in size from 1 to 2, and the y values also increase in size from 1 to 4. As x increases, y increases, and we say that the function $y = x^2$ is strictly increasing over the interval from C to D. The gradient of the line segment CD is positive, and this also is an indication that the function is strictly increasing over this interval. In fact the function $y = x^2$ is strictly increasing for all values of x that are greater than zero.

At the origin, where the gradient is zero, the function $y = x^2$ is neither strictly decreasing nor strictly increasing. For this reason it is called a stationary point.

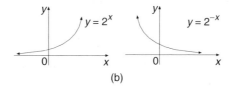

(b)

If a function is strictly increasing throughout its domain, we say it is a strictly increasing function. For example the function $y = 2^x$ (see figure b) is a strictly increasing function, because it is strictly increasing for all values of x. Similarly, the function $y = 2^{-x}$ is a strictly decreasing function.

References: Function, Gradient, Stationary Point.

DEGREE

There are three different meanings for the word degree as generally used in mathematics.

1. A degree is one of the units used for measuring the size of an angle, which is an amount of turning. Other units are radians and gradians. A degree is a very small amount of turning and in size it is 1/360 of a full turn; or complete turn. There are 360 degrees in a full turn; in symbols we write this as $360°$. Angles in degrees are measured with a protractor. A person in the army would soon learn that a half turn is the same as an "about turn" and is a rotation of $180°$. This division of a full turn into 360 equal parts called degrees was devised by the Babylonians about 3000 years ago and is still in use today. 360 is a useful figure to use, because it has so many factors.

2. A degree is a unit for measuring the temperature of something with a thermometer. Two of the units that may be used are degrees Celsius and degrees Fahrenheit.

3. The degree of a polynomial is the highest power of the variable. For example, the polynomial $3x^4 - 2x^2 + 1$ is of degree 4, because the highest power of the variable x is 4. A polynomial does not have negative powers.

References: Acute Angle, Angle, Protractor, Radian.

DELTA

This is the fourth letter of the Greek alphabet; the lower case letter is δ, and the capital letter is Δ. The capital letter Δ is commonly used in mathematics for the discriminant of a quadratic equation or as a very small, but finite, increase of a variable. For example, Δx is a small increase in x.

References: Quadratic Equations, Variable.

DENARY NUMBERS

Reference: Decimal System.

DENOMINATOR

A fraction is made up of two numbers, one above the other. The top number is called the *numerator* of the fraction and the bottom number is called the *denominator* of the fraction. The numerator is sometimes called the *dividend* and the denominator is sometimes called the *divisor*. For example, the numerator of the fraction $\frac{2}{3}$ is 2 and the denominator is 3. The mixed number $2\frac{5}{7}$ must first be written as the improper fraction 19/7, which has a numerator of 19 and a denominator of 7. The algebraic fraction $3ab/4x^2$ has a numerator of $3ab$ and a denominator of $4x^2$.

References: Algebraic Fractions, Fraction, Lowest Common Denominator, Mixed Number.

DEPENDENT VARIABLE

Suppose William has $30 at present, and decides to save $5 each week. After x weeks suppose he has saved $y. The equation connecting the variables x and y is $y = 5x + 30$. In this equation the amount of money saved (y) is dependent on the number of weeks (x) he has been saving, and so y is called the dependent variable. The number of weeks x is called the independent variable. When we graph the equation (see figure) the independent variable x goes on the horizontal axis and the dependent variable goes on the vertical axis.

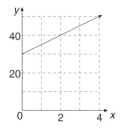

References: Equations, Straight-Line Graph, Variable.

DEPRESSION (ANGLE OF)

Reference: Angle of Depression.

DEPTH

This is the distance from the top to the bottom of an object, whereas the height of an object is the distance from the bottom to the top. So the depth and the height are the same distance, but viewed from different positions. We talk about the depth of a hole or the depth of a swimming pool when viewed from the surface, but we talk about the height of a tower or a power pole when viewed from ground level.

References: Distance, Height.

DEVIATION

This is the difference between one quantity and another quantity. One of the quantities is usually a fixed quantity or a standard quantity.

One of the measures of intelligence is intelligence quotient, which is commonly referred to as IQ. A super brain has an IQ of 140. If Walton has an IQ of 128, the deviation of Walton's brain from a super brain is

$$\text{Deviation} = 140 - 128$$

$$= 12$$

Reference: Standard Deviation.

DIAGONAL

The diagonal of a polygon is a line joining any two vertices, provided that the two vertices are not adjacent to each other on the polygon. For example, in the pentagon *ABCDE* in figure a, the line *AD* is a diagonal, and so is *BD*. The line *AE* is not a diagonal of the pentagon, because *A* and *E* are adjacent vertices on the pentagon. The pentagon has five diagonals, and they are drawn dashed in the figure.

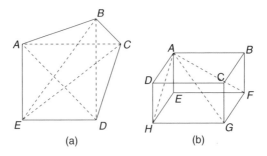

(a)   (b)

If a polygon has n sides, the formula for the number of its diagonals is $d = n(n - 1)/2$, where d is the number of diagonals.

The diagonal of a polyhedron is a line joining any two vertices, provided that the two vertices are not in the same face of the polyhedron. For example, in the cuboid *ABCDEFGH* in figure b, the line *AG* is a diagonal of the cuboid, because the vertices *A* and *G* are not in the same face of the cuboid. *AF* is not a diagonal of the cuboid, because the vertices *A* and *F* are in the same face of the cuboid, which is the face *ABFE*. Similarly, *AH* is not a diagonal. The other three diagonals are *BH*, *CE*, and *DF*.

References: Face, Pentagon, Polygon, Polyhedron, Vertex.

DIAMETER

Reference: Circle.

DICE

Dice is the plural of the word die. A die is a small cube, usually made of plastic, with its faces marked with dots representing the numbers 1, 2, 3, 4, 5, and 6 (see figure a). Since a die is used in many board games and games of chance, it provides intriguing questions in probability. The numbers on its faces are arranged so that 1 is on the opposite face to 6, 2 is opposite to 5, and 3 is opposite to 4. They are arranged so that the numbers on opposite faces add to 7.

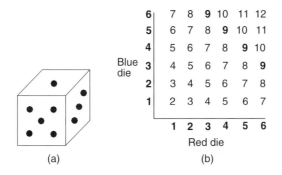

Blue die		Red die					
6	7	8	**9**	10	11	12
5	6	7	8	**9**	10	11
4	5	6	7	8	**9**	10
3	4	5	6	7	8	**9**
2	3	4	5	6	7	8
1	2	3	4	5	6	7
	1	**2**	**3**	**4**	**5**	**6**

(a)   (b)

Example. Two dice, one red and one blue, are rolled together onto a table. The score is obtained by adding the number facing up on the red die to the number facing up on the blue die. What is the probability that the score will be 9? What is the most likely score?

Solution. This problem can be analyzed using a table of all possible outcomes as shown in figure b. When the blue die shows a 2 and the red die shows 4, the score is 6. The total number of ways the dice can score is 36, and this is called the total number of outcomes. The number of ways the dice can score 9 is 4 ways. This is called the number of favorable outcomes. Write

$$\text{Probability(scoring 9)} = \frac{\text{Number of ways of scoring 9}}{\text{Total number of outcomes}}$$

$$= \frac{4}{36}$$

$$\text{Probability(scoring 9)} = \frac{1}{9} \qquad \text{Canceling the fraction}$$

The most likely score is 7, which occurs 6 times, as seen in the table.

References: Outcome, Probability of an Event, Tree Diagram.

DIFFERENCE OF TWO SQUARES

Reference: Factorize.

DIFFERENCE TABLES

These tables are used to find a formula (or general term) for a sequence of numbers. We shall examine only two types of difference tables, the first relating to a linear formula and the second relating to a quadratic formula. They are explained in the following examples.

Example 1. Pat has $5 and saves $4 each week. Write down how much she will have saved after each week for the first 5 weeks, and find a formula for this sequence of numbers.

Solution. The sequence of money saved is written down, and the differences between the terms of the sequence are listed in a table, as follows:

	0 weeks	1st week	2nd week	3rd week	4th week	5th week
Savings in $	5	9	13	17	21	25
Difference		4	4	4	4	4

The differences in the table are between one week's savings and the next. When each difference is the same, as in this sequence of numbers, the formula for the sequence is a linear formula. This formula is expressed in terms of n, the number of weeks, and S, the accumulated savings. The linear formula, in general terms, is $S = an + b$, where a and b are constants which need to be found. The value of the constant a is always equal to the value of the differences in the table, which in this example is 4. Therefore

$a = 4$, and so $S = 4n + b$. The value of b is found by substituting a value for S and a corresponding value for n into the formula $S = 4n + b$. For instance, when $n = 0$ and $S = 5$ we write

$$S = 4n + b$$

$$5 = 4 \times 0 + b \qquad \text{Substituting } S = 5 \text{ and } n = 0$$

$$b = 5 \qquad \text{Solving the equation}$$

The formula is $S = 4n + 5$, where n is the number of weeks and S is the money saved in \$.

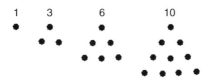

Example 2. If we arrange dots in the shape of triangles we obtain the set of triangle numbers, and the first four are shown in the figure. The next two triangle numbers are 15 and 21. The first five triangle numbers are entered in a table as follows that shows first- and second-order differences:

n	1		2		3		4		5
Triangle numbers	1		3		6		10		15
First-order differences		2		3		4		5	
Second-order differences			1		1		1		1

The first-order differences are not the same, so the formula for the sequence is not linear. The second-order differences are the same and equal to 1, therefore the formula for the sequence is quadratic. The quadratic formula, in general terms, is $T = an^2 + bn + c$, where T is the value of the triangle number and a, b, and c are constants which need to be found. The value of the constant a is always equal to half the value of the second-order difference.

In this example the value of the second-order difference is 1, therefore $a = 1/2$. The formula now becomes $T = \frac{1}{2}n^2 + bn + c$. The values of b and c are found by substituting two pairs of values for T and n into the formula $T = \frac{1}{2}n^2 + bn + c$. For instance, when $n = 1$, $T = 1$, and when $n = 2$, $T = 3$. We write

$$1 = \frac{1}{2} \times 1^2 + b \times 1 + c \qquad \text{Substituting } T = 1 \text{ and } n = 1 \text{ into the quadratic formula}$$

$$b + c = 0.5 \qquad \text{Simplifying the equation}$$

Similarly, by substituting $T = 3$ and $n = 2$ into the equation $T = \frac{1}{2}n^2 + bn + c$ we obtain the simplified equation $2b + c = 1$.

We now have two simultaneous equations:

$$b + c = 0.5$$

$$2b + c = 1$$

Subtracting the first equation from the second equation gives $b = 0.5$. Substituting $b = 0.5$ into the first equation 1 gives $c = 0$.

The formula for the sequence of triangle numbers is $T = \frac{1}{2}n^2 + \frac{1}{2}n$. This can be factorized to $T = \frac{1}{2}n(n + 1)$.

References: Formula, Linear Equation, Patterns, Quadratic Equations, Sequence, Simultaneous Equations, Triangle Numbers.

DIGIT

A digit is any one of the numerals 0, 1, 2, 3, 4, 5, 6, 7, 8, 9. For example, the number 35 is made up of two digits, 3 and 5.

Reference: Numbers.

DIHEDRAL ANGLE

A dihedral angle is the angle between two planes that intersect in a straight line. Suppose two planes intersect in a straight line AB (see figure) and point P is any point on the line AB. A straight line PX is drawn in one plane so that it makes a right angle with the line AB. Similarly, a line PY is drawn in the other plane so that it makes a right angle with the line AB. The dihedral angle is the angle XPY.

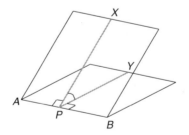

References: Angle between Two Planes, Plane.

DILATION

This is another name for enlargement.

DIRECT PROPORTION

Reference: Proportion.

DIRECT TRANSFORMATION

References: Congruent Figures, Indirect Transformation.

DIRECTED NUMBERS

Reference: Integers.

DISCONTINUOUS

A function is *discontinuous* if its graph is broken into two or more parts that are not connected. If you were drawing a graph of a function, at a point of discontinuity you would have to take your pencil off the paper to continue the drawing of the graph.

The function

$$y = \frac{x^2 - 1}{x - 1}$$

is discontinuous at the point where $x = 1$ (see figure a). At this point its graph is broken into two parts. The point of discontinuity is indicated on the graph by drawing a small circle to highlight the fact that there is a gap in the line at the point $(1, 2)$.

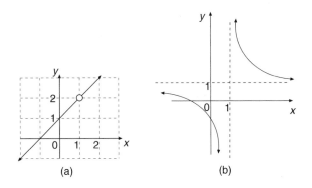

(a)   (b)

The function

$$y = \frac{x + 1}{x - 1}$$

is discontinuous at $x = 1$ (see figure b).

A function is *continuous* if its graph has no break in it.

References: Function, Graphs.

DISCOUNT

Reference: Conversion Factor.

DISCRETE DATA

Reference: Data.

DISCRETE DISTRIBUTION

Reference: Frequency Distribution.

DISPLACEMENT

Before displacement can be adequately described it is better to say what distance is, because displacement and distance are similar. *Distance* is the measure of the change in position of an object or point and can take place in any direction. It is the length of the straight line that can be drawn from the starting point (or position) to the final point (position). The usual units of distance are millimeters (mm), centimeters (cm), meters (m), and kilometers (km).

Displacement is the shift from the starting position of an object to its final position, provided that the distance and the direction are given. Its units are the same as the units of distance, provided that the direction is also stated.

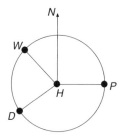

Suppose John walks a distance of 3 km from home, which is at point H in the figure. He could finish up at various places, such as work (W), the dairy (D), or the park (P). All these places are a distance of 3 km from home and lie on a circle of radius 3 km with its center at H. If his wife said he must walk from home 3 km due east so that John finished up at the park, then that instruction would describe a displacement, because the distance and the direction are given. When a displacement is given as an instruction there is only one destination, in this case the park.

DISTANCE

Reference: Displacement.

DISTRIBUTION

This is a term used in statistics. When we collect data, which may be a set of observations or a set of measurements, record the frequency of each item of data, and arrange the data in some readable form, it is called a distribution. The distribution of the data may be displayed graphically.

Example. Thirty families in a village were randomly selected and the data collected from them was regarding the number of children in each family. The distribution of family sizes looked like this:

2, 2, 3, 2, 1, 3, 4, 3, 5, 3, 2, 2, 2, 1, 2, 4, 6, 3, 4, 2, 2, 1, 1, 2, 3, 2, 1, 5, 4, 3

To make these data more manageable, they are arranged as a frequency distribution in the form of a table, and then displayed as a bar graph (see figure):

Number of children	1	2	3	4	5	6
Frequency	5	11	7	4	2	1

Gaps are left between the columns of the bar graph, because the data are discrete.

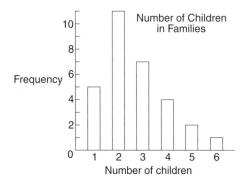

References: Bar Graph, Data, Discrete, Frequency Distribution, Statistics.

DISTRIBUTIVE LAW

References: Addend, Associative Law, Commutative Law.

DIVIDEND

The dividend is the number that is being divided when we divide one number by another. For example, Joanne has 7 Easter eggs for her 3 children. She divides 7 Easter eggs by 3 children to get an answer of 2 eggs for each child and 1 egg left over. In this example 7 is the dividend, 3 is the divisor, 2 is the quotient, and 1 is the remainder. Who gets the egg left over? Dad of course!

The division of 7 by 3 to get an answer of 2 and leave a remainder 1 can be written as a division identity:

$$7 = 3 \times 2 + 1$$

Expressed in words, the division identity is

$$\text{Dividend} = \text{divisor} \times \text{quotient} + \text{remainder}$$

If the remainder is zero, the dividend is exactly divisible by the divisor. This would be the case if Joanne had 6 Easter eggs instead of 7, and we would say that 6 is divisible by 3, because the remainder is zero.

DIVISIBILITY TESTS

These are short cuts to see if one counting number is exactly divisible by another counting number, without actually doing the division. They provide a check to see if one number is a factor or multiple of another number without actually doing the division process. Divisibility tests were more useful some years ago before calculators were freely available, but they still provide a quick test for divisibility.

Test 1. A number is divisible by 10 if the number ends in 0.

Example 1. The number 65**0** is divisible by 10, since it ends in 0, but 50**7** is not, since it ends in 7.

Test 2. A number is divisible by 5 if the number ends in 0 or ends in 5.

Example 2. The numbers 10**5** and 79**0** are both divisible by 5.

Test 3. A number is divisible by 2 if the number ends in 0 or 2, 4, 6, 8. This means it is an even number.

Example 3. The number 73**8** is divisible by 2, since it ends in 8.

Test 4. A number is divisible by 3 if the sum of its digits is divisible by 3.

Example 4. The number 2673 is divisible by 3, because the sum of its digits is $2 + 6 + 7 + 3 = 18$, and 18 is divisible by 3.

Test 5. A number is divisible by 4 if the number formed by the last two digits is divisible by 4.

Example 5. The number 35**24** is divisible by 4, because its last two digits form the number 24, which is divisible by 4.

Test 6. A number is divisible by 8 if the number formed by the last three digits is divisible by 8.

Example 6. The number 933,**248** is divisible by 8, because its last three digits form the number 248, which is divisible by 8. Check: $248 \div 8 = 31$.

Test 7. A number is divisible by 6 if its last digit is even and the sum of its digits is divisible by 3.

Example 7. The number 15,40**8** is divisible by 6 because its last digit, which is 8, is even, and the sum of its digits, which is $1 + 5 + 4 + 0 + 8 = 18$, is divisible by 6.

Test 8. A number is divisible by 9 if the sum of its digits is divisible by 9.

Example 8. The number 67,320 is divisible by 9, because the sum of its digits, which is $6 + 7 + 3 + 2 + 0 = 18$, is divisible by 9.

Test 9. A number is divisible by 11 if the difference and sum alternately of its digits, starting with the last digit and working in order to the first digit, make a number that is either 0 or a number divisible by 11.

Example 9. The number 2783 is divisible by 11, because $3 - 8 + 7 - 2 = 0$.

Example 10. The number 10,857 is divisible by 11, because $7 - 5 + 8 - 0 + 1 = 11$.

The calculator can be used for a divisibility test.

References: Digit, Even, Factor, Integer, Multiple.

DIVISIBLE

Reference: Dividend.

DIVISION IDENTITY

Reference: Dividend.

DIVISOR

Reference: Dividend.

DODECAGON

A dodecagon is a polygon that has 12 sides. To find the angle sum of a dodecagon, search under the entry Polygon. A regular dodecagon is shown in figure a, and each interior angle is equal to 150°.

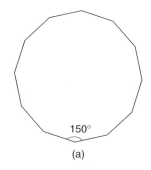

150°

(a)

The dodecagon has 12 axes of symmetry, and its order of rotational symmetry is also 12. The regular dodecagon tessellates with two equilateral triangles and one square; see the entry Tessellations.

The regular dodecagon is made up of 12 isosceles triangles that are all the same size (see figure b).

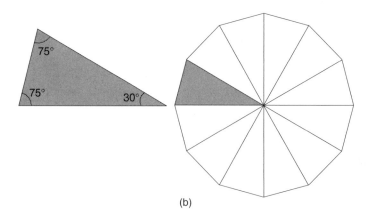

(b)

References: Equilateral Triangle, Isosceles Triangle, Polygon, Regular Polygon, Symmetry, Tessellations.

DODECAHEDRON

This is a regular polyhedron (solid) which has 12 congruent faces and each face is a regular pentagon (see figure). The regular dodecahedron is one of the Platonic solids.

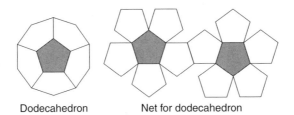

Dodecahedron   Net for dodecahedron

References: Net, Pentagon, Platonic Solids, Polyhedron.

DOMAIN

References: Cartesian Coordinates, Correspondence.

DOMINO

Reference: Polyominoes.

DOT PLOT

This is a graph that is used in statistics and consists of vertical lines of dots to show the frequencies instead of columns, as in the bar graph. An example of a dot plot is included in the following example.

Example. The following frequency table shows how 20 people in Jacob's office travel to work. Draw a dot plot graph to display these data.

Method of travel	Car	Cycle	Train	Walk
Frequency	7	2	5	6

Solution. In the figure each of the methods of travel has a column of dots, and each dot in this example represents one student. This scale needs to be included. The horizontal axis must be labeled and the graph given a title.

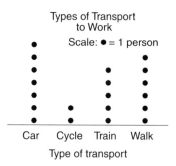

References: Bar Graph, Frequency Table, Statistics.

DUODECIMAL

Sometime called duodenary. Duodecimal is a number system based on 12 and powers of 12. It uses 12 digits for counting, whereas the decimal (denary) system of counting uses 10 digits, because it is based on 10. The digits used in base 12 are $\{0, 1, 2, 3, 4, 5, 6, 7, 8, 9, t, e\}$, where t is ten and e is eleven. Special symbols for ten and eleven are needed, because the units column goes up to eleven, and 10 and 11 cannot be used, because they occupy two columns.

The column headings for the duodecimal system are powers of 12 (compared with the decimal system, which has powers of 10); five of the column headings, expressed in base 10 numbers, look as follows:

$12^4 = 20{,}736$	$12^3 = 1728$	$12^2 = 144$	$12^1 = 12$	Units, 1–e

There are present-day applications of the duodecimal system:

- 12 eggs in one dozen, 12 dozen in one gross
- 12 months in 1 year
- 12 inches in 1 foot
- 12 hours before noon and 12 hours after noon

References: Binary Numbers, Decimal System, Denary Numbers, Digit.

$$E$$

EAST

References: Angle, Bearings.

EDGE

An *edge* is a straight line where two planes of a three-dimensional polyhedron meet. A *plane* of a solid is a flat surface, and a *vertex* is a point where three or more planes meet. The *side* of a polyhedron is sometimes called an edge. The polyhedron shown in figure a is a square-based pyramid. It has eight edges, five planes, and five vertices.

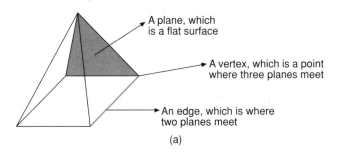

A plane, which is a flat surface

A vertex, which is a point where three planes meet

An edge, which is where two planes meet

(a)

There is a formula connecting the number of edges (E), the number of faces (F), and the number of vertices (V) of three-dimensional polyhedra. This formula is called Euler's formula, named after the man who discovered it. The formula is

$$F + V = E + 2$$

Example. Verify that Euler's formula is true for a three-dimensional triangular prism.

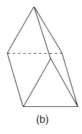

(b)

Solution. Figure b is a sketch of the prism. The formula to verify is

$$F + V = E + 2$$

$$F + V = 5 + 6 \qquad \text{There are 5 faces and 6 vertices}$$

$$F + V = 11$$

$$E + 2 = 9 + 2 \qquad \text{There are 9 edges}$$

$$E + 2 = 11$$

$$\therefore F + V = E + 2 \qquad \text{Each side of the formula is equal to 11}$$
$$\qquad\qquad\qquad\quad \text{(the symbol } \therefore \text{ denotes "therefore")}$$

References: Euler's Formula Plane, Polyhedron, Prism, Pyramid, Vertex.

ELEVATION

An elevation is a view, drawn accurately to scale, from the front, side, or back of an object.

Example. Draw the front, side, and back elevations of the nine blocks in figure a. The blocks are drawn on an isometric grid.

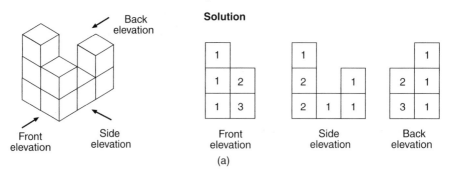

In the solution the squares are numbered to indicate how many blocks are positioned in that particular row. A plan view from above can also be drawn. To draw this view imagine you are positioned at the front of the blocks, looking down on them (see figure b).

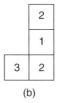

(b)

References: Angle of Elevation, Isometric, Plan.

ELIMINATION

This is a method of solving simultaneous equations.

Reference: Simultaneous Equations.

ELLIPSE

An ellipse looks like a "flattened circle" (see figure a). A common name for an ellipse is an oval, but this is not a word used in mathematics. An ellipse has two axes of symmetry, and the order of rotational symmetry is two. The ellipse is important in science and mathematics because planets move in elliptical orbits around the sun.

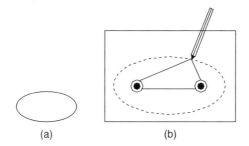

(a)   (b)

One method of drawing an ellipse is to get a length of string and join its ends together to form a loop (see figure b). Place a piece of paper on a drawing board and insert in the paper two thumb tacks placed sufficiently far apart so that when the string is looped around them it is slack. Place a pencil in the loop and with it pull the string taut. Now move the pencil, keeping the string taut, and the shape traced out by the pencil will be an ellipse.

Reference: Conic Sections.

ENDECAGON

This is an 11-sided polygon, which may also be called a hendecagon. The angle sum of the endecagon is given under the entry Polygon.

ENLARGEMENT

This is sometimes called a dilation, or a magnification. An enlargement is one of the geometrical transformations that occurs in everyday life at the movie theater when the film is projected, and enlarged, onto a screen by passing light rays through the film. Using this example, we can say that the source of light in the projector is the center of enlargement, the film is the object, the picture on the screen is the image, and the number of times the image is bigger than the object is the scale factor (k) of the enlargement. The rays of light in this example correspond to rays we can draw with a pencil.

In mathematics when we use the word enlargement we do not necessarily mean the image is larger than the object. For an enlargement of, say, one-half, the dimensions of the image will be only half those of the object. In the work that follows it is assumed that the object shape and the image shape are in the same plane.

Example 1. In figure a, enlarge the flag F (which is the object) with a scale factor of $k = 3$, if the point $P(2, 1)$ is the center of enlargement.

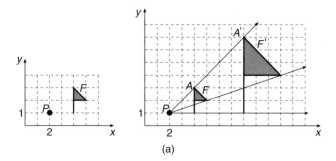

(a)

Solution. Draw rays from P to pass through the points of the flag. A few points on the flag are sufficient in order to be able to draw the image flag F'. Using your compasses (or dividers), step off the distance PA along the ray three times to reach the point A'. Repeat this process along the other rays and build up the image flag. Since the scale factor is $k = 3$, the image lengths of the flag F' will be three times the object lengths of the flag F. For example, the length of the flagpole in the object is 2 units long and the length of its image is 6 units long. For a scale factor of $k = 1$, the object and the image are the same shape in the same place.

Notation: We say that the flag F *maps* onto its image F' under an enlargement center P, and $k = 3$. The symbol for "maps onto" is $\rightarrow$, so we say $F \rightarrow F'$. In all transformations the image shape is denoted by dashes (F') and the object shape is given without dashes (F). Similarly for points A and A'.

Sometimes the center of enlargement is inside the object shape, as in the next example.

Example 2. Draw the image of the shaded triangle ABC in figure b for an enlargement center at the origin, and $k = 2$.

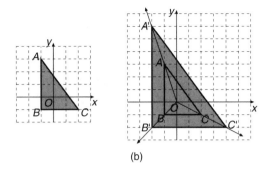

(b)

Solution. The rays are drawn from the origin O through the points A, B, and C, respectively, and the distances stepped out with compasses to get the image points A', B', and C'. The image of triangle ABC is triangle $A'B'C'$.

There is an important rule for enlargements with regard to the ratio of image lengths to object lengths and k, which is stated here:

$$k = \frac{\text{Image length}}{\text{Object length}}$$

In the previous example, when $k = 2$, this rule states that

$$2 = \frac{A'B'}{AB} = \frac{B'C'}{BC} = \frac{A'C'}{AC}$$

Another important rule concerns the scale factor for area. The ratio of the area of the image to the area of the object $= k^2$. This rule is stated here:

$$k^2 = \frac{\text{Area of image}}{\text{Area of object}}$$

In the previous example, when $k = 2$, this rule states that

$$2^2 = \frac{\text{Area of triangle } A'B'C'}{\text{Area of triangle } ABC}$$

$$= \frac{24 \text{ square units}}{6 \text{ square units}}$$

$$= 4, \qquad\qquad\qquad \text{which is true}$$

For negative enlargements the image is upside down in relation to the object, and object and image are on opposite sides of the center of enlargement. The rules about ratio of lengths and about scale factor for area still apply to negative enlargements.

Example 3. Enlarge the arrow in figure c with O as the center of enlargement and with a scale factor $k = -\frac{1}{2}$.

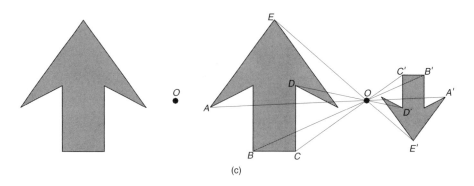

(c)

Solution. It is not necessary to draw rays from the point O through all the points on the arrow, but enough need be chosen to enable the image shape to be constructed. The five points A, B, C, D, and E on the object figure should be sufficient. From the point O draw in the rays OA, OB, OC, OD, and OE, but produce them backward, since it is a negative enlargement and the image is on the opposite side of O to the object. Then halve the distance OA, since $k = -\frac{1}{2}$, and with compasses step off this distance of half OA along OA produced backward and mark the point A'. In a similar way mark the points B', C', D', and E'. Complete the drawing of the arrow. Check that the characteristics of a negative enlargement are demonstrated. The image is upside down, and the object and image are on opposite sides of the center of enlargement. Note also that the image lengths are half the object lengths.

If the object and the image are both drawn, the center of enlargement can be found by drawing the rays $A'A$, $B'B$, etc., which will all meet in the center of enlargement. The scale factor can be found using the formula

$$k = \frac{\text{image length}}{\text{object length}}$$

$$= \frac{A'B'}{AB}$$

Enlargements have the following properties:

- The center of enlargement is an invariant point.
- An enlargement with a scale factor of $k = -1$ is equivalent to a rotation about the center of enlargement of $180°$.
- Each line in the object is parallel to its own image line. For example, AB is parallel to $A'B'$.
- If lines are parallel in the object, they will also be parallel in the image. Parallelism is preserved.
- Angle sizes are invariant.
- The object and image are similar shapes.

References: Invariant Points, Similar Figures, Transformation Geometry.

EQUATIONS

Equations can be formed in order to solve some every-day problems; the process is explained under the entry Abstract. When solving an equation, only use one equals sign per line of working, and keep the equals signs in a straight line underneath each other.

Example. Solve $2x - 3 = 8$.

Solution. Write

$$2x - 3 = 8$$

$$2x = 11 \qquad \text{Adding 3 to both sides of the equation}$$

$$x = 5.5 \qquad \text{Dividing both sides of equation by 2}$$

This alternative setting out is not acceptable:

$$2x - 3 = 8 = 2x = 11 = x = 5.5$$

The following setting out is also unacceptable when an equals sign is used to start each line of working:

$$2x - 3 = 8$$

$$= 2x = 11 \qquad \text{Adding 3 to both sides of the equation}$$

$$= x = 5.5 \qquad \text{Dividing both sides of the equation by 2}$$

References: Balancing an Equation, Graphs, Inequations, Linear Equation, Quadratic Equations, Solving an Equation.

EQUIANGULAR

In mathematics, the prefix equi means equal. Therefore an equiangular polygon means a polygon with all its angles equal in size. If a polygon is equiangular and all its sides are equal in length, it is called a regular polygon. An equiangular triangle is a triangle that has all its angles equal in size. Since the three angles add up to $180°$, each angle is equal to $180 \div 3 = 60°$. Such a triangle is usually called equilateral.

References: Equilateral Triangle, Polygon, Regular Polygon, Square.

EQUIDISTANT

If two or more distances are equidistant, they are equal.

References: Angle Bisector, Locus, Perpendicular Bisector.

EQUILATERAL TRIANGLE

This is a triangle with all three of its angles equal in size to 60° and all three of its sides equal in length (see figure a). An equilateral triangle has three axes of symmetry, and the order of rotational symmetry is three.

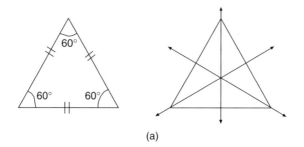

(a)

Example. Fold a piece of paper to make an equilateral triangle.

Solution. Use a circular piece of paper. The paper folding process is easier when the circle is cut out. Fold the paper in half to form a semicircle, and crease it (see figure b). Fold into another semicircle, and crease the paper. Where the two creases meet is the center of the circle. Make a mark at the center and call it point O.

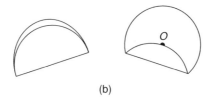

(b)

Now make a fold of the circle so that a point on the circumference of the circle just meets the newly found center of the circle O, and crease it. That crease is one side of the equilateral triangle. Two more creases that are similar are made to form an equilateral triangle.

References: Equiangular, Isometric, Symmetry.

EQUIVALENT EQUATIONS

These are the steps used in solving equations, as set out in the following example.

Example. Solve the equation $3(x + 2) = 1 - x$, showing three equivalent equations as steps of working.

Solution. Write

$$3(x + 2) = 1 - x$$

$$3x + 6 = 1 - x \qquad \text{Expanding the brackets}$$

$$4x = -5 \qquad \text{Adding } x \text{ and } -6 \text{ to both sides of the equation}$$

$$x = -1.25 \qquad \text{Dividing both sides of the equation by 4}$$

This value for x is the solution of the equation.

Reference: Equations.

EQUIVALENT EXPRESSIONS

In algebra two expressions are equivalent if they contain the same information, but expressed in different forms. Equivalent expressions are expressions that are equal to each other. For example, $x + 2x$ and $3x - x$ are equivalent expressions. $2x(x + 3)$ is an equivalent expression to $2x^2 + 6x$, and is obtained by expanding the brackets.

EQUIVALENT FRACTIONS

Fractions are equivalent when they have the same value, but are written in different ways. A convenient way of introducing equivalent fractions is to use areas. Suppose a rectangle has a length of 6 units and a width of 2 units, and is divided up in three different ways.

The rectangle in figure a is divided into 3 equal regions and 1 of them is shaded. This means that $\frac{1}{3}$ of the whole rectangle is shaded. The rectangle in figure b is divided into 6 equal regions and 2 of them are shaded. In this case $\frac{2}{6}$ of the whole rectangle is shaded. In the situation in figure c the same rectangle is divided into 12 equal regions and 4 of them are shaded. We now have $\frac{4}{12}$ of the whole rectangle shaded.

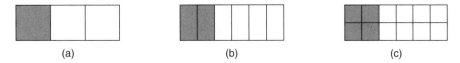

| (a) | (b) | (c) |

Summary. Since equal areas are shaded in each rectangle, it means that the three fractions $\frac{1}{3}$, $\frac{2}{6}$, and $\frac{4}{12}$ are equal, and therefore have the same value:

$$\tfrac{1}{3} = \tfrac{2}{6} = \tfrac{4}{12}$$

When fractions have the same value we say they are equivalent, and the fraction $\frac{1}{3}$ is the simplest form of these three equivalent fractions. Another process for finding equivalent fractions is given in the following example.

Example 1. Write down four fractions equivalent to $\frac{1}{3}$.

Solution. Multiply $\frac{1}{3}$ by $\frac{2}{2}$, by $\frac{3}{3}$, by $\frac{4}{4}$, and by $\frac{5}{5}$ to obtain the four equivalent fractions:

$$\frac{1}{3} \times \frac{2}{2} = \frac{2}{6}, \qquad \frac{1}{3} \times \frac{3}{3} = \frac{3}{9}$$

$$\frac{1}{3} \times \frac{4}{4} = \frac{4}{12}, \qquad \frac{1}{3} \times \frac{5}{5} = \frac{5}{15}$$

Therefore, four equivalent fractions to $\frac{1}{3}$ are $\frac{2}{6}, \frac{3}{9}, \frac{4}{12}$, and $\frac{5}{15}$. Equivalent fractions are used to solve problems.

Example 2. Find x if $\frac{2}{3} = \frac{16}{x}$.

Solution. It can be seen that $2 \times 8 = 16$,

$$\frac{2}{3} \times \frac{8}{8} = \frac{16}{24} \qquad \text{The two fractions } \frac{16}{24} \text{ and } \frac{16}{x} \text{ are equivalent}$$

Therefore $x = 24$.

Reference: Canceling Fractions.

ERATOSTHENES' SIEVE

This is an algorithm for finding prime numbers less than a certain number.

Example. Find all the prime numbers less than or equal to 48.

Solution. Write down, in order, a list of all the counting numbers from 1 to 48. Strike out every second number after 2. This eliminates even numbers. All even numbers, except 2, are not prime. Using the numbers 1–48 again, now strike out every third number, except 3. These are multiples of 3 and are not prime. Repeat the process by striking out every fourth, then every fifth, number, and that will be enough for the prime numbers up to 48. In short, the process is to strike out every multiple of 2, 3, 4, and 5, but not the first number in each case. It should be remembered that 1 is not a prime number and needs to be struck out. As the process develops you will find that some numbers have already been eliminated. This process isolates all the prime numbers, because they are the ones left at the end, just as large stones are left behind in a sieve:

$\{\cancel{1}, 2, 3, \cancel{4}, 5, \cancel{6}, 7, \cancel{8}, \cancel{9}, \cancel{10}, 11, \cancel{12}, 13, \cancel{14}, \cancel{15}, \cancel{16}, 17, \cancel{18}, 19, \cancel{20}, \cancel{21}, \cancel{22}, 23, \cancel{24}, \cancel{25},$
$\cancel{26}, \cancel{27}, \cancel{28}, 29, \cancel{30}, 31, \cancel{32}, \cancel{33}, \cancel{34}, \cancel{35}, \cancel{36}, 37, \cancel{38}, \cancel{39}, \cancel{40}, 41, \cancel{42}, 43, \cancel{44}, \cancel{45}, \cancel{46}, 47, \cancel{48}\}$

References: Multiple, Prime Number.

ERROR

In mathematics error does not necessarily mean that a mistake has been made, but that we may be talking about a quantity that is not expressed exactly. This kind of error may arise due to an estimation or because it is not possible to make measurements with 100% accuracy. Suppose we look at three types of quantities:

1. *Quantities that are counted.* Suppose the number of students in a classroom is known to be 30, and a student counted them and got the answer 32. The difference between the correct answer and the incorrect answer is the error. In this case the error is 2, and in this case is due to a mistake being made in the count.

Sometimes errors are made that are not due to mistakes, but due to errors in estimating and measuring, as identified below.

2. *Quantities that are estimated.* Estimate how many times the letter E appears on the next line, without counting them all:

EEEEEEEEEEEEEEEEEEEEEEEEEEEEEEEEEEEEEEEEEEEEEEEEEEEEEEEEEEEEEEEE

Suppose a person estimated that the letter E appears 60 times, which is quite a good estimate, because by counting them you will find it appears 62 times. The error is $62 - 60 = 2$. This is not a mistake, but an error due to estimating.

3. *Quantities that are measured.* For example, we may be attempting to find the length of a table in the dining room. It can be measured, perhaps to the nearest centimeter, or to the nearest millimeter, but the exact length of the table cannot be measured. This is because there is no instrument or device accurate enough to measure it exactly. The error here is the difference between the true length, if it is known, and the measured length. This is not a mistake, but an error in measuring due to the inaccuracies of the measuring instrument.

This is an opportune point to discuss limits of accuracy. Suppose we are asked to measure the length of a table to the nearest centimeter, and measured it correctly to be 124 cm. Figure a shows an enlarged view of the measuring tape that was used. The actual length of the table, measured to the nearest centimeter, lies somewhere in the interval of 123.5–124.5 cm. The greatest length of the table could not be 124.5 cm, because then it would be rounded up to 125 cm to the nearest centimeter. But any measurement just less than 124.5 cm is OK. The measurement of 124.5 cm is excluded with a hole at the upper end of the range, as shown in figure a. The least length of the table could be 123.5 cm, because that would be rounded up to 124 cm to the nearest centimeter. This measurement is included with a filled hole at the lower end of the interval, as in figure a.

(a)

These two values 123.5 and 124.5 cm are called the limits of accuracy for the length of the table when it has been measured to be 124 cm, to the nearest centimeter.

This means we can assume that the actual length L of the table is between these two extremes. The limits of accuracy of the length L can be expressed as

$$123.5 \leq L < 124.5$$

Example 1. The number N of people in a crowd at a sports event is recorded as 23,000 to the nearest thousand. What are the limits of accuracy?

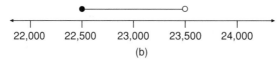

(b)

Solution. The size of the crowd is positioned on a number line as in figure b. The limits of accuracy are

$$22,500 \leq N < 23,500$$

Example 2. The length L of a room is measured as 5.67 meters to two decimal places. What are the limits of accuracy?

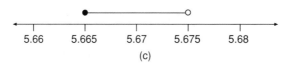

(c)

Solution. The length of the room is positioned on a number line as in figure c. The limits of accuracy are

$$5.665 \leq L < 5.675$$

References: Accuracy, Estimation, Measurement, Percentage Error, Rounding.

ESCHER

Reference: Tessellations.

ESTIMATION

When we estimate we calculate an approximate value for the size of a certain quantity. There are various reasons why we estimate the size of a quantity instead of attempting to find its exact value. Some examples are given here.

Example 1. If we are writing a report on an athletics meet, we only need an approximate value of the size of the crowd, so it is quicker to estimate how many people are there rather than count everyone. Generally, people would not be interested in the exact size of the crowd. This is an example where estimating saves time.

A suitable method of estimating the number of spectators is to count 50 people, and then count roughly how many groups that size there are around the sports arena. If you count 20 such groups, then there are approximately $50 \times 20 = 1000$ spectators.

Example 2. Suppose you are in business manufacturing Chocko chocolates and wonder how many people would buy a new flavor, "crunchy" Chocko bar. Rather than ask everyone in the country, you decide to select a sample of 1000 people, ask them what they think, and then use this result to estimate how many people in the country would buy your new Chocko bar. This is an example where estimating can save the expense, and also the time, of asking everyone in the country.

Out of your sample of 1000 people say there are 40 people who like the new flavor. There are 100 million people in the country, say, which is 100,000 groups of 1000 people. You would expect $100,000 \times 40 = 4,000,000$ people to like your new Chocko bar.

Example 3. In this example suppose you are using your calculator to work out the value of the expression

$$\frac{2.1 + 3.8}{5.9 - 3.2}$$

Using the steps $2.1 + 3.8 \div 5.9 - 3.2$, the answer from the calculator is -0.456, to 3 dp. If you have a suspicion that the answer is incorrect, then estimating the answer using approximate values for the numbers can check it.

The top line is approximated as $2 + 4 = 6$. The bottom line is approximated as $6 - 3 = 3$. The answer is approximately $6 \div 3 = 2$. This confirms that the answer -0.456 could not be correct. The calculator answer isincorrect because brackets need inserting in the calculator steps to make $(2.1 + 3.8) \div (5.9 - 3.2) = 2.185$, to 3 dp.

Example 4. The following example demonstrates how the number of grains of sand on a beach can be estimated. The number of grains of sand on a large beach is finite and not infinite. The number may be estimated in the following way. Count the number of grains of sand in 1 cubic centimeter (cm^3) and say it is 100. There are 1,000,000 cm^3 in 1 cubic meter (m^3) and therefore $1,000,000 \times 100 = 10^8$ grains of sand in 1 m^3. Suppose we measured the length of the beach to be $L = 2$ km, which is 2000 m. The width of the beach is estimated to be $W = 50$ m and its depth is estimated to be $D = 10$ m. Write

$$\text{Volume of beach} = \text{length} \times \text{width} \times \text{depth}$$

$$= 2000 \times 50 \times 10$$

$$= 1,000,000 \ m^3$$

$$= 10^6 \ m^3$$

In each cubic meter there are 10^8 grains of sand. Therefore

$$\text{Number of grains of sand on this beach} = 10^8 \times 10^6$$
$$= 10^{14}$$

10^{14} grains of sand is a finite number.

References: Error, Brackets, Exponent, Sample.

EULER'S FORMULA

This is a formula discovered by Leonhard Euler (1707–1783) that relates the number of vertices V, edges E, and faces F of any simple closed three-dimensional polyhedron. Euler's formula is

$$F + V = E + 2$$

An application of this formula is given in the entry Edge.

There is a similar formula relating the number of nodes N, arcs A, and regions R of a network that is drawn in one plane:

$$N + R = A + 2$$

References: Arc, Edge, Face, Networks, Node, Polyhedron, Regions, Vertex.

EVALUATE

To evaluate an expression means to find the numerical value, which is the number value, of the expression.

Example. The equation of a parabola is $y = x^2 + 2$. Evaluate y when $x = 5$.

Solution. Write

$$y = 5^2 + 2 \qquad \text{Substituting } x = 5 \text{ in the expression } x^2 + 2$$
$$y = 25 + 2 \qquad \text{Squaring 5}$$
$$y = 27$$

By finding a numerical value for $x^2 + 2$ we have evaluated the expression.

References: Formula, Numerical Value, Parabola.

EVEN

A number is even if it is divisible by 2 with zero remainder. For example, 6 is divisible by 2 with zero remainder, so 6 is an even number. Similarly, -4 is an even number, because it is divisible by 2 with zero remainder. The set of even numbers is

$$\{\ldots, -6, -4, -2, 0, 2, 4, 6, \ldots\}$$

Even numbers are multiples of 2, and form an infinite set. The formula for even numbers is $E = 2n$, and the set of even numbers is obtained by substituting each of the integer numbers $I = \{\ldots, -3, -2, -1, 0, 1, 2, 3, \ldots\}$ in turn for n.

A number is odd if it is one more or one less than an even number. The set of odd numbers is

$$\{\ldots, -5, -3, -1, 1, 3, 5, \ldots\}$$

Odd numbers form an infinite set. The formula for odd numbers is $O = 2n + 1$, and the set of odd numbers is obtained by substituting each of the integer numbers $I = \{\ldots, -3, -2, -1, 0, 1, 2, 3, \ldots\}$ in turn for n.

In everyday life, the most frequently used even and odd numbers are those that are greater than zero. The results of adding and multiplying even and odd numbers are given in the two tables below.

+	Even	Odd
Even	Even	Odd
Odd	Odd	Even

×	Even	Odd
Even	Even	Even
Odd	Even	Odd

References: Divisible, Multiple, Quotient, Remainder.

EVEN FUNCTION

A function can be even, odd, or neither even nor odd. The simplest way to recognize whether a function is even or odd is to look at its graph. The graph of an even function is symmetrical about the y-axis. The two examples drawn in figure a are $y = x^2 + 2$ and $y = \cos x$, and both functions are even because their graphs are symmetrical about the y-axis.

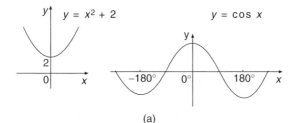

(a)

In a similar way, odd functions can be recognized from their graphs, because an odd function has half-turn rotational symmetry about the origin. In other words, the graph has rotational symmetry of order two, and the center of rotational symmetry is the origin. The two curves drawn in figure b are $y = x^3$ and $y = \sin x$, and both these functions are odd, because their curves have half-turn rotational symmetry about the origin.

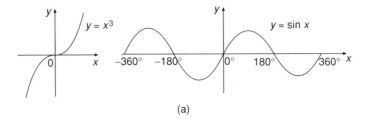

(a)

In addition to the symmetry properties of the graphs of odd and even functions there is an algebraic property of each:

$$\text{If} \quad f(-x) = f(x) \quad \text{the function is even}$$

$$\text{If} \quad f(-x) = -f(x) \quad \text{the function is odd}$$

Example 1. Verify that $f(x) = x^2 + 2$ is an even function.

Solution. Write

$$f(x) = x^2 + 2$$

$$f(-x) = (-x)^2 + 2 \qquad \text{Replacing } x \text{ by } -x$$

$$= x^2 + 2 \qquad (-x)^2 = x^2$$

$$f(x) = f(-x) \qquad \text{They are both equal to } x^2 + 2$$

Therefore $f(x) = x^2 + 2$ is an even function.

Example 2. Verify that $f(x) = x^3$ is an odd function.

Solution. Write

$$f(x) = x^3$$

$$f(-x) = (-x)^3 \qquad \text{Replacing } x \text{ by } -x$$

$$= -x^3 \qquad (-x)^3 = -x^3$$

$$f(-x) = -f(x) \qquad \text{They are both equal to } -x^3$$

Therefore $f(x) = x^3$ is an odd function.

References: Cosine, Cubic Equations, Function, Graphs, Parabola, Sine, Symmetry.

EVENT

There are a number of terms in probability that can be explained together. In this entry the following terms are explained using examples: experiment, outcome, event, sample space.

William tosses two coins together a number of times, and records the results: This process is called an *experiment*. Another example of an experiment might be rolling two dice together, or drawing a card from a pack of 52 playing cards. In an experiment there are a number of possible *outcomes* and we have no way of predicting which outcome is next. The purpose of an experiment is to investigate the truth, or otherwise, of a statement. William is doing his experiment to investigate the truth, or not, that when two coins are tossed together a head (H) and a tail (T) are more likely to turn up than two heads. The sample space is a list of all possible outcomes, and for this particular experiment of tossing two coins together the *sample space* is given in the following table.

First coin	H	H	T	T
Second coin	H	T	H	T

Alternatively, the sample space can be expressed as HH, HT, TH, TT. We say there are four possible outcomes of this experiment.

An *event* is any subset of the sample space. When William tosses two coins together an event may be the coins landing the same side up, which is the subset HH, TT. Another event may be the coins landing differently, which is the subset HT, TH. Both these events are a subset, or part of, the sample space. An event may have only one outcome and then is known as a simple event. In our example, a simple event is tossing two heads, because there is only one possible outcome.

Example. An experiment is tossing three coins together. List the sample space, and list the event of obtaining two heads and one tail.

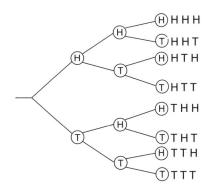

Solution. A convenient way of obtaining a list of all the possible outcomes of this experiment, which is the sample space, is to use a tree diagram (see figure). The sample space is listed at the ends of the branches, and is stated here:

H H H

H H T

H T H

H T T

T H H

T H T

T T H

T T T

The event of tossing two heads and one tail is part of the sample space and is listed here:

H H T

H T H

T H H

References: Complementary Events, Dice, Probability of an Event, Tree Diagram.

EXPANDED FORM OF DECIMALS

Reference: Compact form of Decimals.

EXPANDED FORM OF NUMBERS

Reference: Compact Form of Decimals.

EXPANDING BRACKETS

The process of expanding brackets results in an equivalent expression which contains no brackets. More simply, it is to rewrite an expression in order to remove the brackets. The resulting expression is called the expansion of the brackets, and is usually longer than the original expression. The explanation of expanding brackets will be a practical approach, but when the process is understood a simple algorithm will be used. Before proceeding further the reader should know how to multiply algebraic terms, as explained under the entry Algebra.

Example 1. Andrew has a square sandpit of side 3 m and wishes to extend it in one direction by 2 m (see figure a). Write down an expression for the total area of the new sandpit.

(a)

Solution. Write

$$\text{Area of new sandpit} = \text{width} \times \text{length}$$

$$\text{Area} = 3 \times (3 + 2) \qquad \text{The length is enclosed in brackets}$$

$$\text{Area} = 3 \times 5$$

$$\text{Area} = 15 \text{ m}^2$$

Alternatively, we can consider the new sandpit as being made up of two smaller rectangles:

$$\text{Area of new sandpit} = 3 \times 3 + 3 \times 2$$

$$\text{Area} = 9 + 6$$

$$\text{Area} = 15 \text{ m}^2$$

This example demonstrates that

$$3 \times (3 + 2) = 3 \times 3 + 3 \times 2$$

The expression on the right of the equals sign is written without brackets and is equivalent to the expression on the left of the equals sign. The expression on the right is the expansion of the bracket on the left. This process of expanding a bracket is known as the distributive law.

Brackets can be expanded using algebraic terms, as demonstrated in the next example.

Example 2. Suppose that Andrew did not know that the measurement of the original square sandpit was 3 by 3 and instead used x for its length and width (see figure b). Find an expression for the area of the new sandpit if it is extended by 2 m in one direction.

(b)

Solution. Write

Area of new sandpit = width × length

$$\text{Area} = x \times (x + 2)$$ Replacing the original length of 3 by x

$$\text{Area} = x(x + 2)$$ In algebra the × sign is usually omitted

Alternatively, we can consider the new sandpit as being made up of two smaller rectangles:

$$\text{Area of new sandpit} = x \times x + 2 \times x$$

$$\text{Area} = x^2 + 2x$$ Using algebraic shorthand,
$$x \times x = x^2 \text{ and } 2 \times x = 2x.$$

This example demonstrates that

$$x(x + 2) = x^2 + 2x$$

We write that $x^2 + 2x$ is the expansion of $x(x + 2)$ or $(x + 2)x$, with x written at the end of the brackets.

The rule for expanding a bracket is that both the terms inside the brackets are multiplied by the term outside the brackets, which is the distributive law.

Example 3. Expand the following brackets: (a) $3x(x + 5)$, (b) $4(3x - 5)$, (c) $-3(1 - 2x)$, (d) $x(2x + 3) - 4(x - 6)$.

Solution. For (a), write

$$3x(x + 5) = 3x \times x + 3x \times 5$$ Using the distributive law

$$= 3x^2 + 15x$$ Using algebraic shorthand

For (b),

$$4(3x - 5) = 4 \times 3x - 4 \times 5$$ The subtraction of terms in brackets is preserved

$$= 12x - 20$$

(c) If the term in front of the brackets is negative, the signs of all the terms inside the bracket will be changed when they are multiplied by the term in front. Write

$$-3(1 - 2x) = -3 + 6x \qquad \text{Change in signs: } +1 \to -3, \text{ and } -2x \to +6x$$

For (d), write

$$x(2x + 3) - 4(x - 6) = 2x^2 + 3x - 4x + 24 \qquad \text{Change in signs of the second brackets}$$

$$= 2x^2 - x + 24 \qquad\qquad +3x - 4x = -x$$

The expansion of double brackets will now be explained using a practical example, which will be followed up with a rule.

Example 4. Suppose Andrew has a square sandpit that measures x by x meters, and he decides to extend it 2 m in one direction and 3 m in the other direction. What is the area of the new sandpit?

Solution. The left-hand side of figure c shows the extensions to the sandpit, and the right-hand side shows how the new sandpit is divided up into two rectangles.

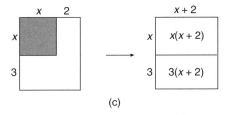

(c)

Using the left-hand side of the figure, write

$$\text{Area of new sandpit} = \text{length} \times \text{width}$$

$$= (x + 3)(x + 2)$$

Using the right-hand side of the figure, write

$$\text{Area of new sandpit} = \text{area of top rectangle} + \text{area of bottom rectangle}$$

$$= x(x + 2) + 3(x + 2)$$

Since both expressions represent the area of the new sandpit, we can write

$$(x + 3)(x + 2) = x(x + 2) + 3(x + 2)$$

$$= x^2 + 2x + 3x + 6 \qquad \text{Using the distributive law for each set of brackets}$$

$$= x^2 + 5x + 6 \qquad\qquad \text{Collecting like terms, } +2x + 3x = +5x$$

The rule for expanding double brackets can now be stated:

The second bracket is multiplied by each term in the first bracket, and then the two brackets are expanded.

Worked examples follow that explain how to solve a variety of problems.

Expand $(x - 3)(x + 5) = x(x + 5) - 3(x + 5)$

$$= x^2 + 5x - 3x - 15 \qquad \text{Note the changes in signs}$$

$$= x^2 + 2x - 15 \qquad +5x - 3x = +2x$$

Expand $(x - 5)(x + 5) = x(x + 5) - 5(x + 5)$

$$= x^2 + 5x - 5x - 25$$

$$= x^2 - 25 \qquad +5x - 5x = 0$$

Expand $(2x - 3)^2 = (2x - 3)(2x - 3)$ Squaring means multiplying by itself

$$= 2x(2x - 3) - 3(2x - 3)$$

$$= 4x^2 - 6x - 6x + 9 \qquad \text{Note the changes in signs}$$

$$= 4x^2 - 12x + 9 \qquad -6x - 6x = -12x$$

Expand $(2x + y)(3a - b) = 2x(3a - b) + y(3a - b)$

$$= 6ax - 2bx + 3ay - by \qquad \text{There are no like terms to collect}$$

References: Algebra, Brackets, Coefficient, Distributive Law, Factorize, Pascal's Triangle, Squaring.

EXPANSION

Reference: Expanding Brackets.

EXPERIMENT

Reference: Event.

EXPERIMENTAL PROBABILITY

Reference: Probability of an Event.

EXPONENT

When a number or an expression is multiplied by itself many times it is simpler to express the result in exponent form. For example, $81 = 3 \times 3 \times 3 \times 3$ and is written as 3^4. We say that 3^4 is the exponent form of $3 \times 3 \times 3 \times 3$ and its value is 81. When 3^4 replaces 81, we say that 3 is the base and 4 is the exponent. Another name for exponent is index. The plural of index is indices.

The power is the number of times the base is multiplied by itself. In the example 3^4 we say that 81 is the fourth power of 3, or 81 is 3 to the power 4. Special names are used for powers of 2 and 3. For example, 4^2 is four squared or four to the power 2, and 5^3 is five cubed or five to the power 3. Another way of expressing exponent form is as follows: 5 is raised to the power 4, which means 5^4.

When numbers or algebraic terms in index form are multiplied or divided certain rules can be applied. The laws of indices are stated here with examples to illustrate their uses. The first law is the exponent form of an expression.

Rule 1:

$$a \times a \times a \times a \times a \times a \ldots \text{ to } n \text{ terms} = a^n$$

Note that 6^n is not the same as $6n$, which means $6 \times n$.

Examples of this rule are

$$2^3 = 2 \times 2 \times 2$$

$$= 8$$

$$y^3 \times y^4 = y \times y \times y \times y \times y \times y \times y$$

$$= y^7$$

$$a^1 = a$$

Rule 2:

$$a^m \times a^n = a^{m+n}$$

When two terms in exponent form are multiplied the exponents are added together, provided they have the same base.

Examples of this rule are

$3^5 \times 3^2 = 3^7$	Both terms are in base 3
$y^5 \times y^4 = y^9$	Both terms are in base y
$(x + 3)^2 \times (x + 3)^4 = (x + 3)^7$	Both terms are in base $(x + 3)$

The rule does not apply when the bases are different. For example, we cannot simplify $3^4 \times 2^6$ using this law of indices, because the bases 3 and 2 are not the same. In the same way $y^2 \times w^3$ cannot be simplified, because the bases y and w are different.

Rule 3:

$$\frac{a^m}{a^n} = a^{m-n} \quad \text{provided } a \neq 0$$

When two terms in exponent form are divided the exponents are subtracted, provided they have the same base. The extra provision of $a \neq 0$ is needed because when $a = 0$, the denominator $a^n = 0$, and a term cannot be divided by zero.
    Examples of this rule are

$$\frac{3^8}{3^2} = 3^6 \qquad \text{Subtracting the indices.}$$

$$y^3 \div y^7 = y^{-4} \qquad \text{Exponents can be negative, and this is explained later}$$

$$\frac{a^n}{a^n} = a^{n-n}$$

$$= a^0$$

$$= 1 \qquad \text{Since } a^n \div a^n = 1, \text{ provided } a^n \neq 0$$

This last result is quoted as the following rule.

Rule 4:

$$a^0 = 1 \quad \text{provided } a \neq 0$$

The extra provision of $a \neq 0$ is needed, because 0^0 is undefined.
    Examples of this rule are

$$10^0 = 1$$

$$(-4)^0 = 1$$

$$(a + b)^0 = 1 \qquad \text{Provided } (a + b) \neq 0$$

Rule 5:

$$(a^m)^n = a^{mn}$$

According to this rule, when a term in exponent form is itself expressed to an exponent, then the exponents are multiplied.
    Examples of this rule are

$$(5^4)^3 = 5^{12}$$

$$(a^4)^3 = a^{12}$$

Note. Confusion may arise, because sometimes we add exponents and at other times we multiply exponents, and the examples above make it clear when to add and when to multiply.

Rule 6:

$$a^{-n} = \frac{1}{a^n}$$

This rule enables negative exponents to be expressed as positive exponents.
   Examples of this rule are

$$6^{-2} = \frac{1}{6^2}$$

$$= \frac{1}{36}$$

$$\frac{3}{4^{-2}} = 3 \times 4^2$$

$$= 48$$

$$\frac{3}{4} y^{-2} = \frac{3}{4y^2}$$

Rule 7:

$$\sqrt{a} = a^{1/2}$$

This rule expresses the square root of a term as an exponent. It also expresses surds as exponents.
   Examples of this rule are

$$9^{1/2} = \sqrt{9}$$

$$= 3$$

$$\sqrt{x^{16}} = \left(x^{16}\right)^{1/2}$$

$$= x^8 \qquad \text{When taking the square root of a term in exponent form the exponent is halved}$$

The bases of some numbers can be changed, and this is explained in the following example.

Example. Simplify the expression $(2^n \times 4^{2n})/8^n$.

Solution. Each of the bases 4 and 8 can be expressed in base 2:

$$\frac{2^n \times 4^{2n}}{8^n} = \frac{2^n \times (2^2)^n}{(2^3)^n} \qquad 2^2 = 4 \text{ and } 2^3 = 8$$

$$= \frac{2^n \times 2^{2n}}{2^{3n}} \qquad \text{Using Rule 5}$$

$$= \frac{2^{3n}}{2^{3n}} \qquad \text{Using Rule 2}$$

$$= 2^{3n-3n} \qquad \text{Using Rule 3}$$

$$= 2^0$$

$$= 1 \qquad \text{Using Rule 4}$$

Calculators are programmed to work out exponents; check with your calculator handbook.

Reference: Square Root.

EXPONENTIAL CURVE

The exponential curve is sometimes called a growth curve, because it often models the way populations grow. This is a curve whose equation is $y = a^x$, where a is any number greater than zero. For example, $y = 2^x$ is the equation of the exponential curve in figure a. The curve has a horizontal asymptote that is the x-axis.

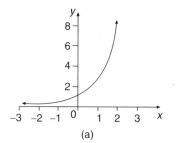

(a)

The curve crosses the y-axis at $y = 1$, which is true for all curves of the type $y = a^x$. A characteristic of this type of curve is that when $a > 1$ the curve becomes very steep. The function $y = 2^x$ is an increasing function for all values of x, which means its gradient is always positive.

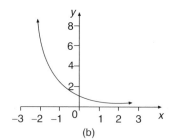

(b)

When a is a fraction, say $a = \frac{1}{2}$, the graph of $y = \left(\frac{1}{2}\right)^x$ is as shown in figure b. This is also the graph of $y = 2^{-x}$. This kind of curve is used to model radioactive decay. The function $y = 2^{-x}$ is a decreasing function for all values of x, which means its gradient is always negative.

References: Asymptote, Exponential Decay, Gradient, Graphs, Increasing Function, Logarithmic Curve.

EXPONENTIAL DECAY

The mass of a radioactive element decays over a period of time according to an exponential rule. The time taken to decay to half of its original mass is called the half-life of the element. The mass m of an iodine element has a fast rate of decay, as shown in the following table. The original mass of the element is 80 grams and in 8 days it decays to 40 grams, which is half its mass, so the element's half-life is 8 days. The graph of this function is an exponential curve, as shown in figure a.

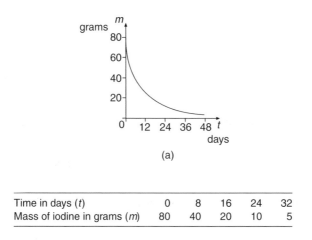

(a)

Time in days (t)	0	8	16	24	32
Mass of iodine in grams (m)	80	40	20	10	5

Exponential Growth The number n of cells of yeast, used to make bread, doubles every hour, and the data showing how the cells multiply is contained in the following table. The exponential graph showing the growth of this population is shown in figure b.

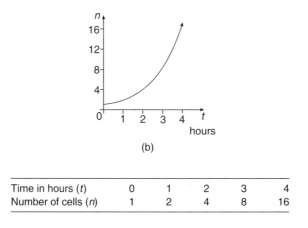

(b)

Time in hours (t)	0	1	2	3	4
Number of cells (n)	1	2	4	8	16

References: Graphs, Exponential Curve, Logarithmic Curve.

EXTERIOR ANGLE OF A CYCLIC QUADRILATERAL

Reference: Cyclic Quadrilateral.

EXTERIOR ANGLE OF A POLYGON

This is the angle between one side of a polygon and the extension of the next side. In the pentagon in figure a it is the shaded angle. A pentagon has five exterior angles.

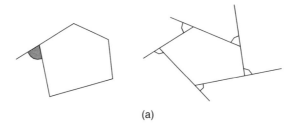

(a)

The geometry theorem about the exterior angles of a polygon states:

The sum of all the exterior angles of a polygon is 360°.

This theorem is true for all polygons whether it has 3 sides, 5 sides, or 100 sides.

Example. One of the exterior angles of a regular polygon is 40°. How many sides has it?

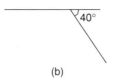

40°

(b)

Solution. Two sides of the polygon are drawn in figure b. Each exterior angle is 40°. Dividing 360° by 40° will give the number of sides of the polygon:

$$\text{Number of sides} = \frac{360}{40}$$

The number of sides of the polygon is 9, which means it is a nonagon.

References: Adjacent Angles, Angle, Nonagon, Pentagon, Polygon.

EXTERIOR ANGLE OF A TRIANGLE

This geometry theorem states that (see figure a):

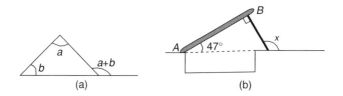

(a)   (b)

The exterior angle of a triangle is equal to the sum of the two interior opposite angles.

Example. A trapdoor AB covers a hole in the ground and is hinged at A (see figure b). The trapdoor is propped open at an angle of of $47°$ by a support which makes a right angle with the trapdoor. Find the angle x that the support makes with the ground.

Solution. Write

$x = 47° + 90°$ Exterior angle of triangle = sum of interior opposite angles

$x = 137°$

The obtuse angle the support makes with the ground is $137°$.

References: Exterior Angle of a Polygon, Obtuse Angle.

EXTRAPOLATION

This is a method of estimating the value of a function beyond the values that are already known. In order to extrapolate, we assume that the function will continue in the same pattern it has already followed. It is because we make this assumption that extrapolation, and interpolation, are not absolutely reliable unless one is certain that the pattern of the function will not change.

Example. The table shows the equivalent ages of dogs and men.

Age of dog in years	1	2	3	4	8	12	15	20
Equivalent age of man in years	15	22	28	34	49	64	76	96

Estimate the equivalent age of a man when a dog is (a) 26 years old, (b) 10 years old.

Solution. To estimate these values, we will need a graph of the data, which is drawn in the figure. Interpolation is a method of estimating the value of a function between the values that are already known, so the age of a man equivalent to a dog's age of 10 years old is read off the graph. The table does not extend as far as a dog of 26 years, so the graph is extended, continuing the line that is drawn so far, which appears to have straightened. This process is called extrapolation. Since we do not know for certain that the relationship between the ages of dogs and men continues beyond the values given in the table, the result is not completely reliable.

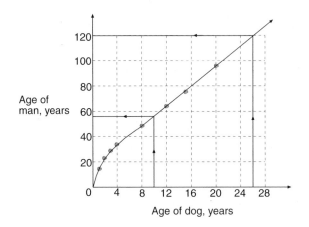

The answer to (a) is found using extrapolation, and the answer to (b) using interpolation:

(a) 26 → 120 years.

(b) 10 → 56 years.

The ages of the men are estimated to be 120 and 56 years, respectively.

References: Estimation, Graphs.

F

FACE

Reference: Edge.

FACTOR

A positive integer that divides exactly into another positive integer, without a remainder, is a factor of that number. The integer 1 is a factor of all positive integers, and every positive integer is a factor of itself. For example, 8 is a factor of 24, because 8 divides exactly into 24, without a remainder. The integer 24 is a factor of 24, and 1 is also a factor of 24. A complete list of the eight factors of 24 is {1, 2, 3, 4, 6, 8, 12, 24}.

A prime factor is a factor that is a prime number. The prime factors of 24 are {2, 3}. The integer 1 is a factor of 24, but it is not a prime number.

A number that is a factor of two numbers is a common factor of those two numbers. For example, 3 is a factor of 6 and of 9, so 3 is a common factor of 6 and 9.

Example 1. Find all the common factors of the two numbers 30 and 36; what is their highest common factor?

Solution. The factors of 30 are {**1**, **2**, **3**, 5, **6**, 10, 15, 30}, and the factors of 36 are {**1**, **2**, **3**, 4, **6**, 9, 12, 18, 36}. The common factors of 30 and 36 are numbers that are highlighted in both lists, which are {1, 2, 3, 6}. The highest common factor (HCF) of 30 and 36 is the largest number in the list of common factors. The HCF of 30 and 36 is 6.

When numbers are factorized they are written as the products of their prime factors. This is also explained under the entry Factor Tree.

Example 2. Factorize 36.

Solution. Write

$$36 = 4 \times 9 \qquad \text{Breaking down 36 into any two products}$$
$$= 2 \times 2 \times 3 \times 3 \qquad \text{Factorizing 4 and 9 into prime numbers}$$
$$= 2^2 \times 3^2. \qquad \text{Using indices}$$

36 is factorized as $2^2 \times 3^2$.

A term often confused with factors is multiples. The multiples of a number are obtained by multiplying that number by 1, 2, 3, 4, ... in turn.

Example 3. Find the first five multiples of 6.

Solution. The numbers 1, 2, 3, 4, 5 multiply the number 6 in turn as follows:

$$6 \times 1 = 6, \quad 6 \times 2 = 12, \quad 6 \times 3 = 18, \quad 6 \times 4 = 24, \quad 6 \times 5 = 30.$$
$$\text{Note that 6 is regarded as a multiple of 6}$$

The first five multiples of 6 are $\{6, 12, 18, 24, 30\}$.

Example 4. Find the lowest common multiple of the two numbers 6 and 8.

Solution. The lowest common multiple (LCM) of 6 and 8 is the smallest number that is a multiple of both of them. Multiples of $6 = \{6, 12, 18, \mathbf{24}, 30, \ldots\}$ and multiples of $8 = \{8, 16, \mathbf{24}, 32, \ldots\}$.

The LCM of 6 and 8 is 24.

References: Factor tree, Integers, Product.

FACTOR TREE

The factor tree is used to factorize numbers, which means write them as the products of their prime factors.

Example. Write 840 as the product of primes.

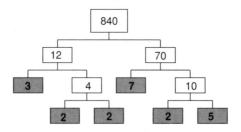

Solution. Find any two numbers that multiply together to give 840, say 12×70. Split 12 into any two numbers which multiply together to give 12, say 3×4, and 70 into 7×10, an so on. All the prime factors 3, 2, 2, 2, 5, 7 that multiply to make 840 are highlighted in the figure. Arranging them in order of size gives

$$840 = 2 \times 2 \times 2 \times 3 \times 5 \times 7$$
$$= 2^3 \times 3 \times 5 \times 7 \qquad \text{Using exponent form } 2^3 \text{ for } 2 \times 2 \times 2$$

The number 840 is now expressed as the product of prime numbers.

References: Factor, Prime Factor, Product.

FACTORIAL

Factorials are used in combinations and permutations. The factorial of a counting number n is the product of the first n counting numbers. The symbol for factorial n is $n!$ By definition $0! = 1$, and also $1! = 1$.

Example. Find the value of factorial 5.

Solution. Write

$$5! = 1 \times 2 \times 3 \times 4 \times 5. \qquad \text{It is the product of the first five counting numbers}$$

Factorial 5 is 120.
    Factorials can be worked out using a scientific calculator.

References: Combinations, Permutations.

FACTORIZE

When we factorize an expression we write the expression as a product of its factors using sets of brackets. Factorizing is the reverse of expanding brackets, and expanding should be well understood before factorizing is studied. There are three kinds of factorizing explained under this entry:

- Common factors, type 1 and type 2
- Quadratic factors
- Difference of two squares

We will also look at combinations of these kinds of factorizing.

Common Factors. Type 1 The factorized answer will contain one set of brackets.

Example 1. Factorize $2x^2 + 6xy$.

Solution.

$$2x^2 + 6xy = 2 \cdot x \cdot x + 2 \cdot 3 \cdot x \cdot y \qquad \text{Writing each term in the form of products.}$$

$$= \mathbf{2 \cdot x} \cdot x + \mathbf{2 \cdot 3 \cdot x} \cdot y \qquad 2x \text{ is common to both terms}$$

$$= \mathbf{2x}(x + 3y) \qquad \text{The common factor } 2x \text{ is written at the front of the brackets, and what is left is written inside the brackets}$$

Expanding the brackets can check the answer. Always make sure that the common factor is as large as possible. For example, using a common factor of only 2 or a common factor of only x is not satisfactory in this example.

Example 2. Factorize $2x^4 - 8x^2 + 2x$.

Solution. Write

$$2x^4 - 8x^2 + 2x = \mathbf{2 \cdot x} \cdot x \cdot x \cdot x - \mathbf{2} \cdot 2 \cdot 2 \cdot \mathbf{x} \cdot x + \mathbf{2 \cdot x}$$

$$= 2x(x \cdot x \cdot x - 2 \cdot 2 \cdot x + 1) \qquad \begin{array}{l}\text{If the common factor} \\ \text{is all of one term,} \\ \text{we must leave a} \\ 1 \text{ in the brackets}\end{array}$$

$$= 2x(x^3 - 4x + 1)$$

Common Factors. Type 2 The common factor may be a term in brackets instead of a single term.

Example. Factorize $2(x + y) + y(x + y)$.

Solution. The principle for factorizing this expression is the same as for factorizing $2b + yb = b(2 + y)$, where b is the term $(x + y)$. Write

$$2(\mathbf{x + y}) + y(\mathbf{x + y}) = (x + y)(2 + y) \qquad (x + y) \text{ is the common factor}$$

In harder problems it may be necessary to group the terms into pairs and then apply common factors to each group. This method is sometimes called "factors by grouping." For example,

$$2x + 2y + xy + y^2 = 2(\mathbf{x + y}) + y(\mathbf{x + y}) \qquad \begin{array}{l}\text{Using common factors on each} \\ \text{pair of terms } 2x + 2y \text{ and } xy + yy\end{array}$$

$$= (x + y)(2 + y) \qquad (x + y) \text{ is a common factor}$$

Note. When factorizing four terms, as in the example above, it may be necessary to regroup them into different pairs at the start if there are no obvious common factors.

Quadratic Factors The rules for factorizing quadratic expressions are easily noticed when we expand pairs of brackets in the following examples. Look for the way the two numbers in the brackets are related to the numbers in the expansion. Write

$$(x + 3)(x + 7) = x^2 + 10x + 21 \qquad 3 + 7 = 10 \text{ and } 3 \times 7 = 21$$

$$(x - 4)(x + 2) = x^2 - 2x - 8 \qquad -4 + 2 = -2 \text{ and } -4 \times 2 = -8$$

$$(x - 7)(x - 1) = x^2 - 8x + 7 \qquad -7 + -1 = -8 \text{ and } -7 \times -1 = 7$$

Having established a pattern in expanding brackets, we can reverse the process of expanding brackets and factorize the expansions.

Example 1. Factorize $x^2 + 7x + 6$.

Solution. Write

$$x^2 + 7x + 6 = (x \quad)(x \quad) \qquad$$ There are two sets of brackets, each containing x, and spaces are left in which to insert the two numbers

The two numbers that go into the brackets must multiply to make $+6$ and add to make $+7$. There is only one possible combination that meets both criteria, and that is $+6$ and $+1$:

$$x^2 + 7x + 6 = (x + 6)(x + 1) \qquad$$ Expanding the two sets of brackets can check the answer

The factors may be written the other way around as $(x + 1)(x + 6)$.

Example 2. Factorize $x^2 - 5x - 14$.

Solution. Write

$$x^2 - 5x - 14 = (x \quad)(x \quad)$$

The two numbers that go into the brackets must multiply to make -14 and add to make -5. One of the numbers must be positive and one negative to obtain a product that is negative. There is only one possible combination that meets both criteria, and that is -7 and $+2$:

$$x^2 - 5x - 14 = (x - 7)(x + 2)$$

With practice the two numbers that go into the brackets can be quickly identified.

Difference of Two Squares The expressions we factorize are of the form $a^2 - b^2$, which factorizes into the two sets of brackets $(a - b)(a + b)$. This can be expressed as a formula:

$$a^2 - b^2 = (a - b)(a + b)$$

The reason for the name "difference of two squares" is because difference means minus, and the two terms a^2 and b^2 are squared terms.

Example 1. Factorize $9 - x^2$.

Solution. Write

$9 - x^2 = 3^2 - x^2$ In the formula a is 3 and b is x

$\quad = (3 - x)(3 + x)$ Substituting $a = 3$ and $b = x$ into $(a - b)(a + b)$

Example 2. Factorize $16x^2 - 25$.

Solution. Write

$16x^2 - 25 = (4x)^2 - 5^2$ In the formula a is $4x$ and b is 5

$\quad = (4x - 5)(4x + 5)$ Substituting $a = 4x$ and $b = 5$ into $(a - b)(a + b)$

Some factorization involves combinations of two types of factors, one of which is a common factor, and is done first.

Example 3. Factorize $2x^2 - 4x - 6$.

Solution. Write

$2x^2 - 4x - 6 = 2(x^2 - 2x - 3)$ The common factor is 2

$\quad = 2(x - 3)(x + 1)$ Factorizing the quadratic $x^2 - 2x - 3$

Example 4. Factorize $\pi R^2 - \pi r^2$.

Solution. Write

$\pi R^2 - \pi r^2 = \pi(R^2 - r^2)$ The common factor is π

$\quad = \pi(R - r)(R + r)$ Using the difference of two squares

Note. It is not possible to factorize all expressions, but when it is possible the process is to proceed as in the examples above. For example, $x^2 + 1$ and $x^2 + x - 1$ are expressions that do not factorize.

References: Brackets, Expanding Brackets.

FIBONACCI SEQUENCE

The Fibonacci sequence of numbers is one of the most famous sequences in mathematics. It was named after the Italian Fibonacci (1180–1250), who was also known as Leonardo of Pisa. The sequence is the set of numbers $\{1, 1, 2, 3, 5, 8, 13, 21, 34, \ldots\}$. The next term is found by adding together the two previous terms. The next term after 34 is 55, because $55 = 21 + 34$.

Fibonacci's sequence of numbers originated with the following problem that he was solving. "How many pairs of rabbits will be produced in a year, beginning with a single pair, if in every month each pair bears a new pair which becomes productive from the second month on?"

References: Golden Ratio, Sequence.

FINITE DECIMALS

Reference: Decimal.

FIXED POINTS

This is another name for invariant points.

Reference: Invariant Points.

FOOT

Reference: Imperial System of Units.

FORMULA

The plural of formula is formulae or formulas. A formula is an equation between two or more quantities which shows how changes in one quantity will affect the other(s). The formula $A = \pi R^2$ is for finding the area of a circle that has a radius of R. This formula represents a relationship between two quantities A and R. As R changes, there are corresponding changes in A.

The formula is $A = \frac{1}{2}BH$ is for finding the area of a triangle that has a base of length B and a perpendicular height of H. This formula represents a relationship between three quantities A, B, and H. Changes in B and H produce a corresponding change in A.

FOUR-COLOR PROBLEM

Suppose you wish to color the different regions on a map, or any similar figure which has many regions. If regions with a common edge are to have different colors, you will never need more than four colors to color the whole map.

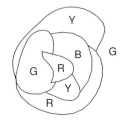

The map shown in the figure has been colored using four colors, red (R), yellow (Y), blue (B), and green (G). The outside region also has to be colored. Regions that meet at a point may have the same color, but those that meet on an edge must be colored differently.

FOUR RULES

The four rules are adding, subtracting, multiplying, and dividing. If you need to learn the four rules of decimals, you will need to know how to add, subtract, multiply, and divide decimals.

FRACTIONS

Only numerical fractions will be studied under this entry. For information about algebraic fractions see the entries Algebraic Fractions and Canceling.

A fraction is a number that is made up of the ratio of two integers. If n and d are two integers then $\frac{n}{d}$ is a fraction, provided that $d \neq 0$, otherwise the fraction is undefined.

Fractions can be positive or negative. Examples of fractions are $\frac{3}{4}$, $-\frac{1}{2}$, 3 (because it can be written as $\frac{3}{1}$), $\frac{12}{7}$, -5, and $2\frac{5}{8}$ (because it can be written as $\frac{21}{8}$). The top number in a fraction is called the numerator and the bottom number is called the denominator. If the numerator is larger than the denominator, we say that the fraction is improper. For example, $\frac{12}{7}$ is an improper fraction. If a fraction is joined to a whole number by addition we say it is a mixed number. For example, $2\frac{5}{8}$ is a mixed number. A proper fraction is one with the denominator larger than the numerator, and an example is $\frac{3}{4}$. Other names for proper fractions are common fractions and simple fractions. In practical terms a proper fraction is part of a whole, as illustrated in these examples:

- Nathan scored 16 out of 20 in a math test, and his score can be written as the fraction 16/20. In turn this fraction can be written as the percentage 80%.
- It takes Nathan and Jacob 5 hours to paint a fence for their dad. Nathan works for 3 hours and Jacob works for 2 hours, and their dad gives them $10. If they divide the $10 into 5 parts, then Nathan should get 3 parts of the money and Jacob 2 parts. Using fractions, we write that Nathan gets $\frac{3}{5}$ of $10 = $6 and Jacob gets $\frac{2}{5}$ of $10 = $4.

Example 1. A chessboard is made up of 64 squares; shade $\frac{3}{4}$ of it.

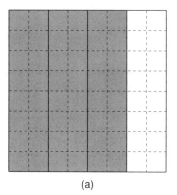

(a)

Solution. The board is divided up into 4 equal strips, and 3 of them are shaded (see figure a). This means $\frac{3}{4}$ of the board is shaded. Alternatively, 48 small squares are also shaded out of a total of 64 small squares, so $\frac{48}{64}$ of the board is shaded. The two fractions $\frac{3}{4}$ and $\frac{48}{64}$ are the same size, and they are called equivalent fractions.

Improper fractions can be converted into mixed numbers, and the process is explained in the following example.

Example 2. Write the improper fraction $\frac{13}{3}$ as a mixed number.

Solution. Write $\frac{13}{3}$ as

$$\frac{3}{3} + \frac{3}{3} + \frac{3}{3} + \frac{3}{3} + \frac{1}{3} = 1 + 1 + 1 + 1 + \frac{1}{3}$$

$$= 4\frac{1}{3} \qquad \text{which is a mixed number}$$

Alternatively, 3 can divide 13, in the way shown in figure b. The remainder 1 is written as a fraction of the divisor 3 to give $4\frac{1}{3}$.

$$\frac{4 \text{ remainder } 1}{3)\overline{13}}$$

(b)

Mixed numbers can also be changed into improper fractions, as explained in the following example.

Example 3. Write the mixed number $3\frac{2}{5}$ as an improper fraction.

Solution. Write $3\frac{2}{5}$ as

$$1 + 1 + 1 + \frac{2}{5} = \frac{5}{5} + \frac{5}{5} + \frac{5}{5} + \frac{2}{5}$$

$$= \frac{17}{5} \qquad \text{which is an improper fraction}$$

Alternatively, a quick way is to multiply the whole number part, 3, by the denominator, 5, to get 15, add the numerator 2 to get 17, and put the 17 over the denominator 5 to get 17/5.

Adding and Subtracting Fractions When fractions have the same denominators they can easily be added or subtracted.

$$\frac{2}{7} + \frac{3}{7} = \frac{2+3}{7} \qquad \text{and} \qquad \frac{7}{8} - \frac{3}{8} = \frac{7-3}{8}$$

$$= \frac{5}{7} \qquad\qquad\qquad = \frac{4}{8}$$

$$= \frac{1}{2} \qquad \text{Canceling the fraction } \frac{4}{8}$$

When fractions have different denominators they cannot be added or subtracted until they have both been rewritten with the same denominator, called a common denominator. To do this we use equivalent fractions (this topic should be understood before proceeding with adding and subtracting fractions).

Example 4. Work out $\frac{5}{6} + \frac{3}{8}$.

Solution. Write down a few equivalent fractions to each of the fractions being added.

$$\frac{5}{6} = \frac{10}{12} = \frac{15}{18} = \frac{20}{24} = \frac{25}{30} = \dots \qquad \text{and} \qquad \frac{3}{8} = \frac{6}{16} = \frac{9}{24} = \frac{12}{32} = \dots$$

From each list select the two fractions that have the same denominator, and then add them together:

$$\frac{5}{6} + \frac{3}{8} = \frac{20}{24} + \frac{9}{24}$$

$$= \frac{29}{24} \qquad \text{Fractions can be added when their denominators are the same.}$$

$$= 1\frac{5}{24} \qquad \text{Writing the improper fraction as a mixed number}$$

This method can be streamlined by identifying the lowest common denominator as the lowest common multiple of the two original denominators, 6 and 8. In this case the lowest common multiple of 6 and 8 is 24. This method is used in the next example.

Example 5. Work out $2\frac{3}{10} - 1\frac{4}{15}$.

Solution. First write each fraction as an improper fraction:

$$\frac{23}{10} - \frac{19}{15}$$

The lowest common denominator of 10 and 15 is 30, and using equivalent fractions, write each fraction with a denominator of 30:

$$\frac{23}{10} - \frac{19}{15} = \frac{23}{10} \times \frac{3}{3} - \frac{19}{15} \times \frac{2}{2}$$

$$= \frac{69}{30} - \frac{38}{30}$$

$$= \frac{69 - 38}{30}$$

$$= \frac{31}{30}$$

$$= 1\frac{1}{30} \qquad \text{Writing the improper fraction as a mixed number}$$

For multiplying and dividing fractions, see the entry Multiplying Fractions.

Some rulers are graduated in fractions, as in the next example.

Example 6. The drawing in figure c is of part of a ruler. Write down the reading to which the arrow points.

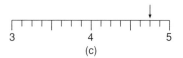

(c)

Solution. Each unit of length from 3 to 4 and from 4 to 5 is divided up into 8 equal parts, so each part is $\frac{1}{8}$. The arrow points to a reading of $4\frac{6}{8}$, which can be written as $4\frac{3}{4}$ by canceling down the $\frac{6}{8}$.

Example 7. What fraction of the figure is shaded?

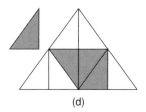

(d)

Solution. The large triangle is made up of eight small triangles like the one pulled out of figure d. Three of these triangles are shaded, so the fraction of the shape that is shaded is $\frac{3}{8}$.

Example 8. David takes his family on a trip. Of the journey, $\frac{3}{4}$ is spent flying and $\frac{1}{6}$ is spent on a train. The rest of the journey is made by cab. What fraction of the journey is made by cab?

Solution. The fraction spent flying and on the train is

$$\frac{3}{4} + \frac{1}{6} = \frac{9}{12} + \frac{2}{12} \qquad \text{Writing each fraction with a lowest common denominator of 12}$$

$$= \frac{9+2}{12}$$

$$= \frac{11}{12}$$

This fraction is now subtracted from 1 to obtain the fraction of the journey spent in the cab:

$$1 - \frac{11}{12} = \frac{12}{12} - \frac{11}{12}$$

$$= \frac{1}{12}$$

The fraction of the journey made by cab is $\frac{1}{12}$.

Fractions can be added, subtracted, multiplied, and divided using a scientific calculator, and you are referred to your calculator handbook.

References: Algebraic Fractions, Canceling, Equivalent Fractions, Integers, Percentage, Ratio.

FREQUENCY

In statistics, frequency refers to the number of times an event occurs in an experiment. When we collect together the frequencies of all the events in an experiment and record them in a table we have a frequency table, or a frequency distribution. These ideas are illustrated in the following example, in which a tally column is used to count the number of times the event occurs.

Example. There are 28 houses on Washington Street and the number of people living in each house is recorded. The results are

0, 2, 1, 1, 4, 2, 3, 3, 2, 3, 0, 4, 2, 0, 3, 3, 3, 2, 3, 1, 2, 3, 5, 3, 4, 0, 3, 3

Draw up a frequency table for these data.

Solution. A frequency table is drawn up and the tally and frequency columns filled in. In the tally column a short 'stick' is drawn to keep a record of each time a number is counted. Each time a fifth number is counted a sloping stick is drawn to form units of five sticks. In this way the tally can be easily counted to get the frequency total. It is usual to call the variable x, in this case the number of people, and the frequency f.

Number of People (x)	Tally	Frequency (f)
0	||||	4
1	|||	3
2	Ⅷ |	6
3	ⅧⅧ |	11
4	|||	3
5	|	1
Total		28

Reference: Statistics.

FREQUENCY CURVE

Reference: Frequency Polygon.

FREQUENCY DISTRIBUTION

References: Frequency, Bar Graph.

FREQUENCY POLYGON

This is a graph that is related to a histogram. The meaning of a frequency polygon will be brought out in the following example.

Nathan collected data about the weight of students in his class of 30 students. He split the weights into class intervals of 5 kg and recorded the data in a frequency table as shown. The class interval 40– means a weight of 40–45 kg, which includes 40 kg, but does not include 45 kg. Nathan used the frequency table to draw a histogram of the results. He joined the midpoints of the tops of the columns with a series of straight-line segments to obtain the frequency polygon shown in the figure.

Weight in kg (x)	Frequency (f)
40–	1
45–	3
50–	4
55–	9
60–	6
65–	5
70–75	2
Total	30

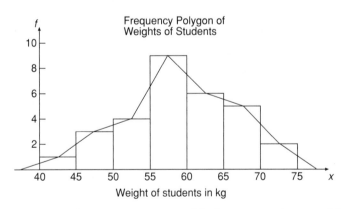

The columns of the histogram are not part of the frequency polygon and they can be removed so that the graph of the straight-line segments is left. The frequency polygon has its ends extended to reach the horizontal axis. To draw a frequency polygon the data should be continuous.

If the midpoints of the tops of the columns are joined with a smooth curve instead of a series of straight-line segments, the graph is called a frequency curve.

References: Class Interval, Continuous Data, Histogram.

FREQUENCY TABLE

References: Arithmetic Mean, Bar Graph, Frequency.

FRUSTUM

A frustum is a solid shape that is left when the top of a cone or a pyramid is sliced off by a plane parallel to the base of the cone. In figure a, the cone is an upright one, called a right cone, and the plane of the slice is a circle. It is sometimes called a truncated cone.

The net of a hollow frustum of a cone, without the two circles for its ends, is shown in figure a. The net is a part of the sector of a circle, and the completion of the sector is shown by dashed lines.

The pattern of a girl's skirt is a net of the frustum of a cone. The surface area of the frustum of the cone (not including the two circular ends) is (see figure b)

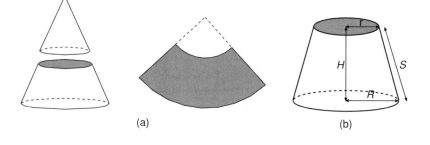

(a)   (b)

$$\text{Area} = \pi S(R + r)$$

The volume of the solid frustum of the cone is

$$\text{Volume} = \frac{1}{3}\pi H(R^2 + Rr + r^2)$$

Example. Helen is cutting out a pattern for a skirt she is making. Her waist measurement is 80 cm, the hem of her skirt is to be 140 cm, and the length of the skirt is 60 cm. What is the area of the pattern for her skirt?

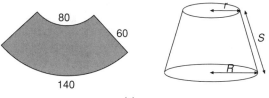

(c)

Solution. The left-hand side of figure c shows the net of the skirt, and the right-hand side is a three-dimensional sketch of the finished skirt. The formula for the area of the skirt is Area $= \pi S(R + r)$. To evaluate this area, we require the values of r and R, neither of which is known. The value of S is 60 cm. The ensuing working shows how to find the values of r and R.

The formula for the circumference of a circle is $C = 2\pi R$. The waist circle in the three-dimensional sketch has a circumference of 80 cm. Write

$80 = 2\pi r$ Substituting $C = 80$ and r is the radius.

$r = \dfrac{80}{2\pi}$ Rearranging the equation to make r the subject

$r = 12.7$ cm   (to 1 dp)   Using the value of π in the calculator

Similarly

$R = \dfrac{140}{2\pi}$ Which is the radius of the hem of the skirt

$R = 22.3$ cm   (to 1 dp)   Using the value of π in the calculator.

Now write

$A = \pi S(R + r)$ Formula for the area of the skirt

$A = \pi \times 60 \times (22.282 + 12.732)$ Substituting $S = 60$, $R = 22.282$, and $r = 12.732$

$A = 6600$ to nearest whole number

The area of the pattern is 6600 cm², or 0.66 m² (1 m² $= 10{,}000$ cm²).

Frustum of a Square-based Pyramid The volume of the frustum is given by (see figure d)

$$\text{Volume} = \tfrac{1}{3}h(a^2 + ab + b^2)$$

where a, b, and h are shown in figure d.

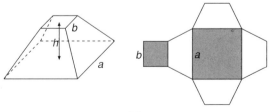

(d)

The right-hand side of the figure is the net of the frustum of the square-based pyramid, which is made up of two squares and four trapeziums.

References: Circle, Cone, Net, Pyramid, Sector of a Circle, Square, Trapezium.

FUNCTION

A full description of function is given in the entry Correspondence. The relation tree in the figure shows all the subsets of relations, and two types of relations are functions

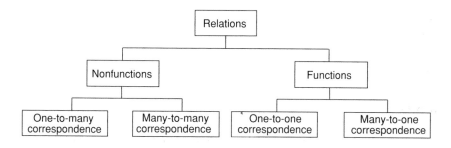

Reference: Correspondence.

FURLONG

Reference: Imperial System of Units.

G

GALLON

Reference: Imperial System of Units.

GEARS

A gear is a wheel with teeth. Two gear wheels are used together to change a turning speed and the direction of the turning. Two gear wheels are used on a bicycle, one at the pedal and the other at the back wheel. A chain connects the gear wheels, which ensures the wheels turn in the same direction.

Example. A chain connects the gear wheels shown in figure a. For one revolution of the larger gear wheel how many times does the smaller gear wheel turn?

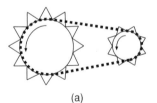

(a)

Solution. The larger gear wheel has 12 teeth and the smaller gear wheel has 8 teeth. Write

$$\text{The ratio of teeth} = 12{:}8 \qquad \text{Larger wheel to smaller wheel}$$

$$= 3{:}2 \qquad \text{Canceling down the ratio}$$

$$= 1.5{:}1 \qquad \text{Dividing each side of the ratio by 2}$$

This means that for every 1 rotation of the larger wheel the smaller wheel rotates 1.5 times. Note the way the ratio is reversed, because the smaller wheel turns faster, so the smaller gear wheel is turning 1.5 times faster than the larger gear wheel.

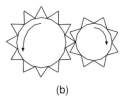

(b)

   If the gear wheels are not connected by a chain, but the teeth mesh together (see figure b), then the ratio of wheel turns is the same as if connected by a chain, but the wheels turn in opposite directions.

Reference: Ratio.

GENERAL TERM

References: Difference Tables, Patterns.

GENERATOR

Reference: Cone.

GEOMETRY THEOREMS

Here is a list of all the geometry theorems, with their references, which are listed in this book. To find them search under their title or their references.

Angles

- Complementary angles add up to 90°
- Supplementary angles add up to 180°
- Conjugate angles add up to 360°

Straight-Line Theorems

- Sum of adjacent angles = 180° (see Adjacent Angles)
- Alternate angles are equal.
- Sum of angles at a point = 360° (see Angles at a Point)
- Sum of angles of a triangle = 180° (see Angle Sum of a Triangle)
- Base angles of an isosceles triangle are equal (see Isosceles Triangle)
- Sum of cointerior angles = 180° (see Alternate Angles)

- Corresponding angles are equal (see Alternate Angles)
- Exterior angle of a triangle = sum of interior opposite angles (see Exterior Angle of a Triangle)
- Sum of exterior angles of a polygon is $360°$ (see Exterior Angle of a Polygon)
- Vertically opposite angles are equal.

Circle Theorems

- Angle at the center is twice the angle at the circumference
- Angles on the same arc are equal
- Angle in the alternate segment
- Sum of opposite angles of a cyclic quadrilateral $= 180°$ (see Cyclic Quadrilateral)
- Angle in a semicircle is a right angle
- Exterior angle of a cyclic quadrilateral = interior opposite angle (see Cyclic Quadrilateral)
- Radius is perpendicular to tangent (see Tangent and Radius Theorem)
- Tangents from a common point are equal

Tests for Concyclic Points

- Converse of the angle in a semicircle is a right angle (see Angle in a Semicircle)
- Converse of sum of opposite angles of cyclic quadrilateral $= 180°$ (see Cyclic Quadrilateral)
- Converse of angles on the same arc are equal (see Cyclic Quadrilateral)

Chords and Tangents

- Intersecting chords inside a circle
- Intersecting chords outside a circle
- Tangent secant theorem (see Intersecting Chords)

GOLDEN RATIO

This is also known as the golden rectangle. Suppose a line segment AB is divided up into two unequal parts by the point C (see figure a). The point C divides the line segment in the golden ratio if the ratio of the complete segment AB to the larger segment AC is equal to the ratio of the larger segment AC to the smaller segment CB. This statement of the golden ratio can be expressed as

$$\frac{AB}{AC} = \frac{AC}{CB}$$

A   1   C   x   B

(a)

Suppose the longer part AC is of length 1 unit, and the shorter part CD is of length x units. The golden ratio can now be expressed as follows:

$$\frac{1+x}{1} = \frac{1}{x} \qquad AB = 1+x, \ AC = 1, \ CB = x$$

$$x(1+x) = 1 \qquad \text{Cross multiplying}$$

$$x + x^2 = 1 \qquad \text{Expanding the brackets}$$

$$x^2 + x - 1 = 0 \qquad \text{Rearranging into quadratic form}$$

$$x = \frac{-b \pm \sqrt{b^2 - 4ac}}{2a} \qquad \text{Formula for solving a quadratic equation}$$

$$x = \frac{-1 \pm \sqrt{1^2 - 4(1)(-1)}}{2 \times 1} \qquad \text{Substituting } a = 1, b = 1, c = -1$$

$$x = \frac{-1 \pm \sqrt{5}}{2}$$

$$x = 0.618 \quad \text{(to 3 dp)} \qquad \text{Discarding the negative solution of the equation}$$

What this value for x means is that if a rectangle has a length of 1 unit and a width of 0.618 unit, then it is a golden rectangle and its sides are in the golden ratio (see figure b). The value $x = 0.618$ is approximately equal to the fraction 21/34.

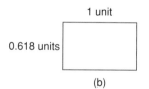

(b)

An enlargement of a golden rectangle is also a golden rectangle. For example if the rectangle of length 1 and width 0.618 is enlarged three times to obtain a rectangle of length 3 units and width 1.854 units, then this rectangle is also a golden rectangle.

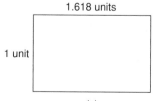

(c)

It is also true that a golden rectangle can have a length of 1.618 units and a width of 1 unit (see figure c), because when the width 1 is divided by 1.618 the result is 0.618.

Example. Test the dimensions of an ordinary playing card to see if it is a golden rectangle (see figure d).

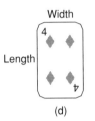

(d)

Solution. The length of the playing card is 8.85 cm and the width is 5.75 cm. Write

$$\frac{\text{Width}}{\text{Length}} = \frac{5.75}{8.85} \qquad \text{Finding the ratio width:length}$$

$$= 0.649 \quad \text{(to 3 dp)}$$

The playing card is not quite a golden rectangle, because the golden ratio of width: length should be 0.618, to 3 dp.

Another property of the golden rectangle is that if the rectangle has a square removed with dimensions equal to the shorter side of the rectangle, then the rectangle that remains is itself a golden rectangle. The rectangle that remains will of course not have lengths of 1 unit and 0.618 unit, but the ratio of length:width will be equal to the ratio 1:0.618. This process can be repeated over and again, as demonstrated in figure e.

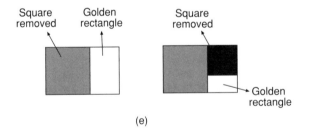

(e)

The ancient Greeks attributed special merits to the golden rectangle, because they believed it was a rectangle in perfect proportion. This made it visually very pleasing to the eye, and they used the golden rectangle in their architecture and designs. It appears in the architecture of the Parthenon, which was built about 400 BC on the Acropolis in Athens, and in many famous paintings, particularly those of Leonardi da Vinci.

The ratio of terms in the Fibonacci sequence is in the golden ratio. The terms of the sequence are 1, 1, 2, 3, 5, 8, 13, 21, 34, 55, 89, 144, 233, ... Suppose we take one term and divide it by the following term:

$$144 \div 233 = 0.618 \quad \text{(to 3 dp)}$$
$$89 \div 144 = 0.618 \quad \text{(to 3 dp)}$$
$$34 \div 55 = 0.618, \quad \text{(to 3 dp)}$$

And so on. To an accuracy of two decimal places the result is the golden ratio. The pattern breaks down for the early terms of the sequence.

References: Enlargement, Fibonacci Sequence, Quadratic Formula, Ratio, Rectangle.

GOLDEN RECTANGLE

Reference: Golden Ratio.

GOODNESS OF FIT

Reference: Line of Best Fit.

GRADIENT

Another name for gradient is slope. The gradient of a straight line is expressed as a fraction, and is a measure of how steep the line is. The gradient may be positive, negative, zero, or undefined. The gradient of a line may be less than 1 or greater than 1. The gradient of a straight line is best described using squares on a grid, or x–y axes.

The symbol for the gradient of a straight line is m. The convention is that m is positive for lines that slope in the direction / and m is negative for lines that slope in the direction \. In other words, if a line is increasing from left to right it has a positive gradient, and if it is decreasing from left to right it has a negative gradient.

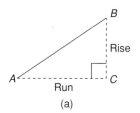

(a)

To find the gradient of the straight line AB in figure a, we first complete a right-angled triangle ACB. The vertical length BC is called the rise and the horizontal length

AC is called the run. The gradient of the line AB is equal to the fraction

$$m = \frac{\text{rise}}{\text{run}}$$

This particular line AB has a positive gradient because it slopes toward the right.

In the examples that follow the gradients of a variety of straight lines will be calculated using the formula $m = \text{rise/run}$. Some of the gradients will be positive and some will be negative according to the direction of the line. The fraction for the gradient is usually canceled down to its simplest form; it may be written as a mixed number or as an improper fraction, but the latter is preferable for ease in drawing the gradient.

(b)

The examples in figure b show that vertical lines have an undefined gradient, which may be referred to as a gradient of infinity, the symbol of which is ∞. Also, horizontal lines have a gradient of zero. It can be seen that a line with a gradient of $\frac{3}{2}$ is steeper than a line with gradient $\frac{1}{2}$, because $\frac{3}{2}$ is greater than $\frac{1}{2}$.

To find the gradient of a longer line segment, we can choose any two points on the line and find the gradient of that line segment, which will be the same as the gradient of the longer line.

Parallel Lines Two lines are parallel if they both have the same gradient. If one line has a gradient of $m_1 = \frac{2}{3}$ and another line has a gradient of $m_2 = \frac{4}{6}$ then the two lines are parallel, because $m_1 = m_2$.

Perpendicular Lines Two lines are perpendicular if the product of their gradients is -1. If one line has a gradient of $m_1 = \frac{3}{4}$ and another line has a gradient of $m_2 = -\frac{4}{3}$, then the two lines are perpendicular, because $m_1 \times m_2 = -1$.

In the examples so far we have counted squares to obtain the gradient of a line, but when different scales are involved, as in graph questions, it is important to measure units rather than count squares.

Example. Harry takes his family out in the car for a picnic. They leave home and travel 80 km to Boulder Bay at a constant speed, and arrive there after $1\frac{1}{2}$ h. Their picnic lasts 2 h. Afterward they travel home in 1 h. On the graph in figure c the journey out to Boulder Bay is marked by A, the picnic by B, and the journey home by C. Find: (a) the speed out, (b) the speed back, (c) the average speed for the whole trip.

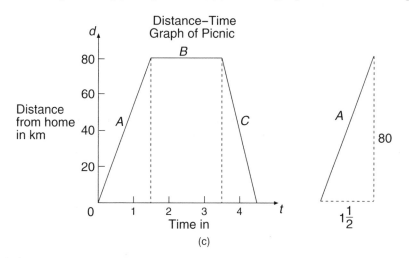

(c)

Solution. (a) The speed for part A is the gradient of the distance–time graph. Write

$$m = \frac{80}{1\frac{1}{2}}$$

Substituting rise = 80 and run = $1\frac{1}{2}$
in the formula for gradient

$$= 53.3 \quad \text{(to 1 dp)}$$

For the journey out the speed is 53.3 km/h.

(b) Similarly, for part C, the gradient is $-80/1 = -80$. For the journey back the speed is 80 km/h. This gradient is of course negative, but since speed cannot be negative (unlike velocity), we ignore the negative sign.

(c) Write

$$\text{Average speed} = \frac{\text{total distance covered}}{\text{total time taken}}$$

$$= \frac{160 \text{ km}}{4\frac{1}{2} \text{ h}}$$

Total time = $1\frac{1}{2} + 2 + 1$ h

$$= 35.6 \text{ km/h} \quad \text{(to 1 dp)}$$

Average speed = 35.6 km/h, including the time for the picnic.

Road Gradients The gradient of a road is defined in the same way as a mathematical gradient. The fraction for the gradient of a road is defined as rise/run. The gradients of roads may be expressed as percentages. For example, if the gradient of a road is $\frac{1}{2}$, the gradient may be expressed as 50%.

Example. Find the average gradient of the viaduct through Otira Gorge from the following data (see figure d). The roadway is 452 meters long. At the start of the viaduct, point A it is 744 m above sea level, and at the end, point B is 797 m above sea level.

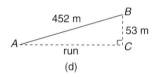

(d)

Solution. The height of point A above sea level is 744 m and the height of B is 797 m, so the length of BC is $797 - 744 = 53$ m. To obtain the gradient of AB the length of AC is required. Write

$AB^2 = BC^2 + AC^2$	Pythagoras' theorem in triangle ABC
$452^2 = 53^2 + AC^2$	Substituting $AB = 452$ and $BC = 53$
$AC^2 = 204{,}304 - 2809$	Rearranging equation, and squaring 452 and 53
$AC = \sqrt{201{,}495}$	Subtracting, and then taking the square root
$AC = 449$ m	To nearest whole number

Now

$$\text{Gradient of the road} = \frac{BC}{AC} \qquad \text{Using the formula for the gradient}$$

$$= \frac{53}{449}$$

The gradient of the Otira viaduct is $53/449$, or

$$\frac{53}{449} \times 100 = 11.8\% \quad \text{(to 1 dp)}$$

Coordinate Geometry Suppose two points A and B have coordinates $A(x_1, y_1)$ and $B(x_2, y_2)$. Then the gradient of the line through the points A and B is given by the formula

$$m = \frac{y_1 - y_2}{x_1 - x_2}$$

Example. Find the gradient of the line through the two points $A(2, -1)$ and $B(-5, 3)$.

Solution. A diagram is not necessary, because there is no need to decide whether or not the gradient is positive or negative, since the formula will take care of it. By comparing A and B with the formula we can see that $x_1 = 2$, $y_1 = -1$ and $x_2 = -5$, $y_2 = 3$. Write

$$m = \frac{y_1 - y_2}{x_1 - x_2}$$ The formula applies equally well with the order reversed,

$$\text{as } m = \frac{y_2 - y_1}{x_2 - x_1}$$

$$m = \frac{-1 - 3}{2 - {}^-5}$$ Substituting values into the formula

$$m = -\frac{4}{7}$$ Simplifying

The gradient of the line through AB is $m = -\frac{4}{7}$.

References: Average Speed, Canceling, Gradient-Intercept Form, Graphs, Parallel, Perpendicular Lines, Trigonometry.

GRADIENT-INTERCEPT FORM

This entry refers to a method of drawing straight-line graphs without using a table of values. The equations of straight-line graphs can be written in the form $y = mx + c$, and this is called the gradient-intercept form of the straight-line graph. The graph of a straight line is often called a linear graph.

In the equation $y = mx + c$, the gradient of the straight-line graph is m and its y intercept is c. When we draw the graph of a straight line we first need to make sure that its equation is written in the form $y = mx + c$ in order to identify the values of m and of c. The graph can be drawn if the values of m and c are known. The whole process is explained in the following examples.

Example 1. Draw the graph of the straight line whose equation is $y = 2x - 3$.

Solution. Compare the equation $y = 2x - 3$ with the formula $y = mx + c$. We can see that the y intercept is $c = -3$, and $m = 2$ (see figure a). The gradient should be expressed as a fraction $m = 2/1$ so that the gradient can be drawn. Always start with fixing the point $c = -3$ on the y-axis, which tells us that the graph crosses this axis at the point $y = -3$. From this point on the y-axis we draw a straight line with a gradient of $m = 2/1$. Then we continue the straight line with a series of gradients $m = 2/1$ until we have an accurate graph of the straight line $y = 2x - 3$. The line should be extended at each end to show that it continues.

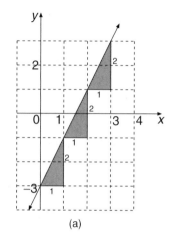

(a)

Example 2. Draw the graph of the straight line $x + 2y = 4$.

Solution. The equation of the straight line must first be rearranged until it is in the form $y = mx + c$. Write

$1x + 2y = 4$	Writing x as $1x$
$2y = -1x + 4$	Subtracting $1x$ from both sides of the equation
$y = -\frac{1}{2}x + 2$	Dividing both sides of equation by 2
$m = -\frac{1}{2}$ and $c = 2$	Comparing with the formula $y = mx + c$.

Start by fixing the point $y = 2$ on the y-axis, and from this point draw a straight line with a gradient of $m = -\frac{1}{2}$ (see figure b). A line with a negative gradient slopes to the left, \. Then continue the straight line with another gradient of $m = -\frac{1}{2}$ until you have an accurate graph of the straight line $x + 2y = 4$.

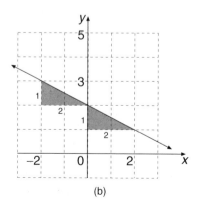

(b)

Special case. A straight line which passes through the origin with a gradient of m is of the form $y = mx$, because $c = 0$.

References: Changing the Subject of a Formula, Gradient, Graphs, Intercepts.

GRADIENT OF A CURVE

The gradient of a curve changes along its length, so a curve does not have one gradient, but a different gradient at each point on the curve. To find the gradient at a point on a curve, we draw a tangent to the curve at that point and find the gradient of the tangent, which is defined to be the gradient of the curve at that point. This process is explained in the following example.

Example. Harry is growing a marrow and hoping to enter it in due course in the garden show. To monitor its growth he decided to weigh it every week, and the first weighing is 50 grams. The marrow's progress is displayed in the following table of values. After 6 weeks the marrow weighed in at 302 grams. Find the rate at which the marrow is growing at the end of week 4.

Weeks (t)	0	1	2	3	4	5	6
Weight in grams (w)	50	68	81	112	151	214	302

Solution. A graph is drawn in the figure showing the weight w of the marrow at the end of week t. To find the rate of increase of the weight of the marrow at the end of week 4, a tangent is drawn to the graph at that point P. The right-angled triangle ABC is completed, with AC being a suitable whole number of units long, which in this case happens to be 4 weeks. The length of BC is now required, and this appears to be 200 grams. Write

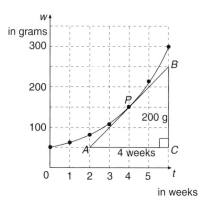

$$\text{Gradient of } AB = \frac{\text{rise}}{\text{run}}$$

$$= \frac{200 \text{ grams}}{4 \text{ weeks}}$$

$$= 50 \text{ grams per week}$$

The rate at which the marrow is growing at the end of day 4 is 50 grams per week.

References: Gradient, Graphs, Rate of Change, Tangent.

GRAM

The gram (abbreviation g) is a small unit of mass; it is one-thousandth of 1 kilogram (kg):

$$1000 \text{ g} = 1 \text{ kg}$$

An important relation is that 1 cubic centimeter of water (1 cm^3) weighs 1 g, provided the water is at a temperature of $4°$ Celsius (1 cm^3 is the same as 1 milliliter). Thus one can also say that 1 cubic meter (1 m^3) of water weighs 1000 kg, or 1 tonne (abbreviation t; note that $1 \, t = 1000 \text{ kg} = 2200$ pounds, whereas 1 U.S. ton $= 2000$ pounds).

References: CGS System of Units, Kilogram, Metric Units.

GRAPHS

For information on statistical graphs see the following entries:

- Bar Graph
- Histogram
- Box and Whisker Graph
- Pie Graph
- Pictogram
- Stem and Leaf Graph
- Frequency Curve
- Line Graph
- Scatter Diagram
- Cumulative Frequency Graph
- Time Series
- Normal Curve

For information on the transformations of the parabola, see the entry Quadratic Graphs. For the following basic curves and how to sketch them, see the following entries:

- Cubic graphs (Quadratic Graphs)
- Distance–time graph (Travel Graphs)
- Ellipse (Ellipse, also Conic Sections)
- Exponential or growth curve (Exponential Curve; see also the present entry)
- Hyperbola (Hyperbola; see also Conic Sections)
- Inequalities of Graphs (Inequality)
- Linear graphs straight-line graphs (Gradient-Intercept Form)
- Logarithmic curve (Logarithmic Curve, also Exponential Curve)
- Parabola (Quadratic Graph; see also Conic Sections)
- Qualitative graphs (Qualitative Graphs)
- Trigonometric curves (Trigonometric Graphs; see also Circular Functions)
- Velocity–time graphs (Acceleration)
- Arrow graphs (Arrow Graph)

In the present entry we shall study the following:

1. How to draw algebraic graphs (which are graphs drawn from algebraic equations) using a table of values.
2. How to draw graphs by sketching.
3. Conversion graphs (see also the entries Conversion and Extrapolation)
4. Circle graphs (see also the entry Conic Sections)
5. Using graphs to solve algebraic equations.

How to Draw Algebraic Graphs An algebraic graph is made up of a set of coordinates plotted on a Cartesian plane and joined by a straight line or by a curve. The set of coordinates is usually written as a table of values, and the coordinates are obtained by substituting into an algebraic equation. This method of drawing graphs is the most basic method and may be used whenever we have little or no idea of what kind of graph will result from the given algebraic equation.

Example 1. Draw the graph of the equation $y = (x - 3)^2$, using values of x from 0 to 6.

Solution. The values of $x = 0, 1, 2, 3, 4, 5, 6$ are entered in a table of values as shown. It is convenient to break down the expression $(x - 3)^2$ into $(x - 3)$ and then square it.

The table of values is now completed.

x	0	1	2	3	4	5	6
$(x-3)$	-3	-2	-1	0	1	2	3
$y = (x-3)^2$	9	4	1	0	1	4	9

The coordinates (x, y) to be plotted on the axes are $(0, 9)$, $(1, 4)$, $(2, 1)$, $(3, 0)$, $(4, 1)$, $(5, 4)$, and $(6, 9)$, which are taken from the table of values. The x-axis is drawn from 0 to 6 and the y-axis from 0 to 9 (see figure a). These two axes form the Cartesian plane. Since the range of y numbers is quite large compared to the range of x numbers, it is convenient to choose a scale of one square to 2 units for y. The graph then takes up less space. It is only necessary to label the y-axis every two units, 0, 2, 4, 6, 8. When the points are plotted they are joined with a smooth curve, using a pencil. This curve is the graph of the algebraic equation $y = (x - 3)^2$, and is called a parabola, or quadratic graph.

Example 2. Sketch the graph of the equation $y = 2^{-x}$ for values of x from -3 to 3.

Solution. The values of x we use are $x = -3, -2, -1, 0, 1, 2, 3$ in the following table of values. The y values can be worked out using the exponent key y^x on the calculator.

x	-3	-2	-1	0	1	2	3
$y = 2^{-x}$ (2 dp)	8	4	2	1	0.5	0.25	0.13

Plotting the coordinates $(-3, 8)$, $(-2, 4)$, $(-1, 2)$, $(0, 1)$, $(1, 0.5)$, $(2, 0.25)$, and $(3, 0.13)$ on the axes gives the exponential graph $y = 2^{-x}$ (see figure b).

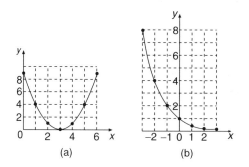

(a)   (b)

How to Draw Graphs by Sketching To learn how to draw sketch graphs of the basic curves, search for the curves by name (see the list at the beginning of this entry). A general description is given here.

We can draw graphs accurately by making a table of values by substituting into an algebraic equation and then plotting the points on a Cartesian plane, as above. Or we can sketch the graphs of algebraic equations by having a good knowledge of their general shape. This may involve finding their intercepts or doing transformations of the basic curves. This sketching of curves is usually a "snapshot" of the general shape and its properties, and usually does not give as accurate a result as drawing them by a table of values. The next example uses intercepts.

Example. Sketch the graph of the parabola $y = (x - 2)(x + 3)$.

Solution. We know that the basic shape of the parabola is as shown in the top part of figure c. A parabola can have up to two x intercepts and one y intercept. To find the x intercepts, substitute $y = 0$ in the equation of the curve, and solve the resulting quadratic equation:

$$0 = (x - 2)(x + 3)$$

$$x = 2 \quad \text{and} \quad x = -3 \qquad \text{See Solving a Quadratic Equation}$$

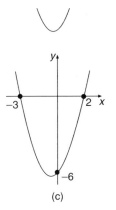

(c)

To find the y intercept, substitute $x = 0$ in the equation of the curve:

$$y = (0 - 2)(0 + 3)$$

$$y = -2 \times 3$$

$$y = -6$$

We fix these three points on the axes, and then sketch the general shape of the parabola through these three points.

The graph shown in figure c is a sketch of the parabola $y = (x - 2)(x + 3)$.

This process is explained more fully in the entry Quadratic Graphs.

Conversion Graphs More information can be found in the entries Conversion and Extrapolation.

Anne's teacher has decided to scale up her students' mathematics marks from a recent examination, because the questions were too hard. The teacher finds the bottom, median, and top marks from the examination and in a table writes the new bottom, median, and top marks she wants her form to have. The original scores in the exam are called raw marks.

	Bottom	Median	Top
Raw marks	20	40	70
Scaled marks	30	60	90

The points (20, 30), (40, 60), and (70, 90) are plotted on a set of axes and joined with two straight-line segments. These lines constitute the conversion graph (see figure d). Anne had a raw mark of 65%, and her scaled mark can be found by locating 65% on the raw marks axis and drawing a vertical line up to the line of the graph. From this point on the graph draw a horizontal line onto the new marks axis, and read off the value 85%. Anne's scaled mark is 85%.

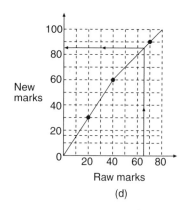

(d)

Circle Graphs More information can be found in the entry Conic Sections. If a circle is drawn that has its center at the origin (0, 0) and with a radius of r, the equation of its graph is $x^2 + y^2 = r^2$. The process of drawing the graph of a circle is demonstrated by the following example.

Example. Draw a circle whose equation is $x^2 + y^2 = 9$.

Solution. Compare the equation $x^2 + y^2 = 9$ with the formula $x^2 + y^2 = r^2$, which is a circle whose center is at the origin (0, 0) and radius r. Write

$$r^2 = 9$$

$$r = \sqrt{9} \qquad \text{Taking the square root}$$

$$r = 3$$

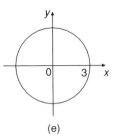

(e)

We now draw the circle of radius 3 and center at (0, 0).

Using Graphs to Solve Algebraic Equations Equations can be solved using algebra; see the entries Linear Equation and Quadratic Equations. We now learn how to solve equations using graphs, but this method is not expected to give an accurate solution to the equation.

The theory for this graphical method is that if we draw two graphs $y = f(x)$ and $y = g(x)$ on the same axes and from the graphs read off the x values where the two graphs intersect , then those values of x are the solutions of the equation $f(x) = g(x)$.

Example. Use a graphical method to solve the equation $2^x = x^2$. This equation can also be written as $2^x - x^2 = 0$.

Solution. The two graphs we need to draw are $y = 2^x$ and $y = x^2$ on the same axes, and the more accurately the graphs are drawn, the more accurate will be the solutions to the equation. We will use a table of values for each graph. First we have

x	-3	-2	-1	0	1	2	3
$y = 2^x$ (2 dp)	0.13	0.25	0.5	1	2	4	8

Plotting the coordinates $(-3, 0.13)$, $(-2, 0.25)$, $(-1, 0.5)$, $(0, 1)$, $(1, 2)$, $(2, 4)$, and $(3, 8)$ on the axes gives the graph of $y = 2^x$:

x	-3	-2	-1	0	1	2	3
$y = x^2$	9	4	1	0	1	4	9

Plotting the coordinates $(-3, 9)$, $(-2, 4)$, $(-1, 1)$, $(0, 0)$, $(1, 1)$, $(2, 4)$, and $(3, 9)$ on the axes gives the graph of $y = x^2$.

The final stage of solving the equation $2^x = x^2$ is to read off from the graph, as accurately as one can, the values of x where the graphs intersect (figure f). There are two values, and the approximate solution of the equation is $x = 2$ and $x = -0.8$.

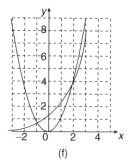

(f)

References: Asymptote, Cartesian Coordinates.

GREATER THAN

The symbol for "greater than" is >. The number 5 is greater than the number −3, since 5 is to the right of −3 on the number line. This is written as 5 > −3.
   The symbol for "less than" is <. The number −3 is less than the number 5, since −3 is to the left of 5 on the number line. This is written as −3 < 5.

Example 1. Find a value for x where $x < -2$.

Solution. We can choose any number to the left of −2 on the number line, say −6 or −2.1. All numbers represented by the continuous line in figure a are solutions. Note there is a "hole" in this line at the point −2 itself, because −2 is not included in the set of solutions.

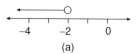

(a)

   The symbol for "less than or equal to" is ≤, and the symbol for "greater than or equal to" is ≥.

Example 2. What are the numbers less than or equal to −2?

Solution. This solution is exactly the same as the solution for "less than −2," except that the number −2 is included in the set of solutions (see figure b). There is no "hole" in the line at the point −2 itself.

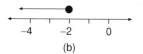

(b)

References: Inequality, Interval, Number Line.

GREATER THAN OR EQUAL TO

References: Greater Than, Inequality, Inequations.

GROSS

This is the number 144. Alternatively, one gross can be described as 12 dozen, where one dozen is 12.

Reference: Duodecimal.

GROUPED DATA

Reference: Arithmetic Mean.

GROWTH CURVE

Reference: Exponential Curve.

H

HALF-LIFE

Reference: Exponential Decay.

HECTARE

This is a unit for measuring land areas such as farms, playing fields, school grounds, parks, etc. The abbreviation for hectare is ha. One hectare is equal in size to 10,000 square meters, which is a square measuring 100 meters by 100 meters. Roughly, two average soccer pitches together have the same area as 1 hectare.

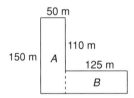

Example. The figure is a sketch of a park, and the measurements are in meters. Find its area in hectares.

Solution. The easiest way to find the area of the park is to split it into two rectangles *A* and *B*, by drawing a dashed line. Write

$$\text{Area of } A = \text{length} \times \text{width}$$

$$= 150 \times 50$$

$$= 7500 \, \text{m}^2$$

$$\text{Area of } B = 125 \times 40 \qquad\qquad \text{The width of } B \text{ is}$$
$$150 - 110 = 40 \, \text{m}$$

$$= 5000 \, \text{m}^2$$

234

Total area of the park = area of A + area of B

$$= 7500 + 5000$$

$$= 12,500 \, \text{m}^2$$

Now write

Area of park in hectares $= 12,500 \div 10,000$ There are $10,000 \, \text{m}^2$ in 1 ha

Area of park $= 1.25$ hectares

One hectare is approximately 2.47 acres. This means that to change hectares into acres we multiply the number of hectares by 2.47 To change acres into hectares we divide number of acres by 2.47.

References: Acre, Area, Metric Units.

HEIGHT

This is the vertical distance from the base of an object to its top. Alternatively, height can be described as the altitude of an object. When finding the areas and volumes of some shapes we need to identify their heights. The height H of a cone is shown in figure a. The slant height, or sloping height, is the length S.

The heights H of two triangles are shown in figure b. The base of each triangle is indicated. Suppose the triangle is twisted and we need to know its height and its base in order to find its area. The height is still the perpendicular distance of the top from the base, as shown in figure c.

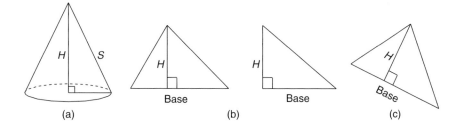

(a)   (b)   (c)

References: Altitude, Cone, Triangle, Vertical.

HELIX

A helix is a three-dimensional curve that is formed when a right-angled triangle is wrapped around a cylinder with no overlapping. Suppose the triangle shown in the figure is very long, so that it can be wrapped around the cylinder a few times. The curve formed by the hypotenuse of the triangle is a helix.

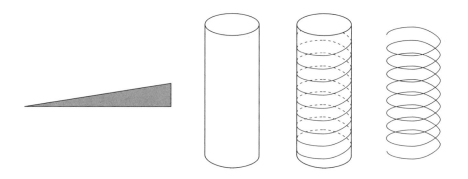

This curve is often confused with a spiral, but they are not the same. Examples of the use of screws in the shape of a helix are glue sticks, lipsticks, and swivel chairs.

References: Cylinder, Right Triangle, Spiral.

HEMISPHERE

A hemisphere is half a sphere and is formed when a sphere is divided equally by a single cut. In figure a, the dome of the cathedral is a hemisphere.

The volume of a hemisphere is half the volume of a whole sphere, but extra care must be taken in finding the surface area of a hemisphere, as shown in the following example.

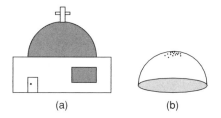

(a)   (b)

Example. An orange with a diameter of 40 cm is cut in half forming two identical hemispheres. Find the surface area of one of the hemispheres of the orange (see figure b).

Solution. The hemisphere is made up of a curved surface area of peel that is half the surface area of a sphere, plus the area of the circular base where the cut was made.

Write

Area of curved surface $= \frac{1}{2} \times 4\pi R^2$ Area curved surface of sphere
$= 4\pi R^2$

$= \frac{1}{2} \times 4 \times \pi \times 20^2$ Substituting radius $= 20$ cm

$= 2513.3$ (to 1 dp)   Using π in the calculator

The area of the curved surface is 2513.3 cm², to 1 dp. Now write

Area of circular base $= \pi R^2$ Formula for the area of a circle

$= \pi \times 20^2$ Substituting radius $= 20$ cm

$= 1256.6$ (to 1 dp)

The area of the circular base is 1256.6 cm².
    The total area of the hemisphere is the sum of the curved surface and the circular base. Thus the surface area of one of the hemispheres of the orange $= 3769.9$ cm².

References: Diameter, Radius, Sphere, Surface Area.

HENDECAGON

This is an 11-sided polygon, also known as an endecagon.

Reference: Polygon.

HENDECAHEDRON

This is a solid with 11 faces.

Reference: Polyhedron.

HEPTAGON

This is a seven-sided polygon and is also called a septagon.

References: Polygon, Septagon.

HEPTOMINO

Reference: Polyominoes.

HERON'S FORMULA

Heron of Alexandria (also known as Hero) lived about the first century AD and derived a formula for the area of any triangle if the length of each side is known. The notation used to describe this formula is the same as the one used in trigonometry, and is briefly explained here.

The vertices of a triangle are labeled A, B, and C, and the lengths of the sides of the triangle opposite those vertices are a, b, and c, respectively. The perimeter of the triangle is $a + b + c$. Half the perimeter of the triangle (semiperimeter) is given by

$$s = \frac{a + b + c}{2}$$

Heron's formula for the area of any triangle is

$$A = \sqrt{s(s - a)(s - b)(s - c)}$$

Example. Jane and her husband, David, are planning to buy a plot of land and build a house on it. The site in which they are particularly interested is triangle-shaped due to a fork in the road (see figure). Jane measures the lengths of the three sides of the triangle section, which are marked on the figure. Use Heron's formula to find the area of the section.

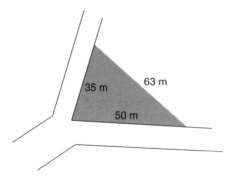

Solution. Write

$$s = \frac{a + b + c}{2}$$

$$s = \frac{35 + 50 + 63}{2} \qquad \begin{array}{l} a = 35, b = 50, c = 63, \text{ but the lengths} \\ \text{can be allocated differently} \end{array}$$

$$s = 74 \, \text{m}$$

Heron's formula for the area of any triangle is

$$A = \sqrt{s(s-a)(s-b)(s-c)}$$

We now substitute for s, a, b, and c:

$$A = \sqrt{74(74-35)(74-50)(74-63)}$$

$A = \sqrt{761904}$ Evaluating with the calculator

$A = 873\,\text{m}^2$ To nearest whole number

The area of the triangular section is $873\,\text{m}^2$.

References: Area, Triangle, Trigonometry.

HEXAGON

A hexagon is a polygon with six sides. Some hexagons are drawn in figure a; the last one is a regular hexagon. The regular hexagon has six sides of equal length, and its six angles are of equal size, which is 120°. It has six axes of symmetry, and the order of rotational symmetry is six (see figure b).

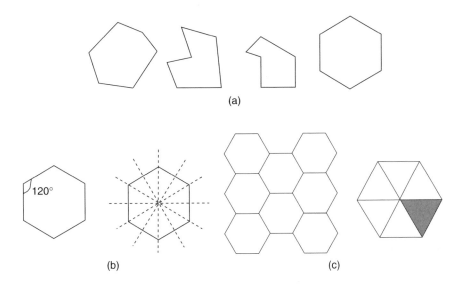

(a)

(b)   (c)

The regular hexagon will tessellate, which means it will form a tiling pattern or mosaic. It is this tiling pattern that bees use to build their honeycomb. The regular hexagon is made up of six equilateral triangles (see figure c).

Thousands of years ago the Babylonians knew how to make a regular hexagon in a circle, in the following way:

Example. Construct a regular hexagon of side 2 cm.

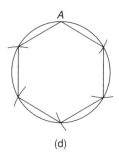

(d)

Solution. Using compasses, draw a circle of radius 2 cm. Starting at a point on the circumference, say point *A*, step off around the circle small arcs with the compasses still of radius 2 cm (see figure d). There will be exactly six steps in getting back to point *A* again. With a ruler, join up the six points where the arcs cut the circle. There are now six chords equal in length to the radius of the circle. This makes a regular hexagon of side 2 cm.

References: Equilateral Triangle, Polygon, Regular Polygon, Symmetry, Tessellations.

HEXAGRAM

If all the sides of a regular hexagon, which is drawn shaded in the figure, are extended, a six-pointed star results which is called a hexagram.

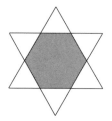

References: Hexagon, Regular Polygon, Star Polygons.

HEXAHEDRON

A hexahedron is a regular polyhedron with six faces, and is another name for a cube.

References: Cube, Polyhedron, Regular Polyhedron.

HEXOMINO

When six squares are joined together edge to edge they form a hexomino. The joining of each pair of squares must be along a complete edge. Altogether there are 35 different combinations of joining together six squares, and each one is called a hexomino. Some of the hexominoes are nets for a cube. Remember that types like the ones in figure a are regarded as the same hexomino.

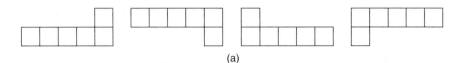

(a)

The 35 hexominoes are shown in figure b.

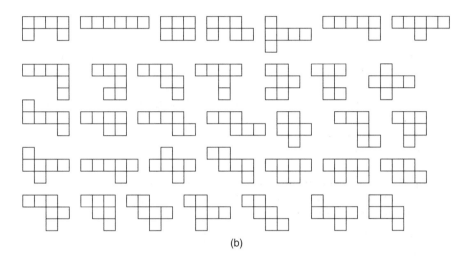

(b)

References: Cube, Pentominoes, Polyominoes, Rectangle, Square.

HIGHEST COMMON FACTOR

Reference: Factor.

HIRE PURCHASE

We use hire purchase (or "buying on time") when we wish to buy an item from a shop, but decide not to pay for it all at once, but pay for it in stages called installments. Provided you pay some of the money to start with, called the deposit, you will be allowed to take the item with you. For example, suppose you wish to buy a CD player for $700. You will be asked to make a payment straightaway of, say, 10% of the price, which is $70. This is called the deposit for the CD player. Then you will be asked to repay the remaining $630 in stages every month over a period of 2 years, say. But the retailer will add some interest to the $630, because he has to wait 2 years before all his money is paid up. That interest may be 15% of the balance owing. Now, 15% of $630 = $94.50, which is the interest the retailer charges.

So the amount you now owe, and have to pay by installments, is $630 + $94.50 = $724.50. If the payments are monthly, over 2 years there will be 24 payments. Each payment will be $724.50 ÷ 24 = $30.1875, which can be rounded off to $30.20.

Summary. You pay a deposit of $70 and then $724.80 by 24 monthly repayments of $30.20. The total cost of the CD player is now $794.80

Reference: Percentage.

HISTOGRAM

A histogram is a statistical graph of a frequency distribution. it resembles a bar graph, but there are differences between them. In a histogram there are no spaces between the columns, because it is a graph of continuous data. In theory the area of the column represents the frequency and so the columns do not have to be of equal widths, but in practice the columns of the histogram are usually equal in width, and the frequency is then represented by the height of the column. The following example illustrates how to draw a histogram.

Example. Madge collected data about the weights of 30 colleagues in her office. She split the weights into class intervals of 5 kg and recorded the data in a frequency table as shown. Draw a histogram of the results.

Solution. The class interval 40– means a weight of 40–45 kg that includes 40 kg, but does not include 45 kg. The use of class intervals is common for continuous data. Madge drew a histogram of the frequency table, ensuring the axes were clearly labeled, there was no space between the columns, and the graph had a title.

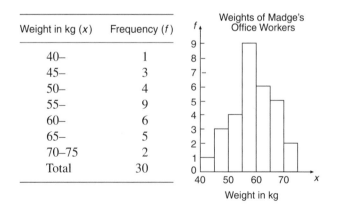

Weight in kg (x)	Frequency (f)
40–	1
45–	3
50–	4
55–	9
60–	6
65–	5
70–75	2
Total	30

References: Bar Graph, Class Interval, Continuous Data, Frequency Distribution.

HORIZONTAL

Horizontal, which is also known as level, is at right angles to the vertical. To define vertical, we can use a device called a plumb line. This is a length of string fastened at one end to a small weight called a plumb bob. The other end of the string is fixed, say to a beam, and if the small weight is allowed to hang freely without swinging, the string will be vertical (see figure). If the beam is at right angles to the string, it will be a horizontal beam.

Vertical
line

Reference: Right Angle.

HORIZONTAL PLANE

Reference: Inclined Plane.

HYPERBOLA

This is a curve that consists of two separate branches, which are sometimes called arms. The equation of a hyperbola is of the type $y = a/x$, or its equation may be written as $xy = a$, where a is a positive or negative number. A hyperbola has asymptotes. They are not part of the graph, but are usually drawn to accompany the graph. Some of the properties of a hyperbola are covered in the example.

Example. Use a table of values to draw the graph of the hyperbola which has the equation $y = 12/x$.

Solution. Suppose we use x values between -12 and 12, which will provide plenty of points to plot. We can reduce the amount of calculation by being selective in our choice of x values. The y value corresponding to each x value is calculated by dividing the x value into 12. The results are recorded in the table of values. When $x = 0$, y is undefined, because a number cannot be divided by zero.

x	-12	-10	-8	-6	-4	-2	-1	0	1	2	4	6	8	10	12
y	-1	-1.2	-1.5	-2	-3	-6	-12	undefined	12	6	3	2	1.5	1.2	1

From the table of values, it can be seen that as the x values get bigger and bigger, the y values get closer and closer to zero. In this way the x-axis, whose equation is

$y = 0$, becomes the horizontal asymptote. In a similar way, the y-axis is the vertical asymptote. More information is provided in the entry Asymptote.

The coordinates are plotted on a graph (see the figure). The graph has half-turn rotational symmetry about the origin. Since the two asymptotes are at right angles, like the sides of a rectangle, the graph is often called a rectangular hyperbola. The gradient of this curve is always negative.

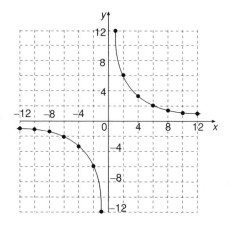

The graph of $y = 3/x$ would be similar to the graph shown and have the same asymptotes, but it would be closer to the axes. The graph of $y = -4/x$ would be drawn in the other two quadrants, and would be further out from the axes than $y = 3/x$, but closer to the axes than $y = 12/x$, and have the same asymptotes. The gradient of $y = -4/x$ is always positive.

When one quantity varies inversely as another quantity, the graph of their relationship is a hyperbola. We say that the two quantities are inversely proportional to each other. For example, suppose you are building a house and deciding how many workers to employ. The more workers you employ, the shorter is the time to complete the building of the house. Time (t) varies inversely with the number of workers (m), and the graph of t against m is a hyperbola. Of course, the graph will only be in the positive quadrant, because you cannot have a negative number of workers!

References: Asymptote, Conic Sections, Gradient, Graphs, Proportion, Quadrants, Table of Values.

HYPOTENUSE

In a right-angled triangle the hypotenuse is the name of the side that is opposite to the right angle. In a right-angled triangle the hypotenuse is always the longest side.

References: Pythagoras' Theorem, Right Angle, Symmetry, Trigonometry.

I

ICOSAHEDRON

This is a polyhedron that has 20 faces; the prefix *icosa* means 20. The regular icosahedron has 20 congruent faces that are equilateral triangles. A regular icosahedron is one of the Platonic solids. The figure shows a regular icosahedron with its net, which is made up of 20 equilateral triangles.

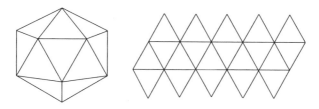

References: Congruent Figures, Equilateral Triangle, Face, Net, Platonic Solids, Polyhedron, Regular Polyhedron.

IMAGE

An image is the figure obtained after applying a transformation to a shape, called an object. The transformations studied in this book are found under the following entries: Enlargement, Reflection, Rotation, Translation.

Reference: Transformation Geometry.

IMPERIAL SYSTEM OF UNITS

This system of units is based on the yard for length and the pound for weight. The main units of length are inch, foot, yard, and mile. The relationships between the units of length are as follows:

$$12 \text{ inches} = 1 \text{ foot}$$

$$3 \text{ feet} = 1 \text{ yard}$$

$$22 \text{ yards} = 1 \text{ chain}$$

$$10 \text{ chains} = 1 \text{ furlong}$$

$$22 \text{ furlongs} = 1 \text{ mile}$$

$$1760 \text{ yards} = 1 \text{ mile}$$

The main units of weight are ounce (oz), pound (lb), hundredweight (cwt), and ton:

$$16 \text{ oz} = 1 \text{ lb}$$

$$112 \text{ lb} = 1 \text{ cwt}$$

$$20 \text{ cwt} = 1 \text{ ton}$$

$$2240 \text{ lb} = 1 \text{ ton}$$

The units for capacity of liquids are pint, quart, and gallon:

$$2 \text{ pints} = 1 \text{ quart}$$

$$4 \text{ quarts} = 1 \text{ gallon}$$

$$8 \text{ pints} = 1 \text{ gallon}$$

For area, there are 4840 square yards in 1 acre.

Reference: Metric Units.

IMPROPER FRACTION

Reference: Fraction.

INCENTER

The incenter is the center of the inscribed circle of a triangle, which is also known as the incircle of a triangle. In the figure the point is the incenter of the triangle. The inscribed circle of a triangle is a circle that lies inside a triangle and touches each of the three sides of the triangle. The radius of the incircle is called the inradius. Each triangle has only one incircle and therefore only one incenter. The three sides of the triangle are tangents to its incircle.

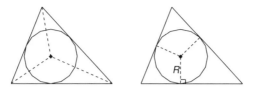

On the left-hand side of the figure the incenter of the triangle is found by constructing the angle bisectors (drawn dashed in the figure) of each of the angles of the triangle. In practice, only two angle bisectors will be needed to find the incenter of a triangle, because the third angle bisector will also pass through the incenter. On the right-hand side of the figure the radius of the incircle is R, and the three radii drawn dashed are at right angles to the sides of the triangle.

References: Angle Bisector, Radius, Tangent.

INCIRCLE

Reference: Incenter.

INCLINED PLANE

This is a plane that is not horizontal or vertical. In the figure the roof of the house is an inclined plane, because it is sloping. The wall of the house is a vertical plane and the floor of the house is a horizontal plane.

Reference: Plane.

INCONSISTENT EQUATIONS

Two equations are inconsistent if they cannot both be true together. We also can say that the two equations are incompatible. For example, the two equations

$$2x + 3y = 5 \quad \text{and} \quad 2x + 3y = 6$$

are inconsistent equations, because they both cannot be true together: $2x + 3y$ cannot be equal to two different quantities at the same time. For example, suppose Nathan and Jacob went to the cafe for lunch. Nathan had two cakes and three buns and his bill was $5. Jacob had exactly the same food, two cakes and three buns, and his bill was $6. If the buns and cakes were the same type for each person, then the results are inconsistent.

Reference: Equations.

INCREASING FUNCTION

Reference: Decreasing Function.

INDEPENDENT EVENTS

Reference: Complementary Events.

INDEPENDENT VARIABLE

Reference: Dependent Variable.

INDEX

The plural of index is indices.

Reference: Exponent.

INDICES

Reference: Exponent.

INDIRECT TRANSFORMATION

Under the entry Congruent Figures, two kinds of congruences are explained:

Directly congruent $\rightarrow$ direct transformations

Oppositely congruent $\rightarrow$ indirect transformations

For a direct transformation the object and the image are directly congruent, and for an indirect transformation the object and the image are oppositely congruent. Reflection is an indirect transformation and direct transformations are rotation, translation, and enlargement.

Another way of recognizing when a transformation is direct or indirect is to check the object and the image in the following way: In the figure the triangle *ABC* is reflected in the mirror line *m* and its image is triangle *A'B'C'*. If you move your finger around the triangle *ABC* in the alphabetical order of the letters you will move in a counterclockwise direction. But if you move your finger around the image triangle *A'B'C'* in alphabetical order you will move in a clockwise direction. This is a property of all indirect transformations. For all direct transformations the alphabetical direction described here is not reversed.

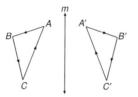

If you place together a pair of your shoes, one of them is an indirect transformation of the other. In this example the transformation is a reflection. If you moved only one shoe to another place, the transformation could never be a single reflection.

References: Congruent Figures, Enlargement, Image, Object, Rotation, Transformation Geometry, Translation.

INEQUALITY

Under this entry we study the graphs of the following inequalities:

- Greater than ($>$)
- Greater than or equal to ($\geq$)
- Less than ($<$)
- Less than or equal to ($\leq$)

We will graph four inequations, and then show how these four inequations can be used to solve a problem. First be certain you know how to draw the graphs of straight lines. Refer to the Gradient-Intercept Form of a straight line.

Example 1. Sketch the following inequations on separate axes:
(a) $x \geq 3$.
(b) $x \leq 5$.
(c) $x + y > 5$.
(d) $x + y \leq 8$.

Solution. To graph an inequation, we first draw the line graph and then shade the region that represents the inequality.

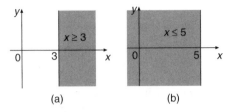

(a)   (b)

(a) Draw the graph of the straight line $x = 3$ and shade all the region to the right of the line, and include the line itself (see figure a). This shaded region represents the inequality x greater than or equal to 3. The line itself is included in the shading by drawing the line in full and not dashed.

(b) Draw the graph of the straight line $x = 5$ and shade all the region which is to the left of the line, including the line itself (see figure b). This shaded region represents the inequality x less than or equal to 5.

(c) Draw the graph of the straight line $x + y = 5$, but draw it dashed, because the line itself is not included in the region, since the inequality is $>$ and not $\geq$. Shade the "greater than" region, which is the entire region above the line, not including the line (see figure c).

(d) Draw the graph of the straight line $x + y = 8$, and draw it in full. Shade the "less than or equal to" region, which is the entire region below the line, including the line itself (see figure d).

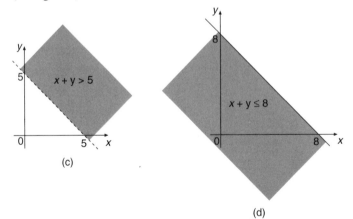

(c)

(d)

These four graphs of inequations will be combined onto a single set of axes to solve the next problem.

Example 2. The Crabtrees family is planning a holiday at Hotel Cheapo, which offers discounts for groups of people. The rules laid down by the hotel, in order to qualify for a discount, are these:

1. There must be more than five people in the group, but no more than eight.
2. There must be at least three adults, but no more than five.

What size groups can go to Hotel Cheapo for a holiday at a discount price?

Solution. Let the number of adults in the group be x and the number of children be y. The inequalities that fit the information in this problem are the same as the inequalities that have just been drawn:

$x \geq 3$	There must be at least three adults
$x \leq 5$	There must be no more than five adults
$x + y > 5$	There must be more than five people
$x + y \leq 8$	There must be no more than eight people

These four graphs will all be drawn on one set of axes. To enable us to concentrate on the intersections of the four inequalities, we will shade the "non" region for each inequality, which leaves the region we want unshaded (see figure e).

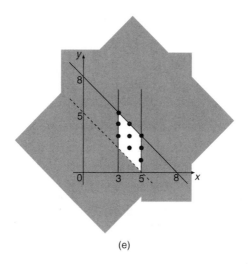

(e)

There are nine points in the unshaded region, and any of those points represents a group of people that qualify for the discount. The coordinates of the nine points are

$$(3, 5), (3, 4), (3, 3), (4, 4), (4, 3), (4, 2), (5, 3), (5, 2), (5, 1)$$

The first coordinate represents the number of adults in the group and the second coordinate represents the number of children in the group. For example, the point $(3, 5)$ is a group of three adults and five children that qualifies for the discount. Note that points on the dashed line are not in the region. This process of solving a problem by graphs of inequalities is called linear programming.

References: Gradient-Intercept Form, Graphs, Greater Than, Inequations, Parabola, Quadratic Graphs.

INEQUATIONS

An inequation is similar to an equation, except that instead of an equal sign joining two expressions, an inequality sign joins them. The equation $2x + 5 = 3$ can be written as an inequation by replacing $=$ by $\leq$ to give $2x + 5 \leq 3$. Inequations can be solved using skills similar to those for solving equations (refer to the entry Balancing an Equation). There is one important difference between balancing equations and balancing inequations. When multiplying (or dividing) both sides of an inequation by a negative number it is necessary to turn around the inequality sign before continuing with the solution. This process is demonstrated in the second example. You may also be interested in graphing the solution to an inequation on a number line, and this is demonstrated in the first example.

Example 1. Solve $\frac{2}{3}x - 2 > -1$.

Solution. Write

$$\frac{2}{3}x - 2 > -1$$

$$\frac{2}{3}x > 1 \qquad \text{Adding 2 to both sides of the inequation}$$

$$x > 1.5 \qquad \text{Dividing both sides by } \frac{2}{3}$$

The solution is graphed on the number line as in the accompanying figure. Note the open circle above 1.5, which indicates that 1.5 is not included in the graph of the answer. Refer to the entry Greater Than.

Example 2. Solve $1 - 2x > 5$.

Solution. Write

$$1 - 2x > 5$$

$$-2x > 4 \qquad \text{Subtracting 1 from both sides of inequation}$$

The next step is to divide both sides of the inequation by -2, and this process includes turning around the inequality. Write

$$x < -2 \qquad \text{The sign } > \text{ is turned around and becomes } <$$

References: Balancing an Equation, Greater Than, Inequality, Number Line.

INFINITE

If something is infinite, then it is not finite. Infinity is a word used to describe the size of a quantity (which may be a set, sequence, or group) which has no limit. Its size has no restriction, it is boundless. For example, the set of positive even numbers $\{2, 4, 6, 8, \ldots\}$ is infinite, because there is no last number in the sequence. In geometry we say that a line has infinite length, whereas a line segment has a finite length.

References: Finite, Line, Line Segment.

INFINITE DECIMALS

These are decimals that have no end to the number of decimal places. There are two kinds of infinite decimals:

    1. *Recurring decimals,* like $\frac{1}{3} = 0.33333333\ldots$, which goes on forever. We usually exclude decimals with recurring zeros, because they are regarded as terminating decimals. For example, 0.3 could be written as $0.300000000\ldots$, but it is not normally regarded as a recurring decimal.

    2. Irrational numbers, like $\pi = 3.14159265\ldots$, which go on forever, but there is no recurring pattern to the numbers.

References: Decimal, Finite Decimals, Irrational Numbers, Pi, Recurring Decimal.

INFINITY

This is the name given to a value that is infinite, too large to be counted, and cannot be calculated. The symbol for infinity is ∞.

References: Infinite.

INFLECTION, POINT OF

Reference: Concave.

INRADIUS

Reference: Incircle.

INSCRIBE

References: Circumscribe, Incenter.

INSTALLMENT

Reference: Hire Purchase.

INTEGERS

The following numbers are explained under this entry:

- Natural numbers
- Whole numbers
- Integers
- Rational numbers
- Irrational numbers
- Real numbers

Natural Numbers The symbol for the natural numbers is N. Another name for them is counting numbers. N is an infinite set of numbers; some of them are listed here:

$$N = \{1, 2, 3, 4, 5, \ldots\}$$

For example, natural numbers are used to count the chapters in a book. There is not a zero chapter.

Whole Numbers The symbol for the whole numbers is W. This set is exactly the same as N, except that it includes zero. W is an infinite set of numbers; some of them are listed here:

$$W = \{0, 1, 2, 3, 4, 5, \ldots\}$$

For example, suppose you were recording the number of people in a room. There could be one, two, three, and so on, but there could also be no people in the room. Whole numbers would be required for this situation.

Integers The symbol for the integers is Z or I. It is an infinite set of numbers; some of them are listed here:

$$I = \{\ldots, -3, -2, -1, 0, 1, 2, 3, \ldots\}$$

These numbers can be described as positive and negative whole numbers, and are sometimes called directed numbers. For each positive number there is an equal and opposite negative number. In real life, directed numbers are used to show that some quantities can fall below zero. A temperature of 10 degrees below zero is written as -10 degrees, and a temperature of 5 degrees above zero is written as $+5$ degrees, or more simply as 5 degrees.

Rational Numbers The symbol for the rational numbers is Q, from the word quotient, which means the result of dividing one number by another. A rational number can be expressed as a quotient of two integers, with the denominator not zero. Q is an infinite set of numbers. In addition to containing all the integers, it also includes all the positive and negative fractions. Some examples of rational numbers are listed here:

$$Q = \{\ldots, -4, 1\tfrac{2}{3}, -2, -\tfrac{7}{8}, 0, 1, 2, 3, \tfrac{1}{4}, 100, \ldots\}$$

It is William's birthday and he has three cakes. He is cutting up the cakes into 20 equal pieces, so each piece is 3/20 of the total number of cakes. This is an example where rational numbers are needed.

Irrational Numbers The symbol for the irrational numbers is Q', which means the complement of Q, and is all those numbers that are not rational numbers, but are real numbers. Some examples of this infinite set of irrational numbers are listed here:

$$\{\pi, \text{ and surds like } \sqrt{2}, \sqrt{7}, \sqrt{10}, \sqrt{34}\}$$

For example, if the two equal sides of a right-angled isosceles triangle are of length 1 unit then the hypotenuse is of length $\sqrt{2}$ units. This is an irrational number.

Real Numbers The symbol for the real numbers is R; this is an infinite set of numbers. This is the family name given to all the numbers, both rational and irrational. The family members of real numbers are related to each other in the following way: The natural numbers are a subset of the whole numbers. The whole numbers are a subset of the integers. The integers are a subset of the rational numbers. The rational and irrational numbers are both subsets of the real numbers. This information is put neatly in the number tree drawn here in figure a.

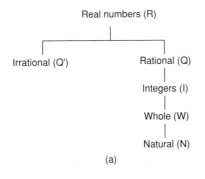

(a)

The rules for adding, subtracting, multiplying, and dividing integers are now explained.

Adding Two Integers Counting "steps" on a number line may used to add integers. If the integer is positive, we count steps to the right, and if it is negative, we count steps to the left.

Example. Work out the answer to $^-3 + {}^+2$.

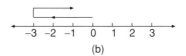

(b)

Solution. Draw a number line (see figure b), or imagine one in your head. Always start at zero and take 3 steps to the left, to represent $^-3$. Then turn and take 2 steps to the right, to represent $^+2$. Where you end up is the answer, which is $^-1$:

$$^-3 + {}^+2 = {}^-1$$

Subtracting Two Integers We can transform the problem from a subtraction into an addition problem by "adding the opposite integer," then doing the addition in the usual way.

Example. Work out $^-3 - {}^+5$.

Solution. Write

$$^-3 - {}^+5 = {}^-3 + {}^-5 \qquad \text{Change subtraction into addition, and replace } {}^+5 \text{ by its}$$
$$\text{opposite, which is } {}^-5$$

$$= {}^-8 \qquad \text{Doing addition in the usual way}$$

Multiplying Two Integers There are two rules to remember for multiplying two integers together:

1. If the signs of the two numbers are the same, either both + or both −, the answer is always positive.
2. If the signs of the two numbers are different, say one is + and one is −, the answer is always negative.

Examples are

$$^-3 \times {}^-4 = {}^+12 \qquad \text{The signs of the two numbers are the same, both negative}$$

$$^-5 \times {}^+3 = {}^-15 \qquad \text{The signs of the two numbers are different, one is negative}$$
$$\text{and the other is positive}$$

If there are more than two numbers being multiplied, multiply them two at a time.

Dividing Two Integers The rules for obtaining the sign of the answer are exactly the same as those for obtaining the sign when multiplying two integers. An example is

$$^+6 \div {}^-4 = {}^-1.5 \qquad \text{The signs of the two numbers are different, and } 6 \div 4 = 1.5$$

It is a straightforward process to use the calculator to add, subtract, multiply, and divide integers.

References: Number Line, Opposite Integers, Pythagoras' Theorem.

INTERCEPT FORM OF A STRAIGHT LINE

Reference: Gradient-Intercept Form

INTERCEPT

An intercept is a point on the x-axis or y-axis where the graph of a straight line or a curve intersects the axis. An intercept can also be the value of the coordinate at that point. More information is given in the entries in the references, especially Graphs.

References: Coordinates, Equations, Gradient-Intercept Form, Graphs.

INTEREST

References: Compound Interest, Simple Interest.

INTERIOR ANGLES

Polygons have interior angles and exterior angles. When we talk about the angles of a polygon we usually mean the interior angles. The figure shown is a pentagon, which has five sides and five angles. The five angles, which are drawn shaded, are all interior angles, because they are inside the polygon. The sum of these five interior angles of the pentagon is 540°.

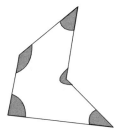

References: Exterior Angle of a Polygon, Polygon.

INTERPOLATION

Reference: Extrapolation.

INTERQUARTILE RANGE

Under this entry we will study the following: median, quartiles (upper and lower), interquartile range, and range.

One set of data is used to explain the different terms. The data we use are the set of test marks, out of 10, scored by Nathan's class of 25 students in a mathematics test. The processes for dealing with a large amount of data are explained in the entry Cumulative Frequency Graph.

The set of 25 marks in the math test, ranked in order of size, is

$$1, 1, 2, 2, 2, 2, 3, 3, 3, 4, 4, 4, 5, 5, 5, 6, 6, 6, 7, 7, 7, 8, 8, 8, 9$$

Median When the marks have been arranged in order of size the median is the middle mark. There are 25 marks, which is an odd number. To find the middle mark, we add 1 to 25 to get 26. Divide 26 by 2 to get 13. The median is the 13th mark, which can be counted from either end of the list. In this example the median mark is 5 (see figure a).

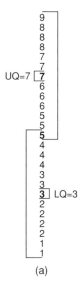

(a)

Lower Quartile (LQ) This is the middle mark of the bottom half of marks. When we are finding the upper and lower quartiles of an odd number of marks the median mark is included in both the bottom half of the list and also in the top half. There are now 13 marks in the bottom half, which is an odd number. To find the middle mark, we add 1 to 13 to get 14. Divide 14 by 2 = 7. The lower quartile is the 7th mark counted from the bottom of the list. The lower quartile mark is LQ = 3.

Upper Quartile (UQ) This is the middle mark of the top half of marks. Using the same process as in finding the LQ gives the upper quartile as the 7th mark counted from the top of the list. The upper quartile mark is UQ = 7.

Interquartile Range (IQR) This is the difference between the UQ and the LQ,

$$IQR = UQ - LQ$$

$$IQR = 7 - 3$$

$$= 4$$

The interquartile range is 4.

The Range This is the difference between the highest mark (9) and the lowest mark (1),

$$Range = 9 - 1$$

$$= 8$$

The range is 8.

It is worth considering how the above calculation would have been affected had there been an even number of students in Nathan's class (say 26 students) instead of 25. The process is slightly different (see figure b). Suppose the 26 marks scored in the test are

1, 1, 2, 2, 2, 2, 3, 3, 3, 4, 4, 4, 4, 5, 5, 5, 6, 6, 6, 7, 7, 7, 8, 8, 9, 9

(b)

Median   There is an even number of marks, so there will be two middle marks, and we have to add the two marks together and divide by 2 to get the median. The two middle marks are found by dividing 26 by 2 to give 13, and counting the 13th and 14th marks from either end of the list. The two middle marks are 4 and 5, so the median will be their mean, which is 4.5.

Quartiles   The median is obtained from the two marks 4 and 5. The lower one, which is 4, is included in the bottom half, and the higher one, which is 5, is included in the top half. Using the same procedures as before, we find that the lower quartile mark is LQ = 3 and the upper quartile mark is UQ = 7.
    The interquartile range = 4 and the range = 8 are calculated as before.

References: Box and Whisker Graph, Central Tendency, Cumulative Frequency Graph, Mean.

INTERSECTING CHORDS

There are three geometry theorems under this heading about the lengths of chords that intersect each other under the following circumstances:

- Inside a circle
- Outside a circle
- When one of the chords is a tangent to the circle

Theorem 1.   When two chords *AB* and *CD* intersect inside a circle at a point *X* (see figure a), the lengths of the chords are related as

$$AX \times XB = CX \times XD$$

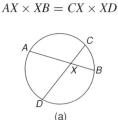

(a)

Example 1.   A knife and a fork are arranged on a dinner plate so that their ends just reach to the edge of the plate, as shown in figure b. The measurements on the figure are in centimeters. Find the length of the fork.

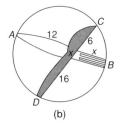

(b)

Solution. Let the length of the prongs of the fork be x cm. Write

$$16 \times 6 = 12 \times x \qquad \text{Intersecting chords inside a circle}$$

$$96 = 12x$$

$$x = 8 \text{ cm} \qquad \text{Dividing both sides of the equation by 12}$$

The length of the fork is 20 cm, because $12 + x = 20$, when $x = 8$.

Theorem 2. For two chords AB and CD that intersect outside a circle at a point X (see figure c; the lines AX and CX are called secants), the lengths of the lines are related as

$$AX \times XB = CX \times XD$$

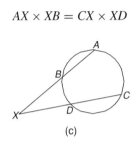

(c)

Example 2. In figure c, $BX = 3$ cm, $AB = 4$ cm, and $XD = 3.5$ cm; find the length of CD.

Solution. Write

$$AX \times XB = CX \times XD \qquad \text{Intersecting chords outside the circle}$$

$$7 \times 3 = CX \times 3.5 \qquad AX = AB + BX$$

$$21 = CX \times 3.5$$

$$CX = 6 \qquad \text{Dividing both sides of the equation by 3.5}$$

$$CD = 2.5 \text{ cm} \qquad CD = CX - XD$$

Theorem 3. In the special case of chords that intersect outside the circle and one of the chords is a tangent to the circle (see figure d), the lengths of a secant and a tangent to the circle are related as

$$AX \times XB = XC^2$$

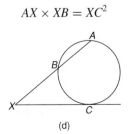

(d)

Example. A watchtower which looks out to sea is of height 100 meters above sea level. The watchman, Bill, is on duty. If the diameter of the earth is 12,800 km, how far can Bill see to the horizon?

Solution. Figure e shows a cross section of the earth. The tower is represented by the line *XB*. Bill can see a distance that is represented by the length of the tangent *XC*. In order to maintain the same units throughout the calculation, the height of the tower must be in kilometers.

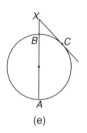

(e)

The height of the tower is given by

$$XB = 0.1 \text{ km} \qquad \text{Divide meters by 1000 to change into kilometers}$$

Now,

$$AX \times XB = XC^2 \qquad \text{Tangent secant theorem}$$
$$12{,}800.1 \times 0.1 = XC^2 \qquad XB = 0.1 \text{ and } XA = 12{,}800 + 0.1$$
$$XC^2 = 1280.01$$
$$XC = 35.78 \quad \text{(to 2 dp)} \qquad \text{Taking square roots}$$

The distance Bill can see to the horizon is 35.78 km.

References: Chord, Circle Geometry Theorems, Quadratic Formula, Secant, Tangent.

INTERVAL

An interval is the set of values between two end points of a line. If the interval includes the end points, it is a closed interval; if it does not include the end points it is an open interval. The interval drawn in figure a is the set of all numbers between 3 and 7, and including 3 and 7. The solid circles at each end of the interval means that those numbers are included. This interval can be written as $3 \leq x \leq 7$ if x is the variable, or as [3, 7]. This is a closed interval.

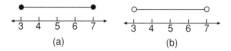

(a)   (b)

The interval drawn in figure b is the set of all numbers between 3 and 7, but excluding 3 and 7. The open circles at each end of the interval means that those numbers are not included. This interval can be written as $3 < x < 7$ if x is the variable, or as (3, 7). This is an open interval.

An interval can be half-closed, which means it is closed at one end and open at the other end. For example, $3 \leq x < 7$ can be written as [3, 7).

Reference: Greater Than.

INVARIANT POINTS

These are points that are not altered by a transformation. They can be described as fixed points under a transformation. For more information see the entries on transformations: Enlargement, Reflection, Rotation, Translation.

Reference: Transformation Geometry.

INVERSE OPERATION

The inverse of an operation is another operation which "undoes" the first operation. This is explained by means of the following example.

Example 1. Think of number: say, 7. The operation on this number is to multiply it by 5: $7 \times 5 = 35$. To undo the operation "multiply by 5" is to do an operation on 35 to take it back to the starting number 7. This operation is "divide by 5." So the inverse operation of "multiply by 5" is "divide by 5," which is the same as "multiply by $\frac{1}{5}$."

In general terms, the inverse operation of multiplying by x is dividing by x or multiplying by $1/x$. Similarly, the inverse operation of adding x is subtracting x or adding $-x$. The inverse operation of squaring is square rooting. The inverse operation in trigonometry is illustrated with the following example.

Example 2. Using a calculator, we can see that $\sin 30° = 0.5$. To undo this operation, we use the inverse sine in the following way:

$$\sin^{-1}(0.5) = 30° \qquad \text{Check this on your calculator}$$

We use inverse operations in solving equations by the "balancing method" and in finding angles in trigonometric problems.

References: Balancing an Equation, Inverse Trigonometric Ratios, Linear Equation, Trigonometry.

INVERSE PROPORTION

Reference: Proportion.

INVERSE RELATIONS

Suppose we have the relation "is older than" between three members of a family: Tom the grandfather, Joanne his daughter, and Luke his grandson. The arrow graph is shown in figure a, and the set of ordered pairs for this relation R is

$$R = \{(\text{Tom, Joanne}), (\text{Joanne, Luke}), (\text{Tom, Luke})\}$$

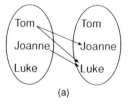

(a)

If the arrow graph and the ordered pairs are reversed, we obtain the inverse relation of R, which has the symbol R^{-1}:

$$R^{-1} = \{(\text{Joanne, Tom}), (\text{Luke, Joanne}), (\text{Luke, Tom})\}$$

In words, the inverse relation is "is younger than."

There is a rule connecting the graph of an algebraic function f with the graph of its inverse f^{-1}, as illustrated in the following example.

Example 1. The graph of the function $f(x) = 2x - 1$ can be drawn using the following ordered pairs (figure b):

$$f = \{(-2, -5), (-1, -3), (0, -1), (1, 1), (2, 3)\}$$

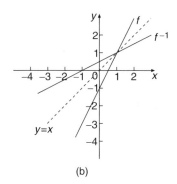

(b)

The graph of the inverse function is drawn by reversing the ordered pairs of f. This is achieved by exchanging x and y values:

$$f^{-1} = (-5, -2), (-3, -1), (-1, 0), (1, 1), (3, 2)\}$$

In figure b the graphs of f and f^{-1} are drawn on the same axes. The rule connecting the graph of f and the graph of f^{-1} is that they are images of each other after reflection in the mirror line whose equation is $y = x$. The mirror line is drawn dashed in the figure. This rule is true for all graphs and their inverses.

The equation of the inverse function f^{-1} can usually be found by rearranging the equation of the function f, as demonstrated in the following example.

Example 2. If the equation of a function is $f(x) = 2x - 1$, find the equation of the inverse function f^{-1}.

Solution. Let the equation of the function be $y = 2x - 1$:

$\quad x = 2y - 1 \qquad$ The rule for finding inverses is to exchange x and y

The next step is to rewrite this equation as $y = ?$, using a process known as changing the subject of a formula:

$\quad x + 1 = 2y \qquad\qquad$ Adding 1 to both sides of the equation

$\quad \dfrac{x + 1}{2} = y \qquad\qquad$ Dividing both sides of the equation by 2

$\quad y = \dfrac{x + 1}{2} \qquad\qquad$ Writing the equation with y on the left-hand side

$\quad f^{-1} = \dfrac{x + 1}{2}$

The inverse of a function f is also a function if it has one-to-one correspondence.

References: Arrow Graph, Correspondence, Function, Mirror Line, Ordered Pairs, Relation.

INVERSE TRIGONOMETRIC RATIOS

References: Inverse Operation, Trigonometry.

INVESTMENT

An investment is lending money to someone, or an institution like a bank, in order to make a profit. For example, Jo invests $1000 in a bank. After 1 year she makes a profit of 10% on her investment: 10% of 1000 = 100, so she has a profit of $100 on her investment.

References: Compound Interest, Interest, Simple Interest.

IRRATIONAL NUMBERS

The symbol for irrational numbers is Q'.

Reference: Integers.

ISOMETRIC

Sometimes we need to draw three-dimensional shapes on two-dimensional paper, but still make them appear as three-dimensional shapes. To do this, we can use isometric paper, which is made up of a regular pattern of dots, or spots. These dots are placed at the vertices of equilateral triangles, as shown in the figure. Isometric means equal measure. This paper is also known as isometric graph paper.

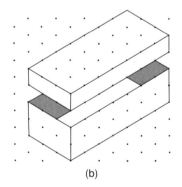

(b)

The drawing shown is of a shoebox with its lid. There are three views of the box that can be seen simultaneously. An isometric drawing does not show perspective. In other words, lines that are the same length do not appear shorter when they are further from your eye.

Reference: Elevation.

ISOMETRIC GRAPH PAPER

Reference: Isometric.

ISOMETRY

References: Congruent Figures, Transformation Geometry.

ISOSCELES TRAPEZIUM

This is a trapezium that has two sides equal in length and two pairs of congruent angles (see figure). It has one axis of symmetry and rotational symmetry of order one. An isosceles trapezium is what is left when an isosceles triangle is removed from a larger isosceles triangle, as shown in the figure. This removal concept is also true for equilateral triangles. In the figure the trapeziums are drawn in an upright position, but they may take up a variety of positions due to rotations and reflections.

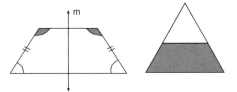

References: Congruent Figures, Equilateral Triangle, Isosceles Triangle, Symmetry, Trapezium.

ISOSCELES TRIANGLE

This is a triangle with two equal sides and two equal angles. The two equal angles are often called the base angles of the isosceles triangle. A geometry theorem is stated as:

Theorem. The base angles of an isosceles triangle are equal.

All isosceles triangles have one axis of symmetry and rotational symmetry of order one. A variety of isosceles triangles are drawn in figure a.

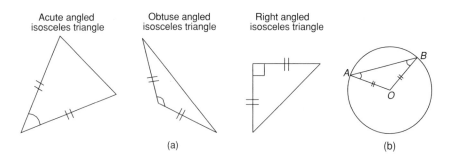

Acute angled
isosceles triangle

Obtuse angled
isosceles triangle

Right angled
isosceles triangle

(a)

(b)

Isosceles triangles frequently occur in circle geometry. In figure b, point O is the center of the circle. Triangle OAB is isosceles because $OA = OB$, since both lengths are radii of the circle. Therefore, angle A = angle B.

Example. Figure c shows a garden roller being pulled along a lawn. The handle makes an angle of $100°$ with the vertical. Find the size of the angle marked x in the figure.

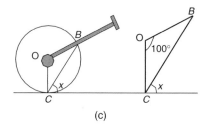

(c)

Solution. Triangle OBC is the important part, and is drawn slightly larger on the right-hand side of the figure. Write

Triangle OBC is isosceles	$OA = OB$, since both lengths are radii of the circle
Angle OCB = angle OBC	Base angles of an isosceles triangle are equal
Angle $OCB = 40°$	Sum of the angles of a triangle = $180°$
$x = 50°$	Radius is perpendicular to the tangent

References: Angle Sum of a Triangle, Angles on the Same Arc, Circle Geometry Theorems, Symmetry, Tangent and Radius Theorem.

K

KILO

Kilo is a prefix meaning 1000. For example, 1 kilometer means 1000 meters. The abbreviation for kilometer is km, where k stands for kilo and m stands for meter. The metric units are mainly based on multiplying some quantity by 1000 (so the unit is preceded by the prefix kilo) or dividing some quantity by 1000 (so the unit is preceded by the prefix milli).

References: CGS System of Units, Metric Units, SI Units.

KILOGRAM

One kilogram means 1000 grams. Kilogram is abbreviated kg, where k stands for kilo and g stands for gram.

References: CGS System of Units, Kilo, Metric Units, SI Units.

KILOLITER

One kiloliter means 1000 liters. Kiloliter is abbreviated kl, where k stands for kilo and l stands for liter.

References: CGS System of Units, Kilo, Metric Units, SI Units.

KITE

A kite is a convex quadrilateral. Convex means that all its angles are less than 180°, and quadrilateral means it has four sides (see figure). A kite has one axis of symmetry, and two pairs of adjacent sides are of equal length. Some kites that people fly at the end of a string on a windy day are kite shaped.

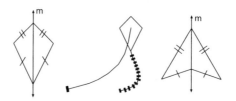

A re-entrant quadrilateral with one axis of symmetry and two pairs of adjacent sides of equal length is called an arrowhead (see far right of figure). An arrowhead has one reflex angle.

References: Adjacent Angles, Convex, Quadrilateral, Reflex Angle, Symmetry.

KONIGSBERG BRIDGE PROBLEM

This is a problem from the history books, which is still popular today. In the year 1736 Leonhard Euler solved a problem which interested the residents of the town of Konigsberg, in Prussia. The town was close to the river Pregel, which divided and flowed around an island. There were seven bridges over the river as shown in the figure. The island is drawn shaded.

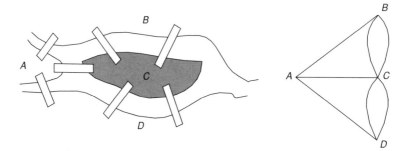

Problem. Start at any point, say point A, and walk over every bridge only once and return to your starting point, without getting your feet wet!

Solution. Euler reduced the bridge problem to a geometry problem. The process of changing a real-life situation into an abstract mathematical problem is called modeling. He constructed a network in which lines represented the routes joining the positions of A, B, C, and D by passing over the bridges, and obtained the right-hand drawing in the figure. In order to understand Euler's explanation, you will need some knowledge of networks, which is found under that entry. The problem now becomes:

Can a network be drawn, without taking the pen from the paper, by starting at point A, traveling over every line once, and finishing up back at the point A? You can pass through a node any number of times, but you cannot redraw a line. In other words, is the figure *traversable*?

A network is traversable if it can be drawn without taking the pen from the paper and traveling over every arc just once. You can pass through a node any number of times, but you cannot redraw a line. Euler discovered the following rule for a network to be traversable, whether you return to your starting point or not:

To be traversable a network must have either two odd nodes or it must have no odd nodes. It may have any number of even nodes.

In Euler's figure representing the bridge problem, the order of the node A is 3, that of node B is 3, that of node C is 5, and that of node D is 3.

This network has four odd nodes, and is therefore not traversable. Euler's conclusion was that the route round the town crossing each of the seven bridges only once and returning to the starting point was impossible.

References: Networks, Traversable Networks.

L

LAWS OF INDICES

Reference: Exponent.

LENGTH

The length of a line segment, which is part of a straight line, is the shortest distance between its end points. A line has one dimension, whereas area is a quantity involving the two dimensions of length and width, and length is usually taken to be greater than width. The principal unit we use for measuring length is the meter. The length of a straight-line segment is measured using a ruler or tape measure. The length of a curved line can be measured by laying a piece of string along it and then measuring the length of string with a ruler, or using a flexible tape measure.

References: Breadth, CGS System of Units, Metric System, SI Units.

LESS THAN (OR EQUAL TO)

References: Greater Than, Inequality.

LEVEL

Reference: Horizontal.

LIKE TERMS

Reference: Algebra.

LIMITS OF ACCURACY

Reference: Error.

LINE

The line segment between A and B, the first line in the figure, is the set of points that move directly between A and B. The line segment AB is written as $\overline{AB}$. The ray AB, the second line in the figure, is the infinite continuation of the straight-line segment between A and B, in the direction of B. It is written as $\overrightarrow{AB}$. The third line in the figure is the infinite continuation in both directions of the straight-line segment between A and B. It is written as $\overleftrightarrow{AB}$.

Summary: The first figure is of a line segment, the second figure is of a ray, and the third figure is of a line.

References: Infinite, Vector.

LINE GRAPH

This entry is confined to the statistical line graph, and is used when the data that are graphed are continuous. The following example explains how to draw and use a statistical line graph. For straight-line algebraic graphs see the entries Gradient-Intercept Form and Graphs.

Example. The temperature, in degrees Celsius, is recorded every 2 hours at East Cape Airport and the data are recorded in a table.

Time of day	2 am	4 am	6 am	8 am	10 am	12 noon	2 pm
Temperature t in degrees C	6	8	12	14	16	18	18

(a) Draw the line graph of these data.
(b) What is the temperature at 7 am?

Solution. (a) The graph is drawn with the time of day as the horizontal axis and the temperature as the vertical axis (see figure). The data from the table are plotted as the coordinates (2, 6), (4, 8), (6, 12), (8, 14), (10, 16), (12, 18), and (14, 18), with 14 representing the time of 2 pm so as not to be confused with 2 am. Then the points are joined up with a series of straight-line segments. The result is a line graph. 0

(b) The temperature at 7 am is found by drawing a vertical line from the time axis to meet the graph, and from that point drawing a horizontal line to the temperature axis. The temperature at 7 am appears to be about 13°C.

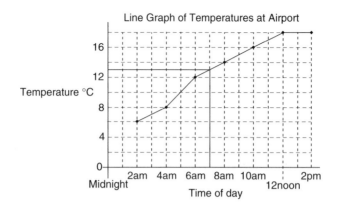

Line Graph of Temperatures at Airport

References: Continuous Data, Gradient-Intercept Form, Graphs, Statistics.

LINE OF BEST FIT

References: Goodness of Fit, Scatter Diagram.

LINE OF SYMMETRY

References: Axis of Symmetry, Mirror Line.

LINE SEGMENT

Reference: Line.

LINEAR EQUATION

This is an equation of degree one, and has one solution. A basic linear equation is of the form $ax + b = c$, where $a, b,$ and c are constants. It may be necessary to read the entries Balancing an Equation and Inverse Operation before continuing. A practical introduction to solving a linear equation is explained in the following demonstration of how an equation is formed.

• Think of a number	x
• Multiply it by 2	$2x$
• Add 3 to the result	$2x + 3$
• If the answer is 8, what is the number?	Solve $2x + 3 = 8$

Starting with the answer, which is 8, we reverse the order of operations, and apply them as the appropriate inverse operations:

- Start with 8
- Subtract 3, which gives 5   "Subtract 3" is the inverse of "add 3"
- Divide by 2, which gives 2.5   "Divide by 2" is the inverse of "multiply by 2"
- The number thought of is 2.5

These ideas are formalized in the following example, using a technique known as "balancing an equation." Whatever operation is done to one side of the equation must be done also to the other side to maintain a "balance." Under this entry we study separately the various methods of solving linear equations using examples, and then solve equations involving more than one of these methods, including equations with brackets.

Method 1. Solve $x - 8 = 2$. Write

$$x - 8 + 8 = 2 + 8$$

Adding 8 to both sides of the equation reduces -8 to zero on the left-hand side. Thus

$$x = 10$$

Method 2. Solve $x + 3 = 2.9$. Write

$$x + 3 - 3 = 2.9 - 3$$

Subtracting 3 from both sides of the equation reduces $+3$ to zero on the left-hand side. Thus

$$x = -0.1$$

Method 3. Solve $7x = 19$. Write

$$\frac{7x}{7} = \frac{19}{7}$$

Dividing both sides of the equation by 7 reduces 7 to 1 on the left-hand side. Thus

$$x = 2.7 \qquad 19 \div 7 = 2.7 \quad \text{(to 2 dp)} \quad \text{or} \quad 2\tfrac{5}{7}$$

Method 4. Solve $x/5 = 0.7$. Note that $x/5$ is the same as $\frac{1}{5}x$. Write

$$5 \times \frac{x}{5} = 0.7 \times 5$$

Multiplying both sides of the equation by 5 reduces $\frac{1}{5}$ to 1 on the left-hand side. Thus

$$x = 3.5 \qquad 0.7 \times 5 = 3.5 \quad \text{or} \quad 3\tfrac{1}{2}$$

We now solve equations that involve more than one of the above methods. When equations have more than one operation, as in those that follow, the order of operations for solving them is the reverse of the order of operations for forming the equations.

In the first example below the equation is formed by multiplying x by 3 and then dividing by 5. To solve the equation, we reverse the order and apply the inverse operations. The inverse of dividing by 5 is multiplying by 5. The inverse of multiplying by 3 is dividing by 3.

Example 1. Solve $\frac{3}{5}x = 4$.

Solution. Write

$$5 \times \tfrac{3}{5}x = 5 \times 4 \qquad \text{Multiply both sides by 5}$$

$$3x = 20$$

$$\frac{3}{3}x = \frac{20}{3} \qquad \text{Divide both sides by 3}$$

$$x = 6.7 \qquad 20 \div 3 = 6.7 \quad (\text{to 1 dp}) \quad \text{or} \quad 6\tfrac{2}{3}$$

Example 2. Solve

$$\frac{x}{3} + 2 = \frac{1}{2} \qquad \text{The two operations are add 2 and divide by 3}$$

Solution. Write

$$\frac{x}{3} + 2 - 2 = \frac{1}{2} - 2 \qquad \text{Subtract 2 from both sides of the equation}$$

$$\frac{x}{3} = -1\tfrac{1}{2}$$

$$3 \times \frac{x}{3} = -1\tfrac{1}{2} \times 3 \qquad \text{Multiply both sides by 3.}$$

$$x = -4\tfrac{1}{2}$$

Example 3. Solve

$$3x + 17 = -6 \qquad \text{The two operations are add 17 and multiply by 3}$$

Solution. Write

$$3x + 17 - 17 = -6 - 17 \qquad \text{Subtracting 17 from both sides of the equation}$$

$$3x = -23$$

$$\frac{3}{3}x = \frac{-23}{3} \qquad \text{Dividing both sides of equation by 3}$$

$$x = -7.7 \qquad -23 \div 3 = -7.7 \quad \text{(to 1 dp)} \quad \text{or} \quad -7\tfrac{2}{3}$$

Example 4. Solve

$$\frac{3x - 5}{7} = 4$$

Solution. The three operations on x, in order to form this equation, are multiply by 3, subtract 5, and divide by 7. In solving the equation this order is reversed, and of course the appropriate inverse operations are applied. Write

$$\frac{(3x - 5)}{7} = 4 \qquad \text{Inserting brackets shows that all of } 3x - 5 \text{ is divided by 7}$$

$$\frac{7(3x - 5)}{7} = 4 \times 7 \qquad \text{Multiplying both sides of the equation by 7}$$

$$(3x - 5) = 28$$

$$3x - 5 = 28 \qquad \text{The brackets are no longer needed}$$

$$3x - 5 + 5 = 28 + 5 \qquad \text{Adding 5 to both sides of the equation}$$

$$3x = 33$$

$$\frac{3}{3}x = \frac{33}{3} \qquad \text{Dividing both sides of the equation by 3}$$

$$x = 11$$

Example 5. Solve $6x - 3 = 2(x + 5)$.

Solution. Write

$$6x - 3 = 2x + 10 \qquad \text{Expanding the brackets}$$

$$6x - 3 + 3 = 2x + 10 + 3 \qquad \text{Adding 3 to both sides of the equation}$$

$$6x = 2x + 13$$

$$6x - 2x = 2x + 13 - 2x \qquad \text{Subtracting } 2x \text{ from both sides to get } x\text{'s on the left side of equation}$$

$$4x = 13 \qquad\qquad 6x - 2x = 4x \text{ and } 2x - 2x = 0$$

$$\frac{4x}{4} = \frac{13}{4} \qquad\qquad \text{Dividing both sides by 4}$$

$$x = 3.25$$

Cross multiplying is a useful way of eliminating fractions in equations, and applies to equations that have one fraction on each side of the equals sign. Cross multiplying is based upon the fact that if $\frac{a}{b} = \frac{c}{d}$, then $ad = bc$.

Example. Solve

$$\frac{3x}{5} = \frac{x - 2}{4}$$

Solution. Write

$$3x \times 4 = 5(x - 2) \qquad \text{Inserting brackets and cross multiplying}$$

$$12x = 5x - 10 \qquad \text{Expanding the brackets}$$

$$12x - 5x = 5x - 10 - 5x \qquad \text{Subtracting } 5x \text{ from both sides}$$

$$7x = -10$$

$$\frac{7x}{7} = \frac{-10}{7} \qquad \text{Dividing both sides by 7}$$

$$x = -1.4 \qquad -10 \div 7 = -1.4 \quad \text{(to 1 dp)} \quad \text{or} \quad -1\tfrac{3}{7}$$

References: Balancing an Equation, Brackets, Degree, Equations, Inverse Operation, Operations.

LINEAR GRAPH

A linear graph is a straight-line graph.

Reference: Gradient-Intercept Form.

LINEAR PROGRAMMING

Reference: Inequality.

LITER

The liter (abbreviation l) is a unit of volume and is used for expressing the volume of liquids and capacity in general. For example, in countries where the metric system is used, milk is often sold in 1-liter quantities. There are 1000 milliliters (abbreviation ml) in 1 liter. The size of 1 liter corresponds to a cube of sides 10 cm. This means that 1 liter $= 1000$ cubic centimeters (cm^3). Other useful equivalent relationships are as follows:

$$1 \text{ milliliter} = 1 \text{ cubic centimeter}$$

$$1000 \text{ liters} = 1 \text{ cubic meter}$$

A volume of 1000 liters of water has a mass of 1 tonne, provided the water is at $4°$ C (for the distinction between the unit tonne and the U.S. ton, see the entry Gram). Rainfall of 1 millimeter is equivalent to 1 liter of water per square meter.

References: CGS Units, Metric Units, SI Units.

LOCUS

The plural of locus is loci. A locus is a path made up of a set of points, and the position of each point in the path obeys a certain rule. This concept is demonstrated in the following example (see figure a). Amanda is training her horse, George, for a show. She stands in one place, at the point A, holding the end of a rope, and the other end of the rope is tied to George (G). As George trots around her, Amanda turns so that she is always facing him, but keeps the rope tight and the same length. Different positions of George make up the set of points, and the path of George is the locus of G. In this example the locus of G is a circle. The rule for this locus is that each point G is the same distance from the fixed point A. This locus is made up of an infinite number of positions of George.

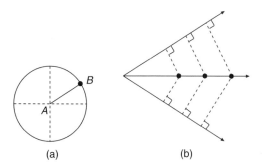

(a)   (b)

Like the circle above, many lines and curves in mathematics can be defined in terms of a locus. Some of them are described here.

The line that is the bisector of an angle can be defined as the locus of points that are equidistant from the "arms" of the angle. In figure b, three of the points of the locus, each at an equal distance from either arm, are marked on the angle bisector.

Example. Jo has a dog, Spot, who is occasionally tied up in the garden with the end of his leash free to pass along a length of wire AB which is fixed to the ground at each end. The length of the leash is 2 meters and the length of the wire is 6 meters. Draw the region in which Spot is free to move, assuming the leash does not get tangled up with the wire.

Solution. This is a locus problem where the set of points covers a region, rather than represents a line. First we draw the locus of Spot at his furthest positions from the wire *AB* (see figure c). The locus of the furthermost position of Spot is made up of two semicircles each of radius 2 meters, and two straight and parallel line segments each of length 6 meters. The region over which Spot is free to move is shaded.

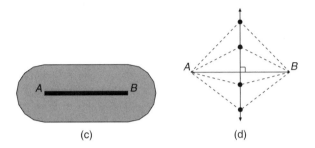

(c)   (d)

The perpendicular bisector of the line segment joining two points *A* and *B* is the locus of points that are equidistant from the points *A* and *B*. This locus is the arrowed line in figure d, and four points of the locus showing the equal distances are marked on the bisector. If there are fires of equal intensity at the points *A* and *B* and you wish to walk between the fires with the least chance of getting burnt, then the route you would take would be the perpendicular bisector of the line segment *AB*.

References: Angle Bisector, Circle, Perpendicular Bisector.

LOGARITHMIC CURVE

Reference: Exponential Curve.

LOWER QUARTILE

References: Cumulative Frequency Graph, Interquartile Range.

LOWEST COMMON DENOMINATOR

Reference: Fractions.

LOWEST COMMON MULTIPLE

Abbreviation LCM. Also known as the least common multiple.

References: Factor, Fractions.

M

MAGIC SQUARE

A magic square is a square grid of counting numbers that is arranged in such a way that the sum of the numbers in each row, or column, or diagonal in the grid is the same total. A 4 by 4 square grid is shown in figure a. If we add up the numbers in each row, or each column, or in each diagonal, we get the same answer, 34. The numbers used in the grid are 1, 2, 3, 4, 5, 6, 7, 8, 9, 10, 11, 12, 13, 14, 15, and 16, since it is a 4 by 4 magic square.

16	3	2	13
5	10	11	8
9	6	7	12
4	15	14	1

(a)

10	1	16	7
15	8	9	2
3	12	5	14
6	13	4	11

(b)

There is more than one way of arranging the numbers 1–16 in a 4 by 4 grid to form a magic square. Another possible arrangement is shown in figure b.

MAGNIFICATION

This is another name for enlargement.

Reference: Enlargement.

MAGNITUDE OF A VECTOR

Reference: Vector.

MAJOR ARC

Reference: Arc.

MAJOR SECTOR

Reference: Sector of a Circle.

MAJOR SEGMENT

Reference: Segment of a Circle.

MANY–MANY CORRESPONDENCE

Reference: Correspondence.

MANY–ONE CORRESPONDENCE

Reference: Correspondence.

MAPPING

Reference: Correspondence.

MAPS ONTO

Reference: Correspondence.

MAXIMUM VALUE

Suppose you are in a car and your speed is increasing to a high speed at point A as shown in the figure, until you approach a bend. Looking ahead, you see the corner and start to slow down from this high speed. As you travel around the corner your speed drops to a minimum, at point B. You accelerate out of the corner, picking up speed as you go. At point C you are traveling fast again. The figure shows a map of the corner, and below that is a velocity–time graph showing how your car speed changes as you travel around the corner. There are three points on the graph corresponding to the three points A, B, and C on the map.

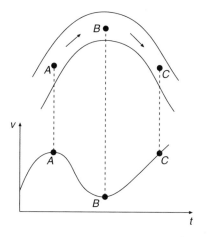

At point *A* the car has a local maximum speed. It is not the overall greatest speed of the car, because further on, after point *C*, it is going faster. That is why we say that at point *A* the car has a local maximum speed. Similarly, at point *B* the car has a local minimum speed, because it could be traveling at a lower speed some other time.

The local maximum and local minimum points on a graph are often simply called maximum and minimum points, respectively.

References: Concave, Turning Points.

MEAN

The mean of a set of data can be found using a scientific calculator. Since calculators vary from one manufacturer to another and may require different instructions, refer to the handbook for instructions. The explanation of mean is found under the entries Arithmetic Mean and Central Tendency.

References: Arithmetic Mean, Central Tendency.

MEAN SQUARE DEVIATION

Reference: Standard Deviation.

MEASUREMENT

A measurement is the size or quantity of something. Here are some examples of measurements and the units used:

- The mass of a car is 850 kilograms.
- The length of a garden is 40 meters.

- The area of a farm is 2450 hectares.
- My lunch break is 50 minutes in duration.
- The temperature of the swimming pool is 21°C.
- The milk jug has a capacity of 800 milliliters.

Reference: Metric Units.

MEDIAN OF A SET OF DATA

References: Central Tendency, Cumulative Frequency Graph, Interquartile Range.

MEDIAN OF A TRIANGLE

The median of a triangle is the line drawn from a vertex of the triangle to the midpoint of the side that is opposite to that vertex. A triangle has three medians, which intersect in a point called the centroid of the triangle, marked G in the figure. The centroid G divides each median of the triangle in the ratio 2:1. In the drawing the line VM is the mediator of the triangle VAB, where M is the midpoint of the side AB which is opposite the vertex V.

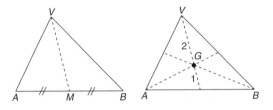

One of the properties of the median of a triangle is that it divides the triangle VAB into two triangles VAM and VBM that are of equal area. If you were asked to divide a triangle into two "equal" parts using a straight line, then the median would be a good way of achieving it.

Reference: Ratio.

MEDIATOR

The mediator of a line segment is the line that bisects the line segment at right angles. In the figure, the mediator of the line segment AB is also the line that is called the axis of symmetry of the line segment AB. The mediator is also known as the perpendicular bisector of the line segment AB.

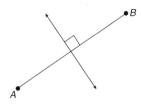

References: Axis of Symmetry, Bisect, Center of a Circle, Line Segment, Locus, Perpendicular Bisector, Right Angle.

METER

The abbreviation of meter is m. The meter is the basic unit of length in the Système International d'Unités, abbreviated SI units. In 1983 the standard for the length of 1 meter was defined to be the length of the path traveled by light in a vacuum during a time interval 1/299,792,458 of a second.

The lengths equivalent to 1 meter in the imperial system of units are 39.3700 inches, 3.2808 feet, and 1.0936 yards, to four decimal places.

In athletic events the lengths of races are measured in meters, and distances thrown and jumped in the field events are measured in meters.

References: CGS System of Units, Imperial System of Units, Metric Units, SI Units.

METRIC UNITS

The metric unit most commonly used to measure length is the meter (m), to measure capacity is the liter (l), and to measure mass is the gram (g) (mass is often incorrectly called weight).

Using the prefixes milli (division by 1000) and kilo (multiplication by 1000), we form other metric units that are related to meter, liter, and gram. There are other prefixes (see SI Units) and one that is frequently used is centi (division by 100). We now look at the relationships among different units of length, area, capacity, volume, and mass.

Length The units of length are:

- Millimeter (mm), which is 1 meter (m) divided by 1000, or 1 centimeter divided by 10
- Centimeter (cm), which is 1 meter divided by 100
- Kilometer (km), which is 1 meter multiplied by 1000

Examples:

- Millimeters are used to measure very small distances, like the length of your fingernail, say 8 mm.
- Centimeters are used to measure small distances, like the length of your arm, say 23 cm.
- Meters are used to measure medium-sized distances, like your height, say 1.7 m.
- Kilometers are used to measure long distances, like that between towns, say 27 km.

Example. Convert the length of a roadway of 3.5 km into centimeters.

Solution. The first step is to change 3.5 km into meters by multiplying the number of kilometers by 1000:

$$3.5 \text{ km} \times 1000 = 3500 \text{ m}$$

The second step is to change 3500 m into centimeters by multiplying the number of meters by 100:

$$3500 \times 100 = 350,000$$

The length of the roadway is 350,000 cm.

Area The units of area are:

- Square millimeter (mm^2), which is a square that measures 1 mm by 1 mm.
- Square centimeter (cm^2), which is a square that measures 1 cm by 1 cm, or 10 mm by 10 mm.
- Square meter (m^2), which is a square that measures 1 m by 1 m, or 100 cm by 100 cm.
- Square kilometer (km^2), which is a square that measures 1 km by 1 km, or 1000 m by 1000 m.
- Hectare (ha), which is equivalent to an area of 10,000 m^2, and is a square that measures 100 m by 100 m. There are 100 ha in 1 km^2.

Examples:

- Square millimeters are used to measure very small areas, like the area of your fingernail, say 40 mm^2.
- Square centimeters are used to measure small areas, like the area of a page of a book, say 300 cm^2.

- Square meters are used to measure medium-sized areas, like the area of a garden, say 950 m^2.
- Square kilometers are used to measure large areas, like the area of a country, for example, 9,363,169 km^2 is the area of the United States.
- Hectares are used to measure fairly large areas, like the area of a farm, 800 ha.

Example. Convert the area of a farm of 800 ha into square kilometers.

Solution. There are 100 ha in 1 km^2. Write

$$800 \div 100 = 8$$

The area of the farm is 8 km^2.

Capacity This is the volume of liquid that a container holds. The units of capacity are:

- Milliliter (ml), which is a cube that measures 1 cm by 1 cm by 1 cm, or 10 mm by 10 mm by 10 mm (remember that a capacity of 1 ml is the same as the volume of 1 cm^3.)
- Liter (l), which is a cube that measures 10 cm by 10 cm by 10 cm.
- Kiloliter (kl), which is a cube that measures 1 m by 1 m by 1 m, or 100 cm by 100 cm by 100 cm.

Examples:

- Milliliters are used to measure very small capacities, for example, 1.5 ml of ink in a ballpoint pen, or 5 ml of medicine in a full teaspoon.
- Liters are used to measure medium capacities, like the quantity of water in a bath, say 830 liters.
- Kiloliters are used to measure large capacities, like the quantity of water in a lake, say 300,000 kiloliters.

The relationships between them are

$$1000 \, ml = 1 \, l$$

$$1000 \, l = 1 \, kl$$

$$1,000,000 \, ml = 1 \, kl$$

Example. A jug holds 2.3 l of milk. Convert this capacity into milliliters.

Solution. We multiply the 2.3 l by 1000 to change it into milliliters:

$$2.3 \times 1000 = 2300$$

The jug holds 2300 ml.

Volume The units of volume are:

- Cubic millimeter (mm^3), which is a cube that measures 1 mm by 1 mm by 1 mm.
- Cubic centimeter (cm^3), which is a cube that measures 1 cm by 1 cm by 1 cm.
- Cubic meter (m^3), which is a cube that measures 1 m by 1 m by 1 m.

Examples:

- Cubic millimeters are used to measure very small volumes, like the volume of a garden pea, say 250 mm^3.
- Cubic centimeters are used to measure medium-sized volumes, like the volume of a tennis ball, say 750 cm^3.
- Cubic meters are used to measure large volumes, like the volume of a bedroom, say 32 m^3.

The relationships between them are

$$1000 \, mm^3 = 1 \, cm^3$$
$$1{,}000{,}000 \, cm^3 = 1 \, m^3$$

The fact that 1 million cubic centimeters = 1 cubic meter is particularly interesting, because it gives us an idea about how big a million is. Another interesting fact is that 1 ml of water weighs 1 g at a temperature of 4°C.

Mass The units of mass are:

- Milligrams (mg), which used to measure the mass of very small quantities, like the mass of a pin.
- Grams (g), which used to measure the mass of small quantities, like the mass of a ballpoint pen, say 25 g.
- Kilograms (kg), which are used to measure the mass of larger quantities, like the mass of a sack of potatoes, say 20 kg.
- Tonnes (t), which are used to measure the mass of very large quantities, like the mass of a truckload of wood, say 2 t (for the relationship between tonnes and U.S. tons, see the entry Gram).

The relationships between them are

$$1000 \text{ mg} = 1 \text{ g}$$
$$1000 \text{ g} = 1 \text{ kg}$$
$$1000 \text{ kg} = 1 \text{ t}$$

Example. Convert the mass of a truckload of wood of 2.4 tonnes to kilograms.

Solution. We multiply the number of tonnes by 1000 to convert to kilograms:

$$2.4 \times 1000 = 2400$$

The mass of the truck of wood is 2400 kg.

References: Système International d'Unités, SI Units.

MIDORDINATE RULE

This is a rule for finding the approximate area between a curve and the x-axis. The process is explained in the example.

Example. Find the approximate area in the figure of the region enclosed by the curve $y = x^2$, the x-axis, and the vertical lines $x = 1$ and $x = 2$.

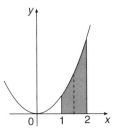

Solution. The region is shaded in the figure. The dashed line is a vertical line through the point $x = 1.5$, which is midway along the width of the shaded strip. The length of the dashed line, called the midordinate, is found by substituting $x = 1.5$ into the equation of the curve $y = x^2$.

$$y = (1.5)^2$$
$$y = 2.25$$

The coordinates of the point where the dashed line meets the curve are (1.5, 2.25). Now write

$$\text{Approximate area of shaded region} = \text{width of region} \times \text{value of midordinate of}$$
$$\text{the region}$$

$$= 1 \times 2.25$$

$$= 2.25$$

The approximate area of the shaded region is 2.3 square units, to 1 dp. A more accurate result can be obtained by splitting the shaded area into two or more strips and using the midordinate rule on each strip. The greater the number of strips, the more accurate is the answer.

A closer look at the process will reveal that the midordinate method is based on regarding the shaded area as a trapezium and then finding its area.

Reference: Trapezium.

MILE

The mile is a measurement of length and is an imperial unit. One mile is approximately 1.609 kilometers, or 1609 meters. An approximate guide for quick conversions between the two units is to say that 5 miles is equivalent to about 8 km.

References: Imperial System of Units, Kilometer, Metric Units.

MILLI

This is a prefix meaning one-thousandth (1/1000) part of some quantity. For example, one-thousandth part of a meter is called a millimeter, which is abbreviated mm. In this abbreviation, the first m stands for milli and the second m stands for meter. The metric units are mainly based on dividing some quantity by 1000 (prefix milli) or multiplying some quantity by 1000 (prefix kilo).

References: Meter, Metric Units, Millimeter, SI Units.

MILLILITER

One milliliter is 1/1000 of a liter, and is abbreviated ml. One milliliter is the same size as 1 cubic centimeter (cm^3) and is a useful unit for measuring medicine. A teaspoon of medicine is about 5 ml.

References: Metric Units, Milli.

MILLIMETER

One millimeter is 1/1000 of a meter, and is abbreviated mm. The first m stands for milli and the second m stands for meter. Builders and architects use millimeters as the unit for measuring distances in preference to centimeters. If you are measuring and cutting wood to build a house, the most accuracy you can expect is to the nearest millimeter.

References: Metric Units, Milli.

MINIMUM VALUE

Reference: Maximum Value.

MINOR ARC

Reference: Arc.

MINOR SECTOR

Reference: Sector of a Circle.

MINOR SEGMENT

Reference: Segment of a Circle.

MINUTE

There are two uses for the unit minute:

1. A minute is an extremely small angle, and is 1/60 of 1 degree. Written as a decimal, 1 minute = 0.017 degree, to 3 dp. The symbol for 1 minute is 1′.

2. A minute is also a measure of time, and is 1/60 of 1 hour. One minute is 0.017 hour, to 3 dp. The minute is subdivided again into seconds, where 1 second is 1/60 of 1 minute. This means that there are 60 seconds in 1 minute, 60 minutes in 1 hour, and 3600 seconds in 1 hour.

Reference: Degree.

MIRROR LINE

If an object is reflected in a mirror to produce an image, then the object and the image are mirror images of each other. In the drawing figure, m is the mirror line, the flag *F* is the object, and the image is the flag *F′*. The mirror line is also known as the axis of symmetry of the whole drawing, and is called the mediator of the two flags.

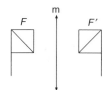

References: Axis of Symmetry, Bisect, Mediator.

MIXED NUMBER

Reference: Fraction.

MKS SYSTEM

This system of units was based upon the meter, the kilogram, and the second. It became the SI system of units.

References: CGS System of Units, SI Units.

MÖBIUS STRIP

The Möbius strip, or band, is a continuous surface that has only one side and only one edge. It is named after A. F. Möbius (1790–1868).

A Möbius strip is extremely simple to make. Take a strip of paper in the shape of a long, thin rectangle, say 30 cm long (see figure a). Give one end a twist through half a turn. Now bring the two ends together and glue them onto each other forming a join at *A*. The resulting shape is a Möbius strip.

Using a crayon, color one side of the paper that makes up the Möbius strip, and you will discovery that you have colored all the paper. If you place your finger at the point *A* and traverse the whole surface of the strip, your finger will arrive back at the same point *A* after traveling a distance of twice 30 cm = 60 cm. This shows that the Möbius strip has only one surface, unlike the original rectangle, which had two surfaces. The Möbius strip also has only one edge.

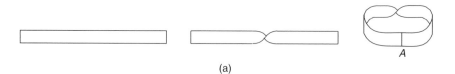

(a)

The property of the Möbius strip of having only one surface can be made use of in machines that are belt-driven. With a normal belt only the inside of the belt is in contact with the drive shaft. If the belt is in the shape of a Möbius strip, the surface in contact with the drive shaft is twice as long and therefore the belt will last longer, because there is less wear. Because of the twist in the belt, the two shafts it connects turn in opposite directions.

(b)

If you cut a Möbius strip into two pieces by cutting with a pair of scissors along the dashed centerline as shown in figure b, you obtain two Möbius strips, which are linked together like a chain.

MODE

References: Bimodal, Central Tendency.

MODELING

Reference: Konigsberg Bridge Problem.

MODULUS

Reference: Absolute Value.

MONOMINO

Reference: Polyominoes.

MOSAICS

Reference: Tessellations.

MOVING AVERAGES

Reference: Time Series.

MULTIPLE

Reference: Factor.

MULTIPLICAND

Suppose two numbers, say 6 and 8, are multiplied together: $6 \times 8 = 48$. The first number (6) is the multiplicand, the second number (8) is the multiplier, and the answer (48) is the product.

MULTIPLIER

Reference: Multiplicand.

MULTIPLYING FRACTIONS

Multiplying fractions is easier than adding and subtracting them. When we multiply two fractions together we multiply the numerators together, multiply the denominators together, and write the result as a single fraction.

Example 1. Multiply the two fractions $\frac{5}{8} \times \frac{4}{15}$.

Solution. Write

$$\frac{5}{8} \times \frac{4}{15} = \frac{5 \times 4}{8 \times 15} \qquad \text{Multiplying the numerators and the denominators}$$

$$= \frac{20}{120} \qquad \text{This fraction needs cancelling down}$$

$$= \frac{1}{6} \times \frac{20}{20} \qquad \text{20 is a common factor of numerator and denominator}$$

$$= \frac{1}{6} \qquad \frac{20}{20} = 1$$

If the two fractions are mixed numbers, it is best to change them into improper fractions before multiplying.

Example 2. Multiply the two mixed numbers $2\frac{3}{4} \times 1\frac{1}{2}$.

Solution. Under the entry Fractions you can see how to change mixed numbers into fractions and back again. These skills are needed to solve this problem. Write

$$2\frac{3}{4} \times 1\frac{1}{2} = \frac{11}{4} \times \frac{3}{2} \qquad \text{Changing mixed numbers into improper fractions}$$

$$= \frac{33}{8}$$

$$= 4\frac{1}{8} \qquad \text{Writing the improper fraction back as a mixed number}$$

Example 3. Uncle Harry leaves $52,000 in his will to be shared between his two nephews, John and Bill. John is to inherit $\frac{2}{3}$ of his uncle's money and Bill is to get the rest. John decides to give $\frac{1}{10}$ of his money to his favorite charity. How much money does John give away?

Solution. John gives away $\frac{1}{10}$ of $\frac{2}{3}$ of his uncle's money. Write

$$\frac{1}{10} \times \frac{2}{3} \times 52{,}000 = \frac{1}{10} \times \frac{2}{3} \times \frac{52{,}000}{1}$$

$$= \frac{1 \times 2 \times 52{,}000}{10 \times 3 \times 1}$$

$$= \frac{104{,}000}{30}$$

$$= 3467 \quad \text{to the nearest dollar}$$

John gives away $3467.

Division of two fractions is done by multiplying the first fraction by the reciprocal of the second fraction.

Example 4. Work out $1\frac{1}{2} \div \frac{3}{5}$.

Solution. Write

$$1\frac{1}{2} \div \frac{3}{5} = \frac{3}{2} \times \frac{5}{3} \qquad \text{Changing } 1\frac{1}{2} \text{ into } \frac{3}{2} \text{ and multiplying by the reciprocal of } \frac{3}{5}$$

$$= \frac{15}{6}$$

$$= 2\frac{1}{2} \qquad \text{Writing the fraction } \frac{15}{6} \text{ as a mixed number}$$

The methods of adding and subtracting fractions are explained in the entry Fractions. Fractions can be added, subtracted, multiplied, and divided using a scientific calculator, and the processes are explained in the calculator handbook.

References: Canceling, Fractions, Reciprocal.

MUTUALLY EXCLUSIVE EVENTS

Reference: Complementary Events.

N

NATURAL NUMBERS

Reference: Integers.

NEGATIVE

This word describes a quantity which is less than zero, whereas positive describes a quantity which is greater than zero. Negative numbers are numbers that are to the left of zero on the number line, and positive numbers are to the right of zero. Negative numbers should not be confused with "minus," which is an instruction to subtract one number from another.

References: Integers, Number Line.

NEGATIVE INTEGERS

The set of integers is $\{\ldots, -3, -2, -1, 0, 1, 2, 3, \ldots\}$. Negative integers are all those integers to the left of zero, which are $\{\ldots, -3, -2, -1\}$. Similarly, positive integers are all those integers to the right of zero, which are $\{1, 2, 3, \ldots\}$. The set of counting numbers is the same as the positive integers. Zero is neither positive nor negative.

References: Integers, Negative.

NET

A net is a two-dimensional drawing which can be cut out and folded up to make a solid shape. The various nets of an open cube are investigated under Pentominoes. The nets of the solid shapes, such as cube, cuboid, cylinder, cone, tetrahedron, etc., are illustrated under their own entries.

NETWORKS

The study of networks, also known as graph theory, is part of a branch of mathematics called topology. A network is made up of a set of points that are joined by lines (see figure a). The lines joining the points are called arcs, and the points are called nodes. The lines do not have to be straight. The enclosed spaces in the network are called regions.

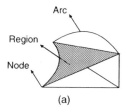

(a)

If the number of nodes in a network is denoted by N, the number of arcs by A, and the number of regions by R, then in the network in figure a, $N = 5$ nodes, $A = 8$ arcs, and $R = 5$ regions. We always count the outside of the network as a region. The data fit Euler's formula for networks, which states that

$$N + R = A + 2$$

We can check that the network drawn in the figure does fit Euler's formula:

$5 + 5 = 8 + 2$ Substituting into Euler's formula $N = 5, A = 8, R = 5$

Which is true.

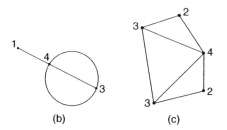

(b)   (c)

A node is a point where arcs meet, or where an arc starts. The order of a node is the number of arcs leaving it. In figure b, the number at the side of each node is its order.

In addition to Euler's formula, another interesting fact about networks is that for a network to exist it must have an even number of odd nodes. The network shown in figure c has two odd nodes, which means that this network complies with the rule, because it has an even number of odd nodes.

A network is *traversable* if it can be drawn without taking the pencil from the paper and traveling over every arc just once. You can pass through a node any number of times, but you cannot redraw a line. Euler discovered the following rule for a network to be traversable, whether you return to your starting point or not:

To be traversable, a network must have either two odd nodes, or it must have no odd nodes. It may have any number of even nodes (see figure d).

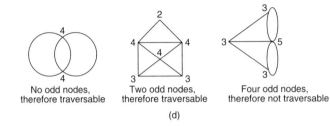

| No odd nodes, therefore traversable | Two odd nodes, therefore traversable | Four odd nodes, therefore not traversable |

(d)

References: Euler's Formula, Konigsberg Bridge Problem.

NODE

References: Konigsberg Bridge Problem, Networks.

NOMINAL DATA

Reference: Data.

NONAGON

A nonagon is a polygon with nine sides. Refer to the entry Polygon for more information on angle sum. Figure a shows three nonagons; the last one is a regular nonagon.

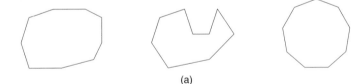

(a)

A regular nonagon has nine sides of equal length, and each of its nine angles is of equal size and equal to 140° (see figure b). The regular nonagon is made up of nine isosceles triangles, and each angle at the center of the nonagon is 40°, since there are 360° in a full turn.

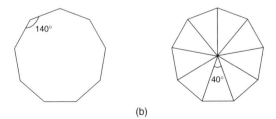

(b)

References: Polygon, Regular Polygon.

NORMAL

In order to describe a *normal*, it is first necessary to know what a *tangent* is. A tangent to a curve is a straight line that touches the curve at a point and has the same gradient, or direction, as the curve at that point. Examples of tangents to curves are drawn in figure a, where the point of contact of the tangent with the curve is indicated by a dot. The drawing on the far right is an example where the tangent to the curve is at a point of inflection. At a point of inflection, the tangent changes from one side of the curve to the other.

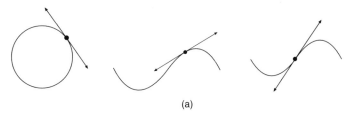

(a)

We can now explain the term normal to a curve in relation to a tangent to a curve. A normal to a curve is a straight line that is at right angles to the tangent to the curve, and passes through the same point of contact with the curve as the tangent. In figure b, the tangents are drawn dashed and the normals are at right angles to the tangents. The normal to a circle passes through the center of the circle.

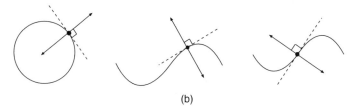

(b)

Gradients of Tangents and Normals If a tangent and normal are drawn to a curve at the same point, the product of their gradients is -1. This rule is true for any two lines in the same plane that are at right angles. If the gradient of the tangent is m_1 and the gradient of the normal is m_2, then

$$m_1 \times m_2 = -1$$

If two lines with gradients m_1 and m_2 are perpendicular, then the quick way to obtain m_2 is to turn m_1 upside down and change its sign.

So far we have discussed the normal to a curve, but there can also be a normal to a plane. A normal to a plane is a straight line that is at right angles to the plane.

References: Gradient; Inflection, Point of; Plane; Slope; Tangent.

NORMAL CURVE

Reference: Normal Distribution.

NORMAL DISTRIBUTION

A normal distribution is part of the study of probability and statistics. It can be best understood using an example, but at this level the example serves only as an introduction to the topic.

William attends a high school and there are 320 students in his year group. The dean of these students weighs them all and enters the data in a frequency distribution table. In this type of assessment, which has a large amount of data, it is usual to place the weights into class intervals, of, say, 5 kilograms. Figure a shows the frequency distribution table and a histogram and frequency polygon of the data.

Weight in kg	Frequency (Number of Students)
40–	10
45–	35
50–	65
55–	90
60–	60
65–	40
70–75	20
Total	320

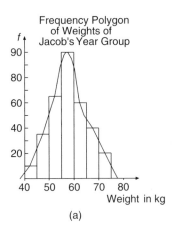

(a)

The weights of the students in William's class show quite a wide variation, but they do follow a pattern. Most of the weights are clustered together in the middle of the graph and very few are at the ends of the graph. Also, the graph has roughly an axis of symmetry.

Some data, like the weights of people, are collected by measuring, and are called continuous data. Other examples of continuous data are the heights of people, the lengths of leaves on a tree, the life of a light bulb, the quantity of paint in a 10-liter can, and examination marks measured in percentages. Examination marks are not really continuous data, but are included here; see the note at the end of this entry. When we measure continuous quantities, which occur naturally in everyday life, and graph the data using frequency polygons, the results usually resemble the graph drawn in figure b. The characteristics of this graph are:

- The shape of the graph resembles a bell, so we say that this kind of curve is bell-shaped.
- If we draw in the axis of symmetry, it passes through the mean of the distribution.
- The mean weight in the example is $\mu = 57.5$ kg. The method of calculating this mean is explained under the entry Mean.
- For a normal curve the mean, mode, and median for the distribution all have approximately the same value, which is at the axis of symmetry of the curve.

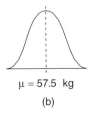

$\mu = 57.5$ kg

(b)

Since this type of bell-shaped curve is usually obtained when we collect and graph continuous data, we say it is a normal curve, and the distribution is called a normal distribution. But a bell-shaped curve will only occur when a large quantity of continuous data are graphed. In our example regarding William's year group, 320 students were weighed, which is probably large enough to obtain only an approximately normal curve. If 1000 students had been in the year group, the curve would have been closer to a normal bell-shaped curve. We can say that "several hundred" is usually a sufficiently large figure.

The mean and the standard deviation of the weights of students in William's year group can be calculated using a scientific calculator. Check the calculator handbook for the method appropriate to your calculator. The mean weight is calculated to be $\mu = 57.5$ kg and the standard deviation is $\sigma = 7.3$ kg. There is a relationship between the standard deviation $\sigma = 7.3$ kg, the mean $\mu = 57.5$ kg, and the number of students in Jacob's year group, which is true for all normal curves. This relationship is described below. Suppose you draw the axis of symmetry on a bell-shaped curve that is at the mean value μ. Measure one standard deviation σ along the horizontal axis and shade in the area, as shown in figure c. The shaded area will represent 34% of all the students in William's year group. Similarly, one standard deviation measured to the left of the mean will also represent 34% of the students. From the symmetry of the bell-shaped curve, we can say that 68% of the data (the students) lie within one standard deviation of the mean for a normal distribution (see figure d). Write

$$\mu + \sigma = 57.5 + 7.3 \qquad \text{Using the data from William's group.}$$

$$= 64.8$$

$$\mu - \sigma = 57.5 - 7.3$$

$$= 50.2$$

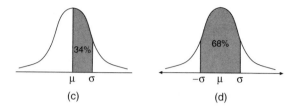

(c)   (d)

This means that 68% of the students lie within the weights 50.2 and 64.8 kg. There are 320 students in Jacob's year group, so we can calculate how many lie within one standard deviation of the mean:

$$68\% \text{ of } 320 = 0.68 \times 320 \qquad \text{Writing 68\% as 0.68}$$
$$= 217.6$$

We can expect that about 218 students weigh between 50.2 and 64.8 kg.

This can be extended further to assert that 95% of the data will lie within two standard deviations of the mean (see figure e):

$$\mu + 2\sigma = 57.5 + 2 \times 7.3$$
$$= 72.1$$
$$\mu - 2\sigma = 57.5 - 2 \times 7.3$$
$$= 42.9$$

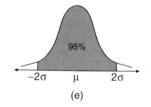

(e)

This means that 95% of 320, or 304, students have weights between 42.9 and 72.1 kg. Further extending this concept, we assert that 99% of the data will lie within three standard deviations of the mean. This means that 99% of 320, or 317, students weigh between 35.6 and 79.4 kg. All the theory outlined above is what we expect to be true for a theoretical normal curve, but everyday life examples do not exactly obey this model. This is certainly true in the example of Jacob's year group, where all the students lie in the range 40–75 kg. To obtain data that will more closely fit the theory of 68%, 95%, and 99%, we would have to consider a much larger group of students of that age. A normal curve containing the standard deviations and relevant percentages is shown in figure f.

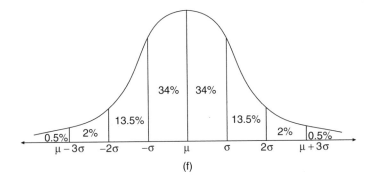

(f)

Useful Descriptors The normal distribution is applied to the study of probability. If a student is selected at random from William's year group, there is a probability of 0.68 (68%) he or she will weigh between 50.2 and 64.8 kg. We say he or she will "probably" or "likely" weigh within one standard deviation of the mean. Similarly, there is a probability of 0.95 (95%) that a randomly selected student will weigh within two standard deviations of the mean. We say he or she will "very probably" or "very likely" lie within this range. There is a probability of 0.99 (99%) that a randomly selected student will weigh within three standard deviations of the mean. We say he or she will "almost certainly" lie within this range.

Example. The owner of the Choco Chocolate Company asked his foreman, John, to do an assurance check on the weights of their medium-size chocolate bars. John randomly selected 500 bars and after weighing them calculated that the sample had a mean weight of 147 grams and a standard deviation of 3 grams.

Assuming that the weights of Choco bars are normally distributed, answer the following:

(a) What is the weight of a bar of Choco very likely to be?

(b) If a bar is randomly selected from a shelf in a supermarket, what is the probability it will weigh more than 150 grams?

(c) The owner of the company decides to print the following guarantee on the wrapper of each Choco bar:

"This bar is guaranteed to weigh 141 grams, or your money back."

How many bars in the sample of 500 will fail the guarantee?

Solution. The first step is to draw a normal curve, complete with all the known information regarding percentages and means and standard deviations (see figure g):

$$\mu = 147, \quad \mu + \sigma = 150, \quad \mu - \sigma = 144, \quad \mu + 2\sigma = 153,$$
$$\mu - 2\sigma = 141, \quad \mu + 3\sigma = 156, \quad \mu - 3\sigma = 138$$

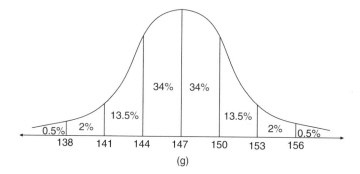

(g)

(a) "Very likely" is within two standard deviations of the mean, which is between 141 and 153 grams. A bar of Choco is very likely to weigh between 141 and 153 grams.

(b) The probability that a bar will weigh more than 150 grams is 13.5% + 2% + 0.5% = 16%. The probability that a Choco bar will weigh more than 150 grams is 16%, which is 0.16.

(c) The percentage of bars that weigh less than 141 grams is 2% + 0.5% = 2.5%. The number in the sample of 500 bars that weigh less than 141 grams is equal to 2.5% of 500: 0.025 × 500 = 12.5. About 13 of 500 Choco bars will be expected to fail the guarantee.

The type of data that result in a normal distribution curve are continuous data, but the properties of a normal curve can sometimes be applied to discrete data. For example, the examination marks of a large number of students display the characteristics of a normal distribution.

The normal curve can be used in quality control, as explained in this next example. Peter is a baker and is concerned with the fluctuations in the weights of his standard loaf of bread. He employs Lisa as a quality control officer to improve the consistency in this area. On a Tuesday, Lisa selects a random sample of 1000 loaves and weighs them. Then she enters the data in a frequency table and draws a normal distribution curve. Lisa implements her quality control system in the bakery to ensure the weights of the standard loaves will be less variable. Three months later, on a Tuesday, she checks the weights of 1000 loaves and draws another normal distribution curve on the same axes as the first batch. It can be seen from the graphs in figure h that the second batch has weights that are less spread out and more closely clustered about the mean weight. Her quality control systems are working.

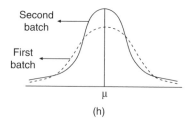

(h)

References: Arithmetic Mean, Axis of Symmetry, Central Tendency, Class Interval, Continuous Data, Frequency Distribution, Frequency Polygon, Histogram, Standard Deviation.

NORTH

Reference: East.

NUMBER BASES

References: Binary Numbers.

NUMBER LINE

Reference: Directed Numbers, Integers.

NUMBER PAIRS

These are sometimes called ordered pairs.

Reference: Cartesian Coordinates.

NUMBER TREE

Reference: Integers.

NUMBERS

Reference: Integers.

NUMERATOR

Reference: Denominator.

NUMERICAL VALUE

The equation $3x + 2 = 8$ is a numerical equation, because the constants are all numbers. The equation $ax^2 + bx + c = 0$ is called a literal equation, because its constants are represented by the letters a, b, and c.

The term numerical value is sometimes used instead of absolute value. For example, the numerical value of -7 and $+7$ is 7.

References: Absolute Value, Equations.

O

OBJECT

Reference: Image.

OBTUSE ANGLE

Reference: Acute Angle.

OCTAGON

An octagon is a polygon with eight sides; the prefix octa means eight. The angle sum of the eight interior angles of any octagon is 1080°. Some octagons are drawn in figure a; the last one is a regular octagon.

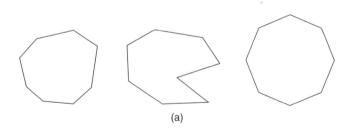

(a)

A regular octagon has eight sides of equal length, and each of its eight angles is of equal size, which is 135° (see figure b). The regular octagon is made up of eight isosceles triangles, and each angle at the center of the octagon is 45°.

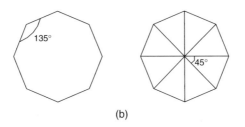

(b)

Tiling Patterns The regular octagon and the square form a tiling pattern (tessellation) as shown in figure c.

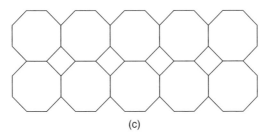

(c)

References: Isosceles Triangle, Regular Polygon, Tessellations, Triangle.

OCTAHEDRON

This is a solid shape (polyhedron) that has eight faces; the prefix octa means eight. All the eight faces of a regular octahedron, which is one of the Platonic Solids, are congruent equilateral triangles. In the figure one can see that the octahedron is made up of two square-based pyramids stuck together. The net of a regular octahedron is made up of eight equilateral triangles.

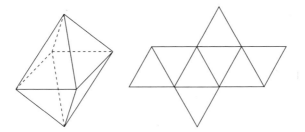

References: Congruent Figures, Equilateral Triangle, Net, Platonic Solids, Polyhedron.

ODD FUNCTION

Reference: Even Function.

ODD NUMBERS

Reference: Even.

ODDS

In order to understand odds it is important first to read the entry Probability of an Event. The odds of an event happening are defined to be the ratio of the probability of the event happening to the probability of the event not happening.

Suppose the probability of an event happening is P. Then the probability of it not happening is $1-P$. Therefore

$$\text{Odds of an event happening} = \frac{P}{1-P} \qquad \text{Using the definition}$$

Example 1. If a die is rolled, what are the odds of it being a 6?

Solution. Write

$$\text{Probability of rolling a six is } P = \tfrac{1}{6}$$

Then

$$\text{Odds of rolling a six} = \frac{\frac{1}{6}}{1 - \frac{1}{6}} \qquad \text{Using } \frac{P}{1-P}$$

$$= \tfrac{1}{5}$$

The odds of rolling a 6 can be referred to as "odds on rolling a 6" and is equal to the ratio of 1:5, or 1 to 5. Similarly, the odds *against* rolling a 6 is equal to the reverse ratio of 5:1, or 5 to 1.

Example 2. If a playing card is drawn from a pack of 52 cards, what are the odds against it being a black king?

Solution. Write

$$\text{Probability of drawing a black king} = \frac{2}{52} \qquad \text{There are 2 black kings out of 52}$$

Thus

$$\text{Odds of drawing a black king} = \frac{\frac{2}{52}}{1 - \frac{2}{52}} \qquad \text{Using } \frac{P}{1-P}$$

$$= \frac{1}{25}$$

The odds of drawing a black king are 25 to 1 against.

When a coin is tossed, the odds of it being a head are 1 to 1 against, or 1 to 1 for. In this case we say the odds are even!

References: Canceling, Probability of an Event.

OGIVE

An ogive is the name of the graph of a cumulative frequency distribution.

Reference: Cumulative Frequency Graph.

ONE–MANY CORRESPONDENCE

Reference: Correspondence.

ONE–ONE CORRESPONDENCE

Reference: Correspondence.

OPERATIONS

An operation is a method of combining numbers. The four basic operations in arithmetic for combining numbers are listed below with their symbol in brackets:

- Addition $(+)$
- Subtraction $(-)$
- Multiplication $(\times)$
- Division $(\div)$

See the entry BEDMAS for the order of operations.

Reference: BEDMAS.

OPPOSITE INTEGERS

One integer is the opposite of another integer when they are positioned on a number line on opposite sides of the origin and at equal distances from the origin. The "opposite of an integer" is a term used to explain the process of subtracting integers, which is done under the entry Integers. For example, the opposite of -2 is $+2$, or just 2. The opposite of 4 is -4.

References: Integers, Number Line.

OPPOSITE TRANSFORMATION

This is the same as an indirect transformation.

References: Congruent Figures, Transformation Geometry.

ORDER OF A NODE

References: Konigsberg Bridge Problem, Networks.

ORDER OF ROTATIONAL SYMMETRY

Rotational symmetry is the symmetry of rotation, or turning, about a fixed point. The order of rotational symmetry is the number of small turns of rotation a shape makes in completing a full turn. The following explanation brings out the meaning of order of rotational symmetry.

Suppose you take grandpa's walking stick and rotate it about its end through four turns of 90° to obtain the complete figure shown here in figure a. This complete figure has four positions of the stick, and each time the stick has been rotated about the same point, through the same angle of 90°. We say the complete figure has rotational symmetry of order four. The end of the stick, about which it is rotated, is called the center of rotation.

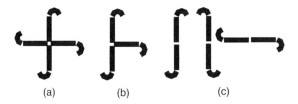

(a)　　　　(b)　　　　　(c)

It should be pointed out that after the stick has been rotated twice, giving three sticks, the complete figure does not appear to have rotational symmetry (see figure b), because the angle between each pair of sticks is not the same size.

On the other hand, the figure made up of two sticks, with 180° between the sticks, has rotational symmetry of order two. This is true in whatever position it is drawn, provided the final figure is obtained by rotation about the same point. Each of the three pairs of walking sticks has rotational symmetry of order two (see figure c).

Note these figures do not have any axes (or lines) of symmetry. A figure can have both rotational and axes of symmetry, as explained below.

Suppose you use a polo stick instead of a walking stick for the rotations (see figure d). The polo stick is rotated about the end of the handle each time, and the size of the angle between each polo stick is 90°. The complete figure has rotational symmetry of order four, and it has four axes of symmetry, which are drawn dashed in the right-hand side of figure d.

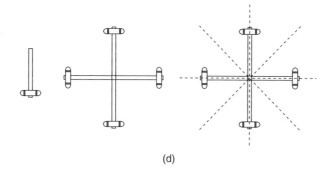

(d)

Some figures do not appear to have rotational symmetry. For example, consider three positions of the walking stick when the angles between the three sticks are 90°, 90°, and 180° (see figure e). In figures like this we say the figure does have rotational symmetry, but its order of rotational symmetry is one. Every figure has at least rotational symmetry of order one.

(e)

Summary 1. If a figure has n axes of symmetry, it will have rotational symmetry of order n, except when $n = 0$, in which case the rule breaks down.

2. If a figure has rotational symmetry of order n, it will have n axes of symmetry or no axes of symmetry.

Example. Madge runs a courier business and the logo is a figure with three running legs (see figure f). What is the order of rotational symmetry of the logo?

(f)

Solution. There are three turns about the point at the center of the figure, and the turn of each leg is through the same angle of 120°. The logo has rotational symmetry of order three, but does not have any axes of symmetry.

Some playing cards have rotational symmetry of order two, as shown by the six of diamonds (see figure g). The center of rotational symmetry is at the center of the playing card. The card has no axes of symmetry. Not all playing cards have rotational symmetry of order two, because the pattern of the shapes has no symmetry, as in the seven of diamonds. Playing cards are designed with this symmetry so they can be easily read from opposite sides of a card table. Many crosswords are designed with a rotational symmetry order of four.

(g)

Total Order of Symmetry This is the result of adding the number of axes of symmetry that a shape has to its order of rotational symmetry. For example, the total order of symmetry of a square is $4 + 4 = 8$. To find the symmetry of well-known polygons search for them under the relevant entry.

References: Axis of Symmetry, Center of Rotation.

ORDERED PAIRS

Reference: Cartesian Coordinates.

ORDINAL DATA

These are data that can be ranked according to size.

Reference: Data.

ORIGIN

Reference: Axes.

ORTHOCENTER

In the accompanying figure the three altitudes of the triangle meet at a point H called the orthocenter of the triangle. In the figure the altitudes of the triangle are *AL*, *BM*, and *CN* and their point of intersection is *H*. We say that the three altitudes of a triangle are concurrent, which means meet at a point.

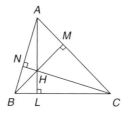

For a right-angled triangle *ABC*, with the right angle at the point *C*, the three altitudes of the triangle intersect at *C*. This means that the point *C* is the orthocenter of the triangle.

If the triangle is obtuse-angled, the orthocenter lies outside the triangle.

References: Altitude Angle in a Semicircle, Collinear, Concurrent.

OUNCE

The ounce (abbreviation oz) is a unit of weight, and is an imperial unit. One ounce is $\frac{1}{16}$ part of a pound, which means 16 oz = 1 pound (lb). One ounce is approximately 28 grams (28.349532 grams, to 6 dp).

References: Gram, Imperial System of Units, Pound.

OUTCOME

Reference: Event.

OVAL

This is another name for ellipse.

Reference: Ellipse.

P

PALINDROMIC NUMBER

This is a number that reads the same backward as forward. For example, the number 1221 is a palindromic number, because when you read it backward you get the same number, 1221. Other examples of palindromic numbers are 12621, 22, and 4884.

Stringing together the last digit of each of the first nine square numbers forms a palindromic number:

$$1^2 = 1, \quad 2^2 = 4, \quad 3^2 = 9, \quad 4^2 = 16, \quad 5^2 = 25, \quad 6^2 = 36,$$
$$7^2 = 49, \quad 8^2 = 64, \quad 9^2 = 81$$

Writing down the last digit of each square number, we obtain the palindromic number 149656941.

Reference: Digit.

PARABOLA

Check the entry Quadratic Graphs for how to draw parabolas. Suppose you throw a stone through the air. Its height h meters after time t seconds may be modeled by the quadratic formula $h = 20t - 5t^2$. The graph of this quadratic formula is the parabola shown in the figure. It was Galileo (1564–1643) who first discovered that when objects are thrown, their path through the air is in the shape of a parabola. This is only true when air resistance is neglected and gravity is uniform. The path of projectiles, fired from a gun or a bow and arrow, or of water coming out of a hosepipe, is a parabola.

One of the properties of the parabolic curve is made use of in car headlights. If a light bulb is placed at the focal point (F) of a parabolic reflector (see right-hand side of figure), the light rays will emerge from the headlight as parallel beams of light. This property is not true for other curves.

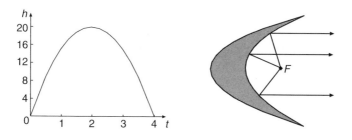

Another application of the parabolic curve is in the telescope and the radiotelescope, which is used in astronomy. A parabolic "dish" is designed to receive parallel radio waves from distant planets and stars and to concentrate those waves at its focus.

References: Conic Sections, Quadratic Graphs.

PARADOX

A paradox is a statement that comes from apparently logical reasoning, but that, on closer inspection, gives an obviously absurd result. Perhaps the most famous paradox is one that was offered by Zeno, a Greek philosopher who lived about 450 BC. He described a race between Achilles and a tortoise. Achilles can run 10 times as fast as the tortoise, so he gave the tortoise a 100-pace head start. At the beginning of the race Achilles is at the point A_0 and the tortoise at the point T_0 and the distance between the two points is 100 paces. The race starts.

When Achilles has run to the point T_0, the starting position of the tortoise, the tortoise has moved to T_1, a distance of 10 paces (see figure). When Achilles arrives at T_1, the tortoise has moved 1 pace to T_2. When Achilles arrives at T_2 the tortoise has moved 0.1 paces to T_3. When Achilles arrives at T_3 the tortoise has moved 0.01 paces to T_4, and so on. In this way, if the reasoning is continued, the tortoise will always be ahead of Achilles (albeit by an ever-decreasing distance) and so can never be caught. Common sense contradicts this conclusion. We know that the winner of the race, if it goes on long enough, will be Achilles, because he can run faster than the tortoise. But it is difficult to find a flaw in Zeno's argument, and so we have a paradox that baffled mathematicians for centuries.

A_0 T_0 T_1 T_2 T_3

100 paces   10 paces   1 pace

Suppose the distances run by Achilles are written down as a sum, given by *D*:

$$D = 100 + 10 + 1 + 0.1 + 0.01 + 0.001 + 0.0001 + \ldots \text{ and so on}$$
$$= 111 + 0.1 + 0.01 + 0.001 + 0.0001 + \ldots$$

$$= 111.11111\ldots$$ Which is $111.\dot{1}$, written as a
recurring decimal.
$$= 111\tfrac{1}{9}$$ $\tfrac{1}{9} = 0.\dot{1}$

The distance run by Achilles when he catches up with the tortoise is $111\tfrac{1}{9}$ paces.

Reference: Decimal.

PARALLEL

Reference: Gradient.

PARALLELOGRAM

A parallelogram is a quadrilateral, which means a four-sided polygon, with its opposite sides parallel and equal in length. In addition, its opposite angles are the same size. The angles a and b in figure a are called cointerior angles and their sum is $180°$, which means $a + b = 180°$.

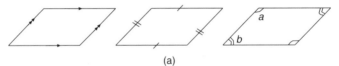

(a)

Each diagonal divides the parallelogram into two congruent triangles (see figure b). The diagonals of the parallelogram bisect each other. It has no axes of symmetry. It has rotational symmetry of order two.

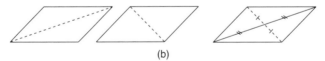

(b)

The shapes rhombus, rectangle, and square are classed as special kinds of parallelograms, but have their own individual properties. More information about these shapes is given under their separate entries.

To find the area of a parallelogram we need to know the lengths of its base (b) and its altitude, that is, its perpendicular height (h). It is not possible to find the area of a parallelogram if we are only given the lengths of its four sides. To find the area of the parallelogram in figure c write

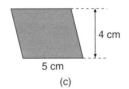

4 cm

5 cm

(c)

$$\text{Area of parallelogram} = \text{base} \times \text{perpendicular height}$$

$$= 5 \times 4$$

$$= 20$$

The area of the parallelogram is 20 cm².

   It is not possible to find the perimeter of a parallelogram if we are only given the lengths of its base and altitude.

   In figure d the rectangle and the parallelogram have the same area because they have the same base and the same perpendicular height, but the perimeter of the parallelogram is greater than that of the rectangle.

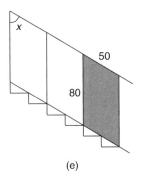

(d)

Example. The handrail on a set of steps on a hillside is shown in figure e. The angle marked *x* is 60°. It is planned to cover each panel (a panel is shaded) with "safety" plastic. Using the measurements marked on the figure, find the area of one panel.

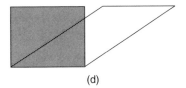

(e)

Solution. A panel is a parallelogram, and to find the area of a parallelogram we need the two dimensions, base (*b*) and perpendicular height (*h*). These dimensions are shown on a panel drawn in the figure.

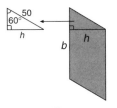

(f)

The base is $b = 80$ cm and the height (h) is found using trigonometry in the small right-angled triangle. Write

$$\frac{h}{50} = \sin 60° \qquad\qquad \text{Sine} = \frac{\text{opposite}}{\text{hypotenuse}}$$

$$h \ = 50 \times \sin 60°$$

$$h \ = 43.3 \text{ cm} \quad \text{(to 3 sf)}$$

Therefore

$$\text{Area of parallelogram panel} = b \times h$$

$$= 80 \times 43.3$$

$$= 3460 \text{ cm}^2 \quad \text{(to 3 sf)}$$

The area of one panel is 3460 cm^2.

References: Area, Base (geometry), Cointerior Angles, Constructions, Perimeter, Polygon, Quadrilateral, Rectangle, Rhombus, Square, Symmetry, Trigonometry.

PARENTHESES

This is another name for round brackets ( ), which are used in algebra to enclose an expression so that it can be treated as a single quantity. Once the brackets have served their purpose, they are often removed by expanding them.

References: Brackets, Expanding Brackets, Rectangle.

PASCAL'S TRIANGLE

The use of Pascal's triangle and its symmetry patterns are explained in the entry Coefficient. Pascal's triangle contains natural numbers, triangle numbers, and powers of 2.

$$
\begin{array}{ccccccc}
 & & & 1 & & & \\
 & & 1 & & 1 & & \\
 & 1 & & 2 & & 1 & \\
1 & & 3 & & 3 & & 1 \\
\end{array}
$$

1	$1 = 2^0$
1 1	$1 + 1 = 2 = 2^1$
1 2 1	$1 + 2 + 1 = 4 = 2^2$
1 3 3 1	$1 + 3 + 3 + 1 = 8 = 2^3$
1 4 6 4 1	$1 + 4 + 6 + 4 + 1 = 16 = 2^4$
1 5 10 10 5 1	$1 + 5 + 10 + 10 + 5 + 1 = 32 = 2^5$

The triangle drawn here contains six rows, but more rows can be added using the patterns of numbers. One of the diagonal rows is the set of counting numbers 1, 2, 3, 4, 5, ..., which are also called natural numbers. Below the natural numbers is a diagonal of triangle numbers, 1, 3, 6, 10, ... Adding the numbers in each row give powers of 2, which are 1, 4, 9, 16, 32, ...

Each number in the table can be expressed as a combination ^nC_r, which is also known as a selection. For example, consider the row of numbers 1, 5, 10, 10, 5, and 1:

$$^5C_0 = 1, \quad ^5C_1 = 5, \quad ^5C_2 = 10, \quad ^5C_3 = 10, \quad ^5C_4 = 5, \quad ^5C_5 = 1$$

The next row of Pascal's triangle is:

$$^6C_0 = 1, \quad ^6C_1 = 6, \quad ^6C_2 = 15, \quad ^6C_3 = 20, \quad ^6C_4 = 15, \quad ^6C_5 = 6, \quad ^6C_6 = 1$$

More information on Pascal's triangle is given in the entry Coefficient.

References: Coefficient, Combinations, Natural Numbers, Powers, Triangle Numbers.

PATTERNS

A set of numbers forms a pattern when it obeys a given rule. A pattern may also exist in a set of drawings or designs, which may also be represented by a given rule. In this entry the aim is to find the rule, or formula, which describes a pattern of numbers or drawings. The rule will be given in terms of n, and the pattern of numbers is obtained by substituting the numbers $n = 1, 2, 3, 4, \ldots$ one at a time into the rule.

Example 1. Harry is a postman and he is delivering letters to the houses with even numbers, which are on one side of the street. The numbers of the houses are 2, 4, 6, 8, ... What is the number of the 20th house, and what is the formula that describes this pattern of house numbers?

Solution. The formula will involve a variable, which we call n, and we can introduce the variable n in a table the following way:

Position of house by counting	1st	2nd	3rd	4th	20th	nth
Number on the house	2	4	6	8		

The variable n is introduced as the nth house in the street of even numbers, and the formula we require is the house number of the nth house. The table contains a pattern that we need to identify. By studying the pattern of numbers in the table we can see that the 5th house is numbered 10, and the 6th house is numbered 12. So the 20th house is numbered 40. The pattern is to multiply n by 2, which equals $2n$, to obtain the number of the nth house. The rule is: House numbers $= 2n$. The pattern of house numbers 2, 4, 6, 8, ... is obtained by substituting $n = 1, 2, 3, 4, \ldots$ in turn into the rule House numbers $= 2n$.

Example 2. What is the formula for the house numbers on the opposite side of the street, which is the set of odd numbers 1, 3, 5, 7, ... ?

Solution. An odd number is one more than an even number, or one less. So the formula is either $2n + 1$ or $2n - 1$. We can check which of these two rules is correct by substituting for n the numbers 1, 2, 3, 4, ...

- First rule: when $n = 1$, $2n + 1 = 3$.
- Second rule: when $n = 1$, $2n - 1 = 1$.

The second rule is the correct formula, because the first house number is 1, and not 3. The formula for the house numbers 1, 3, 5, 7, ... is $2n - 1$, where n is 1, 2, 3, 4, ...

The following table gives rules, in terms of n, for three well-known patterns of numbers:

Square numbers	1	4	9	16	25	...	n^2
Triangle numbers	1	3	6	10	15	...	$\frac{1}{2}n(n+1)$
Cube numbers	1	8	27	64	125	...	n^3

There is not a rule for the set of prime numbers 2, 3, 5, 7, 11, 13, 17, ... Check in the entry Difference Tables for more assistance in finding the rules for patterns of numbers.

The following examples give patterns of figures which will be reduced to patterns of numbers to obtain the rules.

Example 3. Isabel sells square swimming pools, and is interested in knowing how many tiles are needed to tile around the pools. Each tile is 1 meter square. How many tiles are needed for a pool that is n meters square?

Solution. Isabel makes sketches of pools of increasing sizes and counts the tiles needed to fit around them (see figure a). She enters the data into a table in order to recognize the pattern more easily:

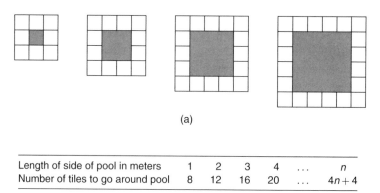

(a)

Length of side of pool in meters	1	2	3	4	...	n
Number of tiles to go around pool	8	12	16	20	...	$4n + 4$

We will study the sketch of (say) the third drawing. There are 5 tiles along the top and 5 along the bottom. The 5 is made up of (3 + 2), where 3 meters is the length of the pool. So for a pool of length n meters, along the top and the bottom there are $(n + 2) + (n + 2) = 2n + 4$ tiles. Going back to the pool of length 3 meters, there are 3 tiles along the two remaining sides, where 3 meters is the length of the pool. So for a pool of length n meters there is an extra $n + n = 2n$ tiles for the two remaining sides. For a square pool of length n meters there are $2n + 4 + 2n = 4n + 4$ tiles needed. The formula can be checked by substituting, in turn, $n = 1, 2, 3, 4, \ldots$ into the formula $4n + 4$.

For a pool of side n meters, $(4n + 4)$ tiles are needed to tile around it.

Example 4. What is the formula for the number of diagonals that can be drawn in a polygon with n sides?

Solution. Drawings of the pattern are shown in figure b. The least number of sides of a polygon is three, which is a triangle. The triangle has no diagonals.

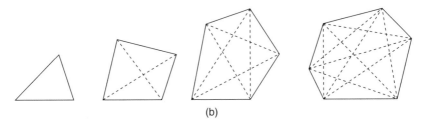

(b)

The number of sides (n) of the polygon is entered in a table and the number of diagonals is counted. It is recognizing the pattern in the table that enables the problem to be solved:

Number of sides (n)	3	4	5	6	...	n
Number of diagonals	0	2	5	9	...	
Pattern	$\dfrac{3 \times 0}{2}$	$\dfrac{4 \times 1}{2}$	$\dfrac{5 \times 2}{2}$	$\dfrac{6 \times 3}{2}$	...	$\dfrac{n(n-3)}{2}$

From the pattern we can deduce the number of diagonals for a polygon with n sides. The number of diagonals that can be drawn in a polygon with n sides is $n(n - 3)/2$.

References: Diagonal, Difference Tables, Formula, Polygon, Variable.

PENTAGON

Check with the entry Polygon for more details on the angle sum. A pentagon is a polygon with five sides. Figure a shows some pentagons; the last one is a regular pentagon.

The regular pentagon has five sides of equal length, and its five angles are of equal size (108°). It has five axes of symmetry, and the order of rotational symmetry is five

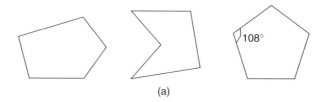

(a)

(see figure b). The regular pentagon is made up of five congruent isosceles triangles whose angles are 54°, 54°, and 72°.

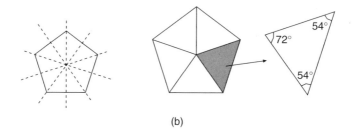

(b)

To construct a regular pentagon, you can use a protractor to measure the 108° angles. The regular pentagon will not tessellate.

References: Axis of Symmetry, Congruent Figures, Isosceles Triangle, Polygon, Protractor, Regular Polygon, Rotational Symmetry.

PENTAGRAM

The pentagram is a star-shaped figure formed by drawing all the diagonals of a regular pentagon (see figure a). The pentagram has five axes of symmetry and it has rotational symmetry of order five, like the regular pentagon.

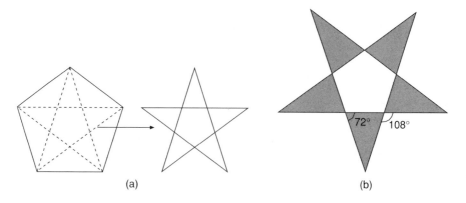

(a)   (b)

Extending, with straight lines, the sides of a regular pentagon can also form a pentagram (see figure b).

References: Axis of Symmetry, Pentagon, Regular Polygon, Star Polygons.

PENTAHEDRON

Reference: Polyhedron.

PENTOMINOES

When five squares are joined together edge to edge they form a pentomino. The joining of each pair of squares must be along a complete edge. Altogether there are 12 different combinations of joining together five squares, and each one is called a pentomino. Eight of the pentominoes are nets for an open box. Remember that types like the three depicted in figure a are regarded as the same pentomino.

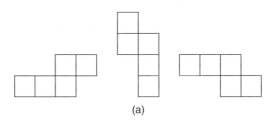

(a)

The 12 pentominoes are drawn in figure b and the squares that are the bases of the open boxes are shaded.

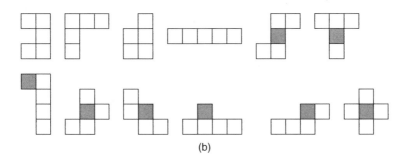

(b)

Jigsaw There are 12 pentominoes each made of 5 squares. Altogether there are $5 \times 12 = 60$ squares. If you imagine a rectangle that is made of 60 squares, then its sides could be 6 by 10, say. It is possible to jigsaw the 12 pentominoes to fit into this rectangle, as shown in figure c. There are many solutions to this jigsaw problem.

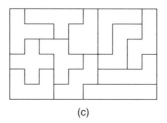

(c)

References: Hexomino, Nets, Polyominoes.

PERCENTAGE

The abbreviation for percentage is percent, or %. This latter symbol is made up by rearranging a 1 and 00. Percentage means "out of 100." For example, 85% means 85 out of 100, which can be expressed as the fraction 85/100, which should then be canceled down to its simplest form as a fraction, which is 17/25.

Summary:

$$85\% = \frac{85}{100} = \frac{17}{20}$$

In the following examples we learn how to convert % to decimals and back again, and how to convert % to fractions and back again.

Example 1. Jacob scored 57% in a math test. Write his mark as a decimal.

Solution. The rule for converting a % to a decimal is to write it over 100, and then express this fraction as a decimal:

$$57\% = \frac{57}{100}$$
$$= 0.57$$

In a similar way,

$$6\% = \frac{6}{100}$$
$$= 0.06$$

Example 2. Of the students in Nathan's class at school, 0.25 are left-handed. Write this as a %.

Solution. The rule for converting a decimal to a % is to multiply the decimal by 100:

$$0.25 \times 100 = 25$$

$$0.25 = 25\%$$

In a similar way, 1.842 when converted to a % is

$$1.842 \times 100 = 184.2$$

$$1.842 = 184.2\%$$

This example demonstrates that a % can be greater than 100%, but a mark in an examination can never be greater than 100%.

Example 3. About 90% of an iceberg is below the surface of the water. Write this as a fraction.

Solution. The rule for converting a % to a fraction is to first write the % over 100, and then cancel down the resulting fraction:

$$90\% = \frac{90}{100} \qquad \text{Which cancels down to } \tfrac{9}{10}$$

$$= \frac{9}{10}$$

Example 4. Anne sells French bread at a profit of 115%. Write this as a fraction.

Solution. Write

$$115\% = \frac{115}{100} \qquad \text{Which cancels down to } \tfrac{23}{20}$$

$$= 1\frac{3}{20} \qquad \text{Writing the improper fraction } \tfrac{23}{20} \text{ as a mixed number}$$

Example 5. Jacob scored 13 out of 15 in an English test. Write this as a %.

Solution. 13 out of 15 is the fraction 13/15. To convert a fraction to % we multiply the fraction by 100, which is the same as the method for converting decimals to %:

$$\frac{13}{15} \times 100 = 86.7 \quad \text{(to 1 dp)} \qquad \text{Using a calculator}$$

13 out of 15 is a mark of 86.7%.

Seen on a wall is the following graffiti: "8 out of 5 Americans have trouble with fractions." Comment: There cannot be more Americans having trouble with fractions than there are Americans! The statement is impossible.

In the following examples we will learn how to find a % of a quantity, and how to increase or decrease a quantity by a %.

Example 6. Bill buys a motorbike for $350. He makes a few improvements to it, and sells it for a profit of 60%. How much profit does he make?

Solution. The profit is 60% of $350. Write

$$60\% \text{ of } 350 = \frac{60}{100} \times 350$$

$$= 210$$

The profit he makes is $210.

Example 7. After the Olympics were over David decided to reduce training by 45%. If he had been running 120 km per week, what would this reduce to?

Solution. To reduce his distance by 45% we multiply his distance by 55% to obtain the new distance he runs (100% − 45%):

$$120 \times 55\% = 120 \times \frac{55}{100} \qquad\qquad 55\% = \frac{55}{100}$$

$$= 66$$

David now runs 66 km per week.

Note. When a quantity is increased by a certain %, and then decreased by the same %, we do not obtain the original quantity. The following example demonstrates this fact:

Example 8. Pat weighs 60 kg. Over Christmas her weight increases by 2%. After a diet her weight then decreases by 2%. Does her weight return to 60 kg?

Solution. After Christmas her weight is

$$60 \times 1.02 = 61.2 \text{ kg} \qquad 102\% = 1.02$$

After the diet her weight is

$$61.2 \times 0.98 = 59.976 \text{ kg} \qquad 98\% = 0.98$$

Her weight does not return to 60 kg.

PERCENTAGE ERROR

This refers to an error written as a % of the correct amount. The formula is

$$\% \text{ error} = \frac{\text{error}}{\text{correct amount}} \times 100$$

Error is the difference between the correct and wrong amounts. The following example brings out the meaning of percentage error.

Example. Tom is having new carpets in his lounge. He asks Pat to measure the length of the lounge, which she measures to be 16.34 meters. On the plans of the house the room is marked as shorter than her measurement, so Tom checks the tape and finds that the first meter has been cut off. The actual length of the lounge is therefore 15.34 meters. Find the % error in Pat's original measurement.

Solution. Write

$$\text{Error} = 16.34 - 15.34$$

$$= 1 \text{ meter}$$

$$= \frac{1}{15.34} \times 100 \qquad \text{Using the formula for \% error}$$

$$= 6.52 \quad \text{(to 2 dp)} \qquad \text{Using a calculator}$$

The % error in Pat's original measurement is 6.52%.

References: Canceling, Conversion Factor, Cost Price, Decimal, Error, Fractions, Percentage.

PERCENTAGE PROFIT

Reference: Cost Price.

PERCENTILES

Percentiles divide a set of data into 100 equal parts, and are often used to give information on how well a person has done in relation to other people in a competition or in an examination. Each of the 100 parts into which the data is divided is called a percentile.

Suppose Luke takes an examination and there are 200 students who take the same examination. Their results are recorded in a cumulative frequency graph, as shown in the figure. Suppose we wish to find out the examination mark of the 90th percentile. There are 200 students in the group, and 90% of 200 students = 180. We can read from the graph that the 180th student from the bottom of the ranking scored 75 in the examination. This mark of 75% is the 90th percentile. It means that 90% of the students, which is 180 students, scored less than 75% in the examination. In addition, it means that 10% of the students scored above a mark of 75%.

Example. In Luke's examination, what examination mark is the 40th percentile?

Solution. Write

$$40\% \text{ of } 200 \text{ students} = 0.4 \times 200 \qquad 40\% = 0.4$$

$$= 80$$

We count 80 students from the bottom of the list and obtain the mark that corresponds to the 80th person. This mark in the examination is 50%.

The 40th percentile is an examination mark of 50%. The 40th percentile means that 40% of the students scored below that examination mark and 60% of them scored above it.

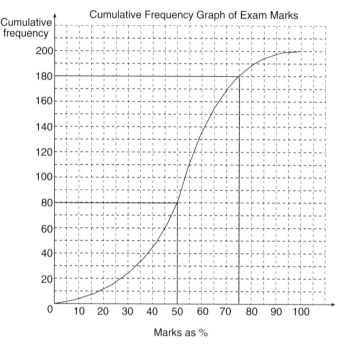

The median mark in the examination is the 50 percentile, which in this example is the mark of the 100th student. From the graph the median examination mark is about 53%. In this example 100 students scored below the median and 100 scored above it. The upper quartile is the 75th percentile, and 75% of the students, which is 150, scored below this mark. The lower quartile is the 25th percentile, and 25% of the students, which is 50, scored below this mark. From the graph, the upper quartile is about 63% and the lower quartile is about 42%.

A decile is a multiple of 10 percentiles. For example, the 20th percentile is the 2nd decile, and the 90th percentile is the 9th decile.

References: Box and Whisker Graph, Cumulative Frequency Graph, Interquartile Range.

PERFECT NUMBER

A perfect number is a whole number that is the sum of all its factors, not including itself. For example, the first perfect number is 6. All the factors of 6 are 1, 2, 3, and 6:

$$6 = 1 + 2 + 3.$$

So 6 is a perfect number, because it is the sum of all its factors except itself.

Example. Is 8 a perfect number?

Solution. The factors of 8 are 1, 2, and 4, not including itself. $1 + 2 + 4 = 7$, so 8 is not a perfect number since it is not the sum of all its factors.

The followers of Pythagoras (who lived about 450 BC) thought that perfect numbers had mystical properties. For example, God created the world in 6 days, and the number 6 is a perfect number. Until the year 1952 there were only 12 known perfect numbers and all of them were even numbers. No one is quite sure whether or not a perfect number can be an odd number. The first three perfect numbers are 6, 28, and 496. At present only 24 perfect numbers have been discovered.

References: Factor, Pythagoras' Theorem, Whole Numbers.

PERFECT SQUARE

Reference: Completing the Square.

PERIMETER

The perimeter of a closed curve or of a polygon is either the boundary line of the closed curve or polygon, or it can mean the total length of the boundary line. In this entry we will discuss the total length of the boundary line. The perimeter of the rectangle in figure a, drawn on a 1-cm grid, is found by adding together the lengths of the four sides, which will be the total length of the boundary line:

$$\text{Perimeter} = 4 + 3 + 4 + 3$$
$$= 14$$

The perimeter of the rectangle is 14 cm.

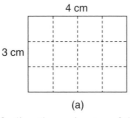

(a)

An alternative method for finding the perimeter of the rectangle is to add together 4 and 3 cm and double the answer.

For curved shapes like the circle it is difficult to find the perimeter by measuring the length of the boundary line, so we use a formula for the perimeter. The perimeter of a circle is usually called the circumference. To find the perimeters of well-known shapes, such as the circle, the parallelogram, the trapezium, the triangle, etc., see the respective entries.

References: Area, Diameter, Pi, Polygon, Radius, Rectangle.

PERIOD

Reference: Cycle.

PERIODIC

Reference: Cycle.

PERMUTATIONS

These are also called arrangements.

Reference: Combinations.

PERPENDICULAR BISECTOR

References: Locus, Mediator.

PERPENDICULAR LINES

Two straight lines are perpendicular if they are at right angles to each other, which means that the angle between them is $90°$. For example, If two lines are perpendicular and one of them is horizontal, then the other line is vertical.

Reference: Gradient.

PI

The symbol for pi is π, which is a letter of the Greek alphabet. The value of π is available in a calculator as $\pi = 3.141592653\ldots$ and cannot be written down as an exact value. π is an irrational number, which means it cannot be written as a fraction, and can never be expressed as a recurring decimal. Sometimes the fraction 22/7 is used as a value for π, but when expressed as a decimal is only accurate to two decimal places. Certain sentences or verses can be used to help remember its value. For example, the value of π to seven decimal places can be remembered according to the number of letters in the respective words of the following sentence:

MAY	I	HAVE	A	LARGE	CONTAINER	OF	COFFEE?
3	1	4	1	5	9	2	6

The origin of π can be explained as follows: Suppose you obtain a circular object, say a dinner plate, and measure its circumference (C) and its diameter (D) using a length of string and a ruler. The definition of π is

$$\pi = \frac{C}{D}$$

You can calculate an approximate value for π using your dinner plate measurements by dividing the circumference C by the diameter D. Using these measurements, provided they are fairly accurate, you can expect your value for π to be accurate to one decimal place. In Biblical times the ancient Hebrews believed that the value for π was equal to 3 (see *2 Chronicles,* Chapter 4, Verse 2). An ancient Chinese ratio for π was 355/113, which is accurate to six decimal places when expressed as a decimal.

Throughout history mathematicians have striven to improve on the accuracy of π, and various formulas have been developed to achieve this. By 300 AD, the mathematician Ptolemy had calculated π to be 3.1416, which is correct to four decimal places. Over 300 years ago a formula for π was known to be given by the following the infinite series, although it converges too slowly to be of practical use:

$$\frac{\pi}{4} = \frac{1}{1} - \frac{1}{3} + \frac{1}{5} - \frac{1}{7} + \frac{1}{9} \cdots$$

Using computers, mathematicians have calculated π to over one billion decimal places. We use π to find, where applicable, the perimeters, areas, and volumes of shapes like the circle, the sphere, the cylinder, and the cone. Explanations are given under the respective entries.

References: Area, Irrational Numbers, Perimeter, Volume.

PICTOGRAM

A pictogram is one of the statistical graphs used for displaying discrete data. The graph is made up of a series of pictures, and each picture represents a certain quantity of data.

Example. Pat is the top salesperson for "We Sell," a business that sells Real Estate. The number of houses sold per year by each member of the "We Sell" team is shown in the table. Draw a pictogram to represent these sales.

Salesperson	Number of Houses Sold
Tom	30
Pat	55
David	40
Joanne	38

Solution. We need to choose a picture, say a small house, to represent a certain number of house sales. One house picture could represent 10 sales or perhaps 5 sales, depending on how much space we wish to use for the graph. Suppose we use one house for 10 sales. We divide by 10 the number of houses sold by each person to obtain the number of pictures for each salesperson. Tom has 3 house pictures, Pat has 5.5, David has 4, and Joanne has 3.8 (see figure). A house picture can be cut in half to represent 0.5 for Pat, but it is difficult to draw 0.8 of a house picture for Joanne, and this is one of the drawbacks of this type of graph. The graph must have a title and a key.

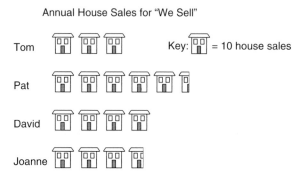

The pictogram has good visual impact, and at a glance it is possible to compare the size of one item of data with another, but it is not always clear what the actual size of each item of data is. The key is helpful for this.

References: Bar Graph, Discrete Data, Statistics.

PIE GRAPH

A pie graph, or pie chart, is one of the statistical graphs used for displaying discrete data. A pie graph does not have axes like a bar graph. It is a circular diagram, and the circle is divided up into sectors. Each item of data is represented by a sector of the circle, whereas in the bar graph a column represents each item of data. The frequency of an item of data is proportional to the angle in the sector that represents it. The frequency is also proportional to the area of the sector. In the bar graph the height of the column is proportional to the frequency. When constructing a pie graph we make use of the fact that in a full turn there are $360°$.

Example 1. Helen earns $600 per week and the table shows her weekly budget. Draw a pie graph of her weekly budget.

Food	$200
Phone	$30
Rent	$120
Clothes	$80
Utilities	$120
Other	$50
Total	$600

Solution. There are 360° in a full circle, so we divide 360° by the total expenses of $600:

$$\frac{360°}{600} = 0.6°$$

This tells us that in the pie graph each dollar is represented by an angle of 0.6°. We now add three extra columns to the table to find the total angle for each item of expenditure and to find the % each item is of the whole budget. For example, food is

$$\frac{200}{600} \times 100 = 33\% \text{ of the weekly budget} \qquad \text{to the nearest whole number}$$

Food	$200	× 0.6° =	120°	33%
Phone	$30	× 0.6° =	18°	5%
Rent	$120	× 0.6° =	72°	20%
Clothes	$80	× 0.6° =	48°	13%
Utilities	$120	× 0.6° =	72°	20%
Other	$50	× 0.6° =	30°	8%
Total	$600	× 0.6° =	360°	99%

The total in the % column should be 100%, but there is a slight discrepancy due to rounding to the nearest whole number. The next stage is to draw a circle. It should be large enough to contain all the information and small enough to easily fit on the page. The angles representing the expenses are measured around the circle with a protractor, starting from a vertical radius at the top of the circle (see figure a). Each sector is labeled with the appropriate expense, or may be color-coded with a key at the side. It is often a good idea to include the percentage for each item of expenses. The pie graph needs a title, like all statistical graphs.

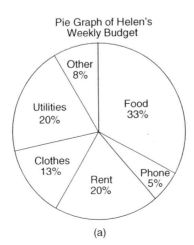

(a)

Example 2. The pie graph in figure b shows how William spends his day (HW= homework). Answer the following questions.

(a) How many hours did William spend at school?

(b) What percentage of his day did he spend on school activities?

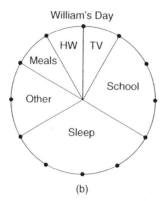

William's Day

(b)

Solution. There are 12 equally spaced marks around the circle. Each mark must represent 2 hours, since there are 24 hours in a day.

(a) William spent 3 markings at school, which is 6 hours. William spent 6 hours at school.

(b) In addition to 6 hours at school he spent 1 marking, or 2 hours, on homework (HW), which is a total of 8 hours on school activities. The fraction of his day spent on school activities is 8/24. To change a fraction into a % we multiply the fraction by 100:

$$\frac{8}{24} \times 100 = 33 \quad \text{to the nearest whole number}$$

William spent 33% of his day on school activities.

A pie graph is mainly used for comparing data and at a glance it is possible to compare the size of one item of data with the others. The drawback is that we do not know the actual size of each item of data as we do in the bar graph. If there are too many categories, the pie graph loses its visual impact, is difficult to label, and may appear quite crowded.

References: Bar Graph, Discrete Data, Frequency, Percentage, Protractor, Rounding, Sector of a Circle, Statistics.

PINT

Reference: Imperial System of Units.

PLACEHOLDER

A placeholder is a zero that is inserted in a number to maintain the value of the number when we are rounding.

Example. The number of spectators at the annual soccer match between Rovers and Wanderers was 54,687. Round this off to three significant figures.

Solution. The figure 54,687 lies between 54,600 and 54,700, which are both three significant figures, but it is closer to 54,700. The number of spectators was 54,700 to 3 sf. The two zeros in the answer are not significant figures, but need to be written to maintain the size of the number. The two zeros are placeholders.

Reference: Accuracy.

PLAN

References: Elevation, Isometric.

PLANE

References: Coplanar, Edge.

PLATONIC SOLIDS

The Platonic solids were named after Plato, who lived about 400 BC. They are also called regular polyhedra, or regular solids. A Platonic solid is one in which every face is the same regular polygon, and as a result all of its edges and vertices are exactly the same. There are only five possible platonic solids, and it was Plato who clearly defined them and showed how to construct the solids from their nets.

The five platonic solids are:

1. *Cube,* made up of six congruent squares.
2. *Regular dodecahedron,* made up of 12 regular congruent pentagons.
3. *Regular icosahedron,* made up of 20 congruent equilateral triangles.
4. *Regular octahedron,* made up of eight congruent equilateral triangles.
5. *Regular tetrahedron,* made up of four congruent equilateral triangles.

For information on these solids, search for them under their names. Crystals grow in the shape of some of the regular polyhedra. For example, the crystals of common salt grow as cubes, and chrome alum crystals grow as regular octahedrons.

References: Coplanar, Edge.

PLOTTING

Reference: Cartesian Coordinates.

POINT

Reference: Edge. .

POINT OF INFLECTION

Reference: Concave.

POLYGON

A polygon is a closed plane figure with straight sides, and no two sides cross over. The sides intersect in points called vertices. When all the sides are the same length, the polygon is said to be regular, and the angles of each regular polygon are the same size. There is a formula for the sum of the interior angles of a polygon, and it is derived in the following way (see the figure).

The three interior angles of a triangle add up to $180°$. This means $a + b + c = 180°$.

A quadrilateral is made up of two triangles that have a common vertex V: $2 \times 180° = 360°$, which is the angle sum of a quadrilateral.

A pentagon is made up of three triangles that have a common vertex V: $3 \times 180° = 540°$, which is the angle sum of a pentagon.

Using the pattern of the triangle, quadrilateral, and the pentagon, we can extend the process to include a polygon with n sides. A polygon with n sides is made up of $(n - 2)$ triangles that have a common vertex.

The angle sum of a polygon with n sides is $(n - 2) \times 180°$. The formula is

Sum of interior angles of a polygon with n sides $= (n - 2) \times 180°$

Example. A decagon is a polygon with 10 sides. Find the interior angle sum of a decagon.

Solution. Write

Angle sum of a decagon $= (10 - 2) \times 180°$ Substituting $n = 10$ into $(n - 2) \times 180°$

$$= 8 \times 180°$$

$$= 1440°$$

The interior angle sum of a decagon is $1440°$.

The table gives information about some of the well-known polygons. For more information search under the name of the given polygon. From the table, you can see that each time the number of sides of the polygon increases by one, the angle sum increases by 180°.

Name of Polygon	Number of Sides	Angle Sum
Triangle	3	180°
Quadrilateral	4	360°
Pentagon	5	540°
Hexagon	6	720°
Heptagon	7	900°
Octagon	8	1080°
Nonagon	9	1260°
Decagon	10	1440°
Hendecagon	11	1620°
Dodecagon	12	1800°

References: Exterior Angle of a Polygon, Regular Polygon.

POLYHEDRON

The plural of polyhedron is polyhedra. A polyhedron is a solid figure with plane faces that are polygons. The faces meet in edges, and the edges meet in vertices. For regular polyhedra see the entry Platonic Solids. Some examples of polyhedra are shown in the figure.

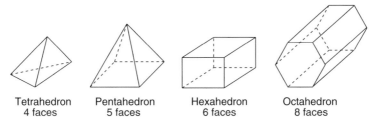

| Tetrahedron | Pentahedron | Hexahedron | Octahedron |
| 4 faces | 5 faces | 6 faces | 8 faces |

The polyhedra drawn in the figure have the following, more commonly known names:

- Tetrahedron, triangle-based pyramid
- Pentahedron, square-based pyramid
- Hexahedron, cuboid
- Octahedron, hexagonal prism

For more information about polyhedra search for them under their respective names.

References: Edge, Plane, Platonic Solids, Prism, Regular, Polyhedron, Vertex.

POLYNOMIAL

Reference: Degree.

POLYOMINOES

When squares are joined together edge to edge they form polyominoes. The various kinds of polyominoes are as follows:

- *Monomino.* This type of polyomino is made up of just one square (see figure a). There is only one type of monomino.

(a)

- *Domino.* Two squares joined together form a domino (see figure b). There is only one type of domino.

(b)

- *Triomino.* This is made up of three squares joined together (see figure c). There are two types of triominoes.

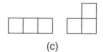

(c)

- *Tetromino.* This is made of four squares joined together (see figure d). There are five types of tetrominoes.

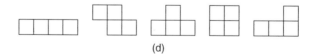

(d)

- *Pentomino.* This is made up of five squares joined together. There are 12 types of pentominoes (see the entry Pentominoes).
- *Hexomino.* This is made of up six squares joined together. There are 35 types of hexominoes (see the entry Hexominoes).

- *Heptomino.* This is made up of seven squares joined together. There are 108 types of heptominoes, including one with a "hole," shown in figure e.

(e)

References: Hexomino, Pentominoes.

POPULATION

Population is a term used in statistics and means the complete set of data being considered when choosing a sample. Before you can choose a sample you must be clear about what population you are going to choose your sample from. For example, if you were choosing a sample of students to answer a questionnaire on how they feel about having a representative on the Board of Trustees, the population will probably be all the students in your school. But if you were working for the Examination Board and were choosing a sample of students to obtain some general information on the examination results, your population would probably be all the students in the country. It would not be a good idea to have all the students in the world as the population, because they would not all have taken the same examination.

Reference: Bias, Sample.

POSITIVE

Reference: Negative.

POSITIVE INTEGERS

Reference: Negative Integers.

POUND

The pound is a unit of weight in the imperial system of units. The abbreviation is lb.

Reference: Imperial System of Units.

POWERS

Reference: Exponent.

PRIME FACTOR

References: Factor, Prime Number.

PRIME NUMBER

A positive integer is prime if it has exactly two factors, which are 1 and itself:

- 3 is a prime number because its only factors are 1 and 3.
- 2 is a prime number because its only factors are 1 and 2.
- 6 is not a prime number because its factors are 1, 2, 3, and 6. Since it has four factors, it is not a prime number.
- 1 is not a prime number because it has only one factor, which is 1.

There are 25 positive prime numbers less than 100:

2, 3, 5, 7, 11, 13, 17, 19, 23, 29, 31, 37, 41, 43, 47, 53, 59, 61, 67, 71, 73, 79, 83, 89, 97

The next prime number is 101.

Through history mathematicians have been interested in prime numbers and in obtaining a formula for them. This has never been achieved, but some interesting attempts are discussed in the entry Counterexample.

References: Counterexample, Eratosthenes' sieve, Factor.

PRINCIPAL

Reference: Capital.

PRISM

A prism is a polyhedron (solid) which has a base and a top that are congruent and parallel, and cross sections which are parallel and identical to the base (see figure a).

The prisms that we will consider are *upright* prisms, which are usually called "right" prisms. The names of some prisms are given according to the shape of their cross section. Some well-known prisms are drawn in figure b. The cross section may be a regular polygon, as in the cube, or it may be a nonregular prism, as in the trapezium prism. Notice that the trapezium prism is the shape of a swimming pool that has a deep end and a shallow end.

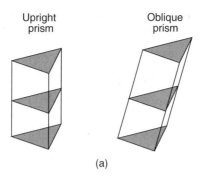

Upright prism   Oblique prism

(a)

Examples of prisms in everyday life are pencils, coins, swimming pools, and boxes.

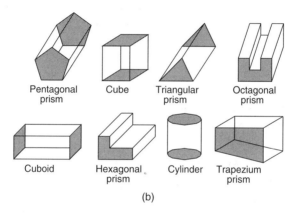

Pentagonal prism   Cube   Triangular prism   Octagonal prism

Cuboid   Hexagonal prism   Cylinder   Trapezium prism

(b)

Volume of a Prism The distance between the cross-sectional ends of a prism is called the length. The volume is given by

$$\text{Volume of prism} = \text{area of cross section} \times \text{length}$$

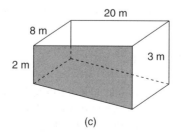

20 m

8 m

2 m

3 m

(c)

Example. The length of the swimming pool is 20 meters, the width is 8 meters, the deep end is 3 meters deep, and the shallow end is 2 meters deep (see figure c). Find the volume of the pool.

Solution. The cross section is the trapezium drawn shaded in figure c. Write

Area of a trapezium $A = \dfrac{(a + b)}{2} \times h$ Here a and b are the lengths of the parallel sides and h is the distance between them

Write

Area of trapezium $= \dfrac{(2 + 3)}{2} \times 20$ Substituting $a = 2, b = 3, h = 20$ in the area formula

$$= 2.5 \times 20$$

$$= 50$$

The area of the trapezium is 50 square meters.
    To find the volume, write

Volume of prism $=$ area of cross section $\times$ length

$\qquad\qquad\quad = $ area of trapezium $\times$ length

$\qquad\qquad\quad = 50 \times 8$ The distance between the parallel sides is 8 meters

$\qquad\qquad\quad = 400$ cubic meters

The volume of the pool is 400 m^3.
    Note: The length of 8 meters in the volume formula is the distance between the parallel faces of the cross section, not the length of the pool.

References: Congruent Figures, Cross Section, Parallel, Trapezium.

PROBABILITY OF AN EVENT

Suppose the event we are going to consider is rolling a die once and obtaining a 3. The die could land in a total of six different ways. We say that the total number of outcomes of rolling the die is six, which means there are six ways it could land. The number of ways of obtaining the particular outcome of rolling a 3 is one. The formula for probability is

$$\text{Probability of a particular event} = \frac{\text{number of ways the event can happen}}{\text{total number of outcomes}}$$

We can apply this formula to our example of the probability of rolling a 3 with a die:

$$\text{Probability of rolling a three} = \tfrac{1}{6}$$

The probability of an event happening is another way of answering the question: "What are the chances or likelihood that an event will happen?" When we roll a die it has an equal chance of landing on any of the six numbers 1, 2, 3, 4, 5, or 6, provided it is not biased. It is certain to land on one of these numbers, but which one? The chances, or probability, of it landing on the number 3 is one chance in six. If you rolled the die six times, in theory, it is expected to land once on a 3 and once on each of the other five numbers. In practice, however, in six rolls the die may not land on a 3 at all, or it could land on a 3 every time! All we can say is this: If we roll a die, the probability of it landing on a 3 is $\tfrac{1}{6}$, provided each of the numbers on the die are equally likely to turn up. Suppose you rolled a die four times and it landed on a 6 every time. The next time you roll the die the probability of it landing on a 6 is still $\tfrac{1}{6}$. But the probability of rolling five 6's in succession is $\left(\tfrac{1}{6}\right)^5 = 1/7776$. When we calculate probabilities we write them as fractions, or may express them as decimals or percentages.

Example 1. A die is rolled once. What is the probability the outcome is an odd number?

Solution. The number of ways of rolling an odd number is three, because the die could land as a 1, 3, or 5. The total number of outcomes is six, because the die could land as a 1, 2, 3, 4, 5, or 6. Therefore,

$$\text{Probability of rolling an odd number} = \tfrac{3}{6}$$

$$= \tfrac{1}{2} \qquad \text{Canceling down the fraction}$$

Example 2. In a bag there are five red marbles, three green and two blue. If John puts his hand in the bag and, without looking, brings out a marble, what is the probability that the marble is:
(a) Red?
(b) Blue?

(c) Red or green?
(d) Red and green?

Solution. There are 10 marbles in the bag and 5 of them are red. Write

$$\text{Probability of choosing a red marble} = \frac{\text{number of ways of choosing a red marble}}{\text{total number of ways of choosing a marble}}$$

$$= \tfrac{5}{10}$$

$$= \tfrac{1}{2} \qquad \text{Canceling down the fraction}$$

(b) There are 10 marbles in the bag and 2 of them are blue. Write

$$\text{Probability of choosing a blue marble} = \tfrac{2}{10} \qquad \text{Using the formula for probability}$$

$$= \tfrac{1}{5} \qquad \text{Canceling down the fraction}$$

(c) There are 10 marbles in the bag and 8 are either red or green. Write

$$\text{Probability of either a red or green} = \tfrac{8}{10}$$

$$= \tfrac{4}{5} \qquad \text{Canceling down the fraction.}$$

(d) It is impossible to have marbles in the bag that are both red and green. Write

$$\text{Probability of a red and green} = \tfrac{0}{10}$$

$$= 0$$

We have seen in the example above that if the probability of an event is zero, then that event is impossible. In addition, if an event is certain to happen, its probability is one. An example of an event that is certain to happen is that every human being will die. The probability that we will die is one.

All probabilities lie between zero and one, inclusive.

There are three kinds of probabilities we now need to study:

- The probability of mutually exclusive events
- The probability of complementary events
- The probability of independent events

Before proceeding it may be a good idea to read the entry Complementary Events, where all these terms are explained.

The probability of mutually exclusive events are explained in the first example.

Example 3. Ann rolls a die once.

(a) What is the probability she rolls a 3 and a 6?

(b) What is the probability she rolls a 3 or a 6?

Solution. (a) When one die is rolled, the event of rolling a 3 and the event of rolling a 6 are events that cannot both happen at the same time, and are called mutually exclusive events. So the probability of rolling a 3 and a 6 is impossible on one roll of a die, and equal to zero.

(b) The probability of rolling a 3 or a 6 is also a mutually exclusive event and is calculated by adding the probability of each event:

$$\text{Probability of rolling a 3 or 6} = \tfrac{1}{6} + \tfrac{1}{6} \qquad \text{Prob}(3) + \text{prob}(6).$$

$$= \tfrac{1}{3} \qquad \text{Canceling the fraction.}$$

Summary. If E_1 and E_2 are mutually exclusive events, then

$$\text{Probability of } E_1 \text{ and } E_2 = 0$$

$$\text{Probability of } E_1 \text{ or } E_2 = \text{ probability of } E_1 \ + \ \text{probability of } E_2$$

The probability of complementary events are explained in the next example.

Example 4. Using the same example of Helen rolling a die once, what is the probability she rolls an even number or an odd number?

Solution. The event of rolling an even number and the event of rolling an odd number are mutually exclusive events, because they both cannot happen at the same time, so we add the probabilities. In addition, these two events make up all the possible outcomes, so they are complementary events. Write

$$\text{Probability of an even number or odd number} = \tfrac{1}{2} + \tfrac{1}{2} \qquad \text{Prob(even)} + \text{prob(odd)}$$

$$= 1$$

For complementary events the sum of the probabilities is always 1.

Summary. If E_1 and E_2 are complementary events, then

$$\text{Probability of } E_1 \text{ or } E_2 = \text{ probability of } E_1 \ + \ \text{probability of } E_2$$

$$= 1$$

When the outcome of one event has no effect on the outcome of another event, we say that the two events are *independent events*. To obtain the probability of independent events we multiply the probabilities of the separate events.

Example 5. A coin is tossed and a die is rolled. What is the probability of obtaining a head and a prime number?

Solution. The result of tossing a coin cannot possibly affect the outcome of rolling a die. In other words, if the coin landed as a head, it would not affect the way the die would land. When there are two trials that have no effect on the outcome of each other, we say the outcomes are independent events.

The probability of tossing a head is $\frac{1}{2}$ and the probability of rolling a prime number with a die is $\frac{3}{6}$, because there are three numbers 2, 3, and 5 that are prime. When two events are independent we multiply the probabilities. Write

Probability of head and prime number $= \frac{1}{2} \times \frac{3}{6}$ Prob(head) × prob(prime)

$$= \frac{1}{4}$$

Summary. If E_1 and E_2 are independent events, then

Probability of E_1 and E_2 = probability of E_1 × probability of E_2

Experimental Probability The probability we have studied so far is theoretical probability, which is what we expect to happen based on a theoretical calculation. We now look briefly at experimental probability. This kind of probability is a prediction based on what has happened previously.

Example 6. Pat's birthday is 29 April. What is the probability it will rain on Pat's next birthday?

Solution. We need some historical data to make a prediction. Records for the past 10 years, say, are obtained from the Weather Office. We find that during the past 10 years it has rained three times on Pat's birthday. Based on these data we predict what will happen this year:

Probability it will rain on Pat's birthday $= \frac{3}{10}$

A better result may be obtained if we go back 20 years or even longer.

Sometimes experimental results are compared with theoretical results. For example we know that if we toss a coin once, the probability of a head is $\frac{1}{2}$. Suppose we tossed a coin 50 times and recorded that 23 heads came up. We would say that the experimental probability of tossing a head is 23/50, which of course is not the same as $\frac{1}{2}$. However, the more times we toss the coin, the closer the experimental probability will be to the theoretical probability.

References: Canceling, Dice, Event, Odds, Prime Number, Tree Diagram.

PROBABILITY TREE

Reference: Tree Diagram.

PRODUCT

Reference: Multiplicand.

PRODUCT OF PRIME FACTORS

Reference: Factor.

PROFIT

Reference: Cost Price.

PROJECTION

Reference: Angle between a Line and a Plane.

PROOF

To prove a statement is true is to present a logical argument that establishes the statement as a fact. The proof of a geometry theorem follows a logical argument that is often based on axioms or fundamental geometry theorems.

Example. Prove the theorem "The sum of the angles of a triangle = $180°$."

Proof. Suppose the three angles of the triangle are referred to as angles A, B, and C (see figure). We have to prove that $A + B + C = 180°$. In this proof it is necessary

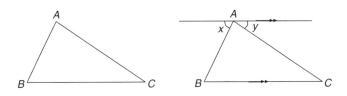

to draw in an extra line through the point A and parallel to the line BC. Write

Angle x = angle B Alternate angles are equal

Angle y = angle C Alternate angles are equal

Angle x + angle A + angle y = $180°$ Sum of adjacent angles = $180°$

Therefore $B + A + C = 180°$ Since $y = C$ and $x = B$

In proving this theorem, it was necessary to make use of two fundamental theorems that are found under the entries Adjacent Angles and Alternate Angles.

References: Adjacent Angles, Alternate Angles, Angles on the Same Arc, Geometry Theorems.

PROPER FRACTION

Reference: Fraction.

PROPORTION

Two terms that are related to each other, but often confused, are *ratio* and *direct proportion*. You may wish to study the entry Ratio before continuing. When two ratios are equal we say they are in direct proportion. A good way of explaining direct proportion is to have a photograph enlarged. In the figure, the length of the smaller photograph is 3 cm and its width is 2 cm:

$$\text{Ratio of } \frac{\text{length}}{\text{width}} = \frac{3}{2}$$

In the enlarged photo, the length is 6 cm and the width is 4 cm:

$$\text{Ratio of } \frac{\text{length}}{\text{width}} = \frac{6}{4}$$

$$= \frac{3}{2} \qquad \text{Canceling the fraction}$$

We say the lengths and widths of the two photographs are in direct proportion because the ratios of length/width for each photograph are equal.

If two figures are in direct proportion, we say they are *similar*. This means that two figures are similar if one figure is an enlargement of the other. Other quantities can be in direct proportion besides similar figures. For example, the extension of a spring is in direct proportion to the weight producing it, which is known as Hooke's law.

Another kind of proportion is *inverse proportion,* which we now compare with direct proportion. An example of direct proportion is extending a spring by adding weights to it. As one quantity (the weight) increases, the other quantity (the extension of the spring) also increases in proportion. Inverse proportion also deals with two quantities, but as one quantity, increases the other quantity decreases.

Example. Suppose Harry has the job of painting a wall on both sides. He asks his friends for help, and five of them paint one side of the wall in 7 hours. Later, three more join him and the eight friends paint the other side of the wall. If all the painters work at the same rate, how long will it take them to paint the other side of the wall?

Solution. With extra help the other side of the wall will be painted more quickly than the front side. As one quantity (the number of painters) increases, the other quantity (time to paint the wall) decreases, and so this is an example of inverse proportion. One method of solving inverse proportion problems is the "unity method" as set out below:

- Five people paint the wall in 7 hours.
- One person paints the wall in 35 hours (Number of people × number of hours = 5 × 7).
- Eight people paint the wall in $4\frac{3}{8}$ hours (number of hours ÷ number of people = 35 ÷ 8).

The eight friends paint the wall in $4\frac{3}{8}$ hours, which is 4 hours, 22 minutes, and 30 seconds.

Take care with some problems in which two quantities may appear to be in direct proportion, but on a closer inspection are not. For example, Luke weighs 20 kg at age 6 years. How much will he weigh at age 9 years? The two quantities weight and age are not in direct proportion, because people do not grow uniformly. This problem cannot be solved from the information given.

References: Cross Multiply, Enlargement, Hyperbola, Rate, Ratio, Similar Figures.

PROTRACTOR

A protractor is a flat, transparent plastic semicircle with two scales on it that are used for measuring angles in degrees (see figure a). Each scale measures from 0° to 180°, and each marking on the scale is 1°. In figure a, the 10° markings are shown but not the 1° markings. The protractor will measure angles to an accuracy of 1°. The diameter of the semicircle is called the straight edge and the center of the semicircle is called the center. A protractor may also be fully circular.

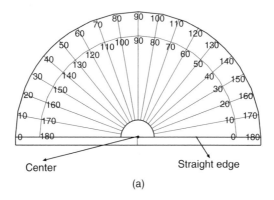

(a)

Each angle has two arms, or rays, which intersect in a point, or vertex. In figure b the vertex of the angle is O and the arms are OA and OB. When we are using the protractor to measure the angle we place the center of the protractor on the vertex O of the angle and the straight edge on one of the arms of the angle as shown in the figure. We place the straight edge of the protractor on the arm OA (or we could use the arm OB).

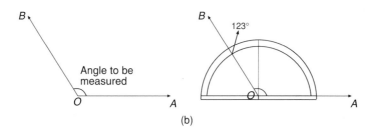

(b)

Starting at the arm of the angle where the straight edge is placed, we examine the two scales on the protractor and notice that the inside scale begins at zero and the outside scale starts at 180°. Using the inside scale that begins at zero, we measure around the scale until we come to the other arm OB. The arm OB crosses the inside scale at 123°. The size of the angle AOB is 123°.

Always use the scale that begins at zero on the arm of the angle on which you place the straight edge. If the arms of the angle are not long enough to reach the scale on the protractor, they will have to be lengthened before using the protractor. It is a good idea to have an approximate estimate of the angle before you measure it in case you use the wrong scale.

Example. Use the protractor to measure the reflex angle shown in figure c.

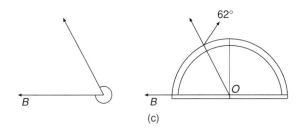

(c)

Solution. Since a protractor can only measure up to $180°$, you cannot measure a reflex angle directly unless you have a circular protractor. The method is to measure the acute angle and then subtract your answer from $360°$ to obtain the desired angle. You will recall there are $360°$ in a full turn. We place the protractor on top of the angle so that the straight edge lies on the arm OB of the angle, as shown. The scale on the protractor that begins at zero on the arm OB is the outside scale. The acute angle measured is $62°$. We now subtract this angle from $360°: 360° - 62° = 298°$. The required angle to be measured is $298°$.

We use a similar method for drawing angles, and it is always necessary to draw one arm of the angle first and measure from that arm the angle you require.

References: Acute Angle, Degree, Ray, Reflex Angle, Vertex.

PYRAMID

This is a solid figure (a polyhedron) that has a base (which can be any polygon) and edges drawn from each vertex of the base that meet at a single point V called the apex of the pyramid (see figure a). The base of a pyramid can be any polygon, but its other faces are always triangles. We will study the right pyramids, and in addition those that have regular polygons for their bases. The shape of its base categorizes a pyramid. The pyramid shown in figure a is a square-based pyramid.

Here V is the vertex, and point X is the center of the square base, which is shaded. VX is the altitude and the pyramid has four sloping triangular faces that meet at V. The triangles are either isosceles or equilateral. The volume of a pyramid is given by

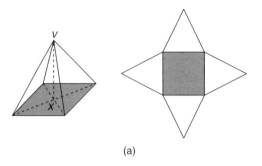

(a)

the formula

$$\text{Volume} = \tfrac{1}{3} \times \text{area of base} \times \text{altitude}$$

The surface area of a pyramid is found using the descriptive formula

$$\text{Surface area} = \text{area of base} + \text{sum of areas of triangular faces}$$

The net of the square-based pyramid is drawn on the right-hand side of figure a. The four triangles are congruent, and may be either isosceles or equilateral, depending on length of the altitude.

Various kinds of pyramids are drawn in figure b.

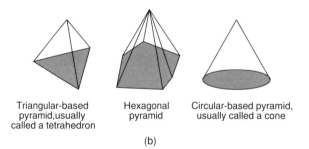

Triangular-based pyramid, usually called a tetrahedron

Hexagonal pyramid

Circular-based pyramid, usually called a cone

(b)

References: Altitude, Cone, Cube, Edge, Equilateral Triangle, Isosceles Triangle, Polyhedron, Regular Polyhedron, Right Pyramid, Vertex.

PYTHAGORAS' THEOREM

This theorem is named after the Greek mathematician Pythagoras, who lived about 500 years BC. The theorem applies to any right-angled triangle, and is a formula connecting the areas of squares that can be drawn on the sides of the triangle.

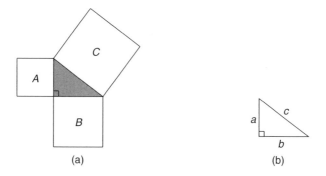

(a)   (b)

In figure a, the shaded triangle is right-angled, and squares A, B, and C are drawn on the sides of the triangle. The theorem of Pythagoras states that

$$\text{Area of square } C = \text{area of square } A + \text{area of square } B$$

The theorem is usually written in the following form:
    If the lengths of the sides of the triangle are a, b, and c units (see figure b), then

$$c^2 = a^2 + b^2$$

In words we say: "The square on the hypotenuse of a right-angled triangle is equal to the sum of the squares of the other two sides." The theorem of Pythagoras is used for finding the length of one side of a right-angled triangle if the other two sides are known.

    Figure c will enable you to demonstrate in a practical way the truth of the theorem of Pythagoras. You will need a copy of the figure and a pair of scissors. The square B has been divided up by drawing two lines each passing through the center of the square, with one of them parallel to the hypotenuse of the triangle. The two lines are perpendicular. Using scissors, remove the three squares from the triangle. Cut up

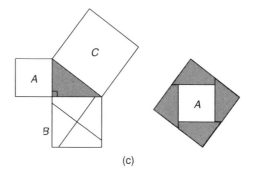

(c)

square B into the four congruent pieces. Place the four pieces from B along with the square A onto the square C. These five pieces should jigsaw together to exactly cover the square C.

This demonstrates that the area of C = area of A + area of B.

A proof of the theorem of Pythagoras is given in the entry Triangle. The three examples that follow illustrate how to use this theorem to solve problems.

Example 1. A ladder rests on horizontal ground and leans against a vertical wall (see figure d). The foot of the ladder is 1.3 meters from the wall and reaches 2.9 meters up the wall. Find the length of the ladder to an accuracy of three decimal places.

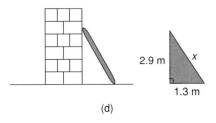

(d)

Solution. We make a simple sketch of the right-angled triangle and its dimensions, calling the length of the ladder x meters. Write

$$x^2 = 2.9^2 + 1.3^2 \qquad \text{Theorem of Pythagoras: } c^2 = a^2 + b^2$$

$$x^2 = 8.41 + 1.69 \qquad \text{Using squares on a calculator}$$

$$x^2 = 10.1$$

$$x = \sqrt{10.1}$$

$$x = 3.178 \quad \text{(to 3 dp)} \qquad \text{Using square roots on a calculator}$$

The length of the ladder is 3.178 meters

Example 2. A "flying fox" consists of a wire stretched between two upright poles (see figure e). The poles are 10 meters apart. If the height of the shorter pole is

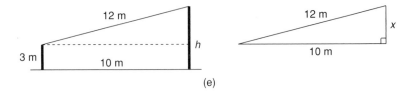

(e)

3 meters and the length of the wire is 12 meters, find the height of the longer pole. Assume that the wire has no "sag" and lies straight.

Solution. In order to use Pythagoras' theorem we need a right-angled triangle, and drawing a horizontal line from the top of the shorter pole to meet the longer pole forms this. Let the unknown length in the triangle be x meters. We can now apply Pythagoras' theorem to this triangle. Write

$$12^2 = x^2 + 10^2 \qquad \text{Pythagoras' theorem}$$

$$144 = x^2 + 100$$

$$144 - 100 = x^2 \qquad \text{Subtracting 100 from both sides of the equation}$$

$$x^2 = 44$$

$$x = \sqrt{44}$$

$$x = 6.633 \quad \text{(to 3 dp)} \qquad \text{Using square roots on a calculator}$$

The distance x must be added to the height of the 3-meter pole to obtain the height of the longer pole. The height of the longer pole is 9.633 meters.

Example 3. Amy works in a carpet warehouse. She stacks one roll of tufted carpet on top of two other rolls so that the rolls of carpet touch each other, as shown in figure f. The diameter of each roll is 1.4 meters. How high is the top of the tufted roll above the floor of the room?

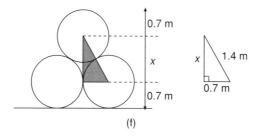

(f)

Solution. We use Pythagoras' theorem in the right-angled triangle shown in figure f. The hypotenuse is of length 1.4 meters, which is equivalent to the diameter of the end of the carpet. The base of the triangle is of length 0.7 meters, which is the radius of the end of the carpet. Let the other side of the triangle be x meters. Write

$$1.4^2 = x^2 + 0.7^2 \qquad \text{Pythagoras' theorem}$$

$$1.96 = x^2 + 0.49$$

$$1.96 - 0.49 = x^2 \qquad \text{Subtracting 1.96 from both sides of the equation}$$

$$x^2 = 1.47$$

$$x = \sqrt{1.47}$$

$$x = 1.212 \quad \text{(to 3 dp)} \qquad \text{Using a calculator for the square root}$$

Therefore

$$\text{Height of the tufted carpet above the floor} = 0.7 + x + 0.7$$

$$= 0.7 + 1.212 + 0.7$$

$$= 2.612 \text{ meters}$$

The converse of the theorem of Pythagoras is as follows:

If the square on the longest side of a triangle is equal to the sum of the squares on the other two sides, then the triangle is right-angled.

Example. Darren is a groundsman and is laying out a new soccer pitch. He needs to make sure that the corners of the pitch are right angles. He gets three tape measures and pulls them out to lengths of 30, 40, and 50 meters. Using two other helpers, he tightly stretches the tapes to form a triangle. This triangle is right-angled, because $50^2 = 30^2 + 40^2$.

References: Converse of a Theorem, Hypotenuse, Pythagorean Triples, Right Angle, Square, Square Root, Vertex.

PYTHAGOREAN TRIPLES

A right-angled triangle has three sides which obey the theorem of Pythagoras, $c^2 = a^2 + b^2$. Any three whole numbers that fit this formula are called Pythagorean triples. For example, the numbers 3, 4, and 5 are Pythagorean triples, because $5^2 = 3^2 + 4^2$. So are any multiples of 3, 4, and 5, like 6, 8, and 10, or 30, 40, and 50.

There is an infinite number of Pythagorean triples, and some well-known ones are as follows:

- 5, 12, 13 (check: $13^2 = 5^2 + 12^2$)
- 7, 24, 25
- 8, 15, 17
- 9, 40, 41
- 11, 60, 61

Any multiples of these triples are also Pythagorean triples.

References: Pythagoras' Theorem, Triangle.

Q

QUADRANTS

Reference: Circular Functions.

QUADRATIC EQUATIONS

A quadratic equation is an equation of the form

$$ax^2 + bx + c = 0$$

where a is not equal to zero, and a, b, and c are real numbers. Examples of quadratic equations are

$$x^2 + 9x + 18 = 0$$

$$x^2 + x - 6 = 0$$

$$x^2 - 4x = -4 \qquad \text{Which is the same as } x^2 - 4x + 4 = 0$$

$$2x^2 - 6x + 4 = 0$$

$$x^2 = 21 \qquad \text{Which is the same as } x^2 + 0x - 21 = 0$$

$$x^2 - 9 = 0 \qquad \text{Which is the same as } x^2 + 0x - 9 = 0$$

$$4x^2 = 10x \qquad \text{Which is the same as } 4x^2 - 10x + 0 = 0$$

Some of the above equations may not appear to be of the form $ax^2 + bx + c = 0$, but they can be rearranged into that form. To solve quadratic equations we first have to factorize the quadratic expression $ax^2 + bx + c$. Before proceeding it may be necessary for you to study the entry Factorize. The theory behind solving quadratic equations is outlined here:

Suppose two terms A and B multiply together to give zero.

$$AB = 0$$

Then it follows that this equation is true if $A = 0$ or if $B = 0$.

If the two terms A and B are brackets, we can apply this theory in the following way:

Example 1. Solve (a) $(x + 1)(x - 5) = 0$, (b) $2x(x + 2) = 0$.

Solution. For (a), write

$$(x + 1)(x - 5) = 0$$

$(x + 1) = 0$ or $(x - 5) = 0$ If $AB = 0$, then either $A = 0$ or $B = 0$

$x + 1 = 0$ or $x - 5 = 0$ We can dispense with the brackets

$x = -1$ or $x = 5$ Solving linear equations

For (b), Write

$$2x(x + 2) = 0$$

$2x = 0$ or $x + 2 = 0$ If $AB = 0$, then either $A = 0$ or $B = 0$

$x = 0$ or $x = -2$

In the next example the first stage is to express the quadratic equation in factorized form.

Example 2. Solve $x^2 + x - 6 = 0$.

Solution. Write

$$x^2 + x - 6 = 0$$

$(x - 2)(x + 3) = 0$ Factorizing the quadratic

$x - 2 = 0$ or $x + 3 = 0$

$x = 2$ or $x = -3$

Example 3. Solve $x^2 - 4x = -4$.

Solution. Write

$$x^2 - 4x = -4$$

$x^2 - 4x + 4 = 0$ Adding 4 to both sides of equation to make it $= 0$

$(x - 2)(x - 2) = 0$ Factorizing the quadratic

$x - 2 = 0$ or $x - 2 = 0$

$x = 2$ (twice)

When the two brackets are the same there will be only one solution, or we could say there are two equal solutions.

Example 4. Solve $2x^2 - 6x + 4 = 0$.

Solution. Write

$$2x^2 - 6x + 4 = 0$$

$$2(x^2 - 3x + 2) = 0 \qquad \text{2 is a common factor; always look for these first}$$

$$2(x - 1)(x - 2) = 0 \qquad \text{Factorizing the quadratic}$$

$$(x - 1)(x - 2) = 0 \qquad \text{Dividing both sides of the equation by 2}$$

$$x - 1 = 0 \quad \text{or} \quad x - 2 = 0$$

$$x = 1 \quad \text{or} \quad x = 2$$

Example 5. Solve $x^2 - 9 = 0$.

Solution. Write

$$x^2 - 9 = 0$$

$$(x - 3)(x + 3) = 0 \qquad \text{Factorizing by difference of two squares}$$

$$x - 3 = 0 \quad \text{or} \quad x + 3 = 0$$

$$x = 3 \quad \text{or} \quad x = -3$$

Example 6. Solve $4x^2 = 10x$.

Solution. Write

$$4x^2 = 10x$$

$$4x^2 - 10x = 0 \qquad \text{Subtracting } 10x \text{ from both sides, to make equation} = 0$$

$$2x(2x - 5) = 0 \qquad 2x \text{ is a common factor}$$

$$2x = 0 \quad \text{or} \quad 2x - 5 = 0$$

$$x = 0 \quad \text{or} \quad x = 2\tfrac{1}{2} \qquad \text{Solving the linear equations}$$

Summary:

• Always make sure the quadratic equation $= 0$.
• Look for common factors first.

- Answers may be left as fractions or decimals, but if decimals are used, some may need rounding off.
- There are three kinds of factors to know: common, quadratic, and difference of two squares.

There are three types of quadratic equations that are not solved by factorizing, and their solutions are explained here.

Example 7. Solve $x^2 = 21$.

Solution. Write

$$x^2 = 21$$

$$\sqrt{x^2} = \pm\sqrt{21}$$ Reduce x^2 to x by taking the square root of both sides of the equation

$$x = \pm 4.58 \quad \text{(to 2 dp)}$$ When we take the square root we get $\pm$

Example 8. Solve $\sqrt{x} = 9$.

Solution. Write

$$\sqrt{x} = 9$$

$$\left(\sqrt{x}\right)^2 = 9^2$$ Make $\sqrt{x}$ into x by squaring both sides of the equation

$$x = 81$$

Example 9. Solve $(x - 3)^2 = 16$.

Solution. Write

$$(x - 3)^2 = 16$$

$$\sqrt{(x - 3)^2} = \pm\sqrt{16}$$ Taking the square root of both sides of the equation

$$x - 3 = \pm 4$$

$$x - 3 = +4 \quad \text{or} \quad x - 3 = -4$$ Separating out into two equations

$$x = 7 \quad \text{or} \quad x = -1$$

To solve other quadratic equations that do not factorize, see the entry Quadratic Formula for the method.

QUADRATIC FACTORS

References: Factor, Quadratic Equations.

QUADRATIC FORMULA

The general form of a quadratic equation is

$$ax^2 + bx + c = 0$$

The coefficient of x^2 is a, the coefficient of x is b, and the constant term is c. The two roots of this quadratic equation are given by the formula

$$x = \frac{-b \pm \sqrt{b^2 - 4ac}}{2a}$$

The example that follows demonstrates how to solve a quadratic equation using this formula.

Example. Solve the equation $2x^2 - 3x - 6 = 0$ to 2 dp.

Solution. Comparing $2x^2 - 3x - 6 = 0$ with $ax^2 + bx + c = 0$, we have $a = 2$, $b = -3$, $c = -6$. Write

$$x = \frac{-(-3) \pm \sqrt{(-3)^2 - 4 \times 2 \times (-6)}}{2 \times 2}$$

Substituting for a, b, c into

$$x = \frac{-b \pm \sqrt{b^2 - 4ac}}{2a}$$

$$x = \frac{3 \pm \sqrt{9 + 48}}{4}$$

$$x = \frac{3 \pm \sqrt{57}}{4}$$

$$x = \frac{3 + \sqrt{57}}{4} \quad \text{or} \quad x = \frac{3 - \sqrt{57}}{4}$$

Separating out into two answers

$$x = 2.64 \quad \text{or} \quad x = -1.14 \quad \text{(to 2 dp)}$$

Reference: Completing the Square, Quadratic Equations.

QUADRATIC GRAPHS

The graph of the quadratic expression $y = ax^2 + bx + c$, where b or c, or both may be zero, is a called a parabola. In this formula $a \neq 0$, and a, b, and c are real numbers. The parabolas we study may be "cup"-shaped, in which case $a > 0$, or "hat"-shaped, in which case $a < 0$ (see figure a).

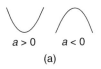

a > 0   a < 0

(a)

The most basic quadratic graph is $y = x^2$, and its graph is drawn in figure b. The points used to draw this graph are

$$(-3, 9), (-2, 4), (-1, 1), (0, 0), (1, 1), (2, 4), (3, 9)$$

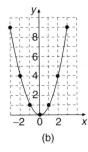

(b)

In fact all the graphs of $y = x^2 + bx + c$, that is, when $a = 1$, are exactly the same as $y = x^2$, except they have been given a translation so that the vertex of each graph is at a different point. The theory of the translation of parabolas is explained here.

The equation of the parabolic graph $y = x^2 + bx + c$ can be expressed as $y = (x + h)^2 + v$, where h and v are constants, using a process called completing the square. When the equation is expressed in this form it is easier to sketch its graph. The rules for doing this are explained in the examples that follow.

Example 1. Sketch the graph of $y = (x - 2)^2$.

Solution. By comparing $y = (x - 2)^2$ with $y = (x + h)^2 + v$, we see that $h = -2$ and $v = 0$. This value of h means the graph of $y = x^2$ is given a horizontal translation of two squares to the right to obtain the graph of $y = (x - 2)^2$. The result of this process is shown in figure c with $y = x^2$ drawn dashed; $y = (x - 2)^2$ has its vertex at the point $(2, 0)$.

(c)

The graph of $y = (x + 2)^2$ would have its vertex at the point $(-2, 0)$. This is because $h = +2$ and the graph of $y = x^2$ is translated two squares to the left. The graph of $y = x^2$ can also be translated vertically, as shown in the next example.

Example 2. Sketch the graph of $y = x^2 - 3$.

Solution. By comparing $y = x^2 - 3$ with $y = (x + h)^2 + v$, we see that $h = 0$ and $v = -3$. This value of v means the graph of $y = x^2$ is given a vertical translation of three squares downward to obtain the graph of $y = x^2 - 3$. The result of this process is shown in figure d with $y = x^2$ drawn dotted; $y = x^2 - 3$ has its vertex at the point $(0, -3)$.

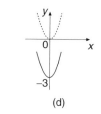

(d)

The graph of $y = x^2 + 3$ would have its vertex at the point $(0, 3)$. This is because $v = 3$ and the graph of $y = x^2$ is translated three squares upward.

Summary. When graphing the equation $y = (x + h)^2 + v$, where one unit on the graph is one square:

- If h is positive and $v = 0$, move $y = x^2$ horizontally to the left through h squares.
- If h is negative and $v = 0$, move $y = x^2$ horizontally to the right through h squares.
- If v is positive and $h = 0$, move $y = x^2$ vertically upward through v squares.
- If v is negative and $h = 0$, move $y = x^2$ vertically downward through v squares.
- When neither h nor v is zero, we sketch a combination of horizontal and vertical translations.

The same processes are applied to sketching the graphs of parabolas with the equation $y = -(x + h)^2 + v$, except the graph is "hat"-shaped instead of "cup"-shaped.

Stretches of $y = x^2$ So far we have taken the value of a in the equation of $y = ax^2 + bx + c$ to be either 1 or -1. For other values for a, we stretch the $y = x^2$ in the horizontal direction and make it "flatter" or stretch $y = x^2$ in a vertical direction and make it "steeper." For example, the graphs in figure e are $y = x^2$, which is shown dotted, $y = 2x^2$, which is steeper, and $y = \frac{1}{2}x^2$, which is flatter.

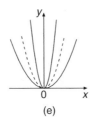

(e)

When the equation of parabolas are given in factorized form we can use the "intercept" method of sketching them. The method is simply to find the points where the parabolas cross the x- and y-axes and then sketch a "hat"- or "cup"-shaped parabola. This method produces a sketch graph that passes through the intercept points, and is explained using the examples that follow.

Example 3. Sketch the graph of the parabola $y = (x - 3)(x + 1)$ clearly showing the curve's axis of symmetry and vertex.

Solution. The equation of the curve after expanding the brackets is $y = x^2 - 2x - 3$. Now that the brackets are expanded we can identify whether a is positive or negative. In this case the value of a is 1, which is positive, so the parabola is "cup"-shaped. The x intercepts are the points where the curve crosses the x-axis. To obtain these points we substitute $y = 0$ into the equation of the curve and solve the resulting equation:

$y = (x - 3)(x + 1)$

$0 = (x - 3)(x + 1)$ Substitute $y = 0$ to obtain the x intercepts

$x = 3$ and $x = -1$ are the x intercepts   Solving the quadratic equation

The y intercept is the point where the curve crosses the y-axis. To obtain this point we substitute $x = 0$ into the equation of the curve and find the value of y:

$y = (x - 3)(x + 1)$

$y = (0 - 3)(0 + 1)$ Substitute $x = 0$ to obtain the y intercept

$y = -3 \times 1$

$y = -3$ is the y intercept

We can now sketch the parabola. We know it is "cup"-shaped and crosses the y-axis at -3 and the x-axis at the two points 3 and -1. The axis of symmetry is the vertical line that passes through the vertex, and is drawn dashed in figure f. It crosses the x-axis at $x = 1$, which is halfway between the x intercepts 3 and -1. We now substitute the value $x = 1$ into the equation of the curve to obtain the y coordinate

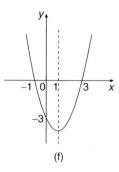

(f)

of the vertex:

$$y = (1 - 3)(1 + 1) \qquad \text{Substituting } x = 1 \text{ into the equation of the curve}$$

$$y = -2 \times 2$$

$$y = -4$$

The coordinates of the vertex are $x = 1$ and $y = -4$, or more briefly $(1, -4)$.

Sketching Cubic Graphs Cubic equations are of the form $y = ax^3 + bx^2 + cx + d$, where b, c, and d may be zero. Under this entry we will study the graphs of cubics in factorized form. Typical examples of cubics in factorized form are

$$y = (x - 2)(x + 2)(x + 3)$$
$$y = 2x(x - 3)(x + 4)$$
$$y = (x - 2)(x + 3)^2$$
$$y = (3 - x)(x + 2)(x - 4)$$

These types of cubics have two basic shapes, depending on whether or not the value of a in $y = ax^3 + bx^2 + cx + d$ is positive or negative (see figure g). A cubic graph is not made up of two parabolas joined together, but has its own characteristics.

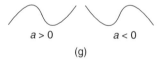

$a > 0$ $\qquad\qquad$ $a < 0$

(g)

Example 4. Sketch the graph of the cubic $y = (x - 2)(x + 2)(x + 4)$.

Solution. When the brackets are expanded the value of a is $+1$, and so its basic shape is of the type in which $a > 0$. The x intercepts are found by substituting $y = 0$

into the equation of the curve:

$$0 = (x - 2)(x + 2)(x + 4) \qquad \text{Substituting } y = 0$$

$$x = 2, \quad x = -2, \quad x = -4 \qquad \text{Solving the equation}$$

The y intercept is found by substituting $x = 0$ into the equation of the curve:

$$y = (0 - 2)(0 + 2)(0 + 4) \qquad \text{Substituting } x = 0$$

$$y = -2 \times 2 \times 4$$

$$y = -16$$

The graph of $y = (x - 2)(x + 2)(x + 4)$ will cross the x-axis at $x = 2$, $x = -2$, $x = -4$ and it will cross the y-axis at $y = -16$ (see figure h). Since the y-axis needs to go down rather a long way, to $y = -16$, it is a good idea to reduce the scale on this axis.

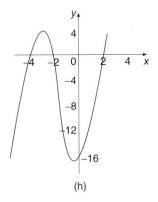

(h)

Without further work, which is beyond the scope of this book, we do not know the coordinates of the turning points, but endeavor to draw a smooth curve through the intercepts.

References: Factorize, Intercept, Parabola, Substitution, Translation, Turning Points, Vertex.

QUADRILATERAL

A quadrilateral is a polygon with four sides. See the entry Polygon for the angle sum of a quadrilateral. Many of the following quadrilaterals are studied separately under their own names:

• Square (this is a regular quadrilateral)
• Rectangle

- Rhombus
- Parallelogram
- Kite
- Trapezium

Figure a shows some quadrilaterals; the last one is a regular quadrilateral, or square.

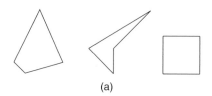

(a)

Quadrilaterals will always tessellate. Suppose we draw any quadrilateral, like the one shaded in figure b. The white tile is obtained by rotating the shaded tile through half a turn, which is 180°. This process is repeated to obtain the tiling pattern drawn here.

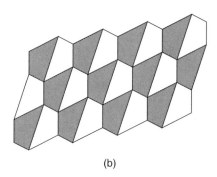

(b)

References: Mosaics, Polygon, Regular Polygon, Rotation, Tessellations, Triangle, Tiling Patterns.

QUADRUPLE

This word means "four times". For example, suppose John invested $100 and in 5 years he quadrupled his investment. This means that his investment increased four times to the value of $4 \times \$100 = \$400$.

QUALITATIVE DATA

Reference: Data.

QUALITATIVE GRAPHS

The data that are used in most graphs are *quantitative* data about numbers and relationships between them. *Qualitative* data are descriptive data, such as eye color, gender, nationality, attitudes, and so on. When we study qualitative graphs we are examining relationships between items of descriptive data—particularly attitudes. For example, the figure shows the attitudes of four students towards their study of English and Mathematics. Boz dislikes English and has a fairly good attitude to Math. Ray hates both subjects. Liz likes English more than Math. Ron likes Math and quite likes English.

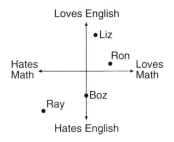

References: Data, Graphs, Qualitative Data, Quantitative Data.

QUANTITATIVE DATA

Reference: Data.

QUARTILES

References: Box and Whisker Graph, Interquartile Range.

QUESTIONNAIRE

A survey is an inquiry into something, such as how shoppers feel about alcohol being sold in supermarkets. A survey may take the form of a questionnaire to be completed by a sample of the population. The information gathered from the sample is used to make predictions and draw conclusions about the whole population. For example, it may be possible from the responses of the questionnaire to conclude that it is not popular with people generally to sell alcohol in supermarkets. The questionnaire will be designed to gather information that truly represents the attitude of the people toward the sale of alcohol, and not to produce a biased view to support the people who own wine shops! The questionnaire is a set of written questions, and is not the same as a verbal interview.

Questions must be chosen with care in order not to get a biased result. For example, if you wanted to prevent supermarkets selling alcohol, you might set a question that may produce a negative answer, such as "Do you think supermarkets should be allowed to sell spirits?" Some people might agree to the sale of beer or wine, but object to whisky and gin. Another negative response might be obtained to the question "Do you think supermarkets should sell alcohol at all times ?" Many people might generally agree to supermarkets selling alcohol, but make an exception for Sunday.

References: Bias, Census, Sample, Survey.

QUOTIENT

Reference: Dividend.

R

RADIAN

A radian (rad) is a unit used for measuring the size of an angle. A radian is defined in the following way (see figure a). Suppose we have a circle of center O and radius one unit. A sector AOB is drawn that has an arc length also of one unit, as shown in the circle on the left-hand side of the figure. The angle AOB is defined to be one radian. Instead of the arc length AB being 1 unit, suppose the chord length AB is 1 unit, as shown in the right-hand side of the figure. The angle AOB in the circle on the right-hand side of the figure is 60° because triangle AOB is an equilateral triangle. Comparing the two figures we can see that one radian is slightly smaller than 60°.

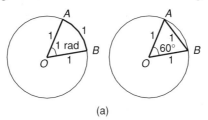

(a)

There are a little over six radians in a full turn of 360°. The exact number of radians in a full turn is obtained by dividing the circumference (C) of a circle of radius one unit by an arc length of one unit:

$C = \pi \times \text{diameter}$ The formula for the circumference of a circle

$C = \pi \times 2$ The diameter of the circle is two units

$2\pi \div 1 = 2\pi$ Circumference $\div$ arc length

There are 2π radians in a full turn of 360°. We also write

1 radian $= 360° \div 2\pi$

     $= 57.29577951\ldots$ degrees   This number cannot be written as an exact decimal

1 radian $= 57.3°$ (to 1 dp)

Conversions To convert degrees to radians multiply the number of degrees by $\pi/180$. To convert radians to degrees multiply the number of radians by $180/\pi$.

Example 1. (a) Convert 3.2 radians to degrees. (b) Convert 94.6° to radians.

Solution. For (a), write

$$3.2 \times \frac{180}{\pi} = 183.3° \quad \text{(to 1 dp)} \qquad \text{Using conversion formula and then the calculator}$$

For (b), write

$$94.6 \times \frac{\pi}{180} = 1.65 \text{ rad} \quad \text{(to 2 dp)} \qquad \text{Using the conversion formula and then the calculator}$$

Radians are used to solve problems concerning arc length (s) and area of a sector (A); see figure b, where the appropriate formulas are stated.

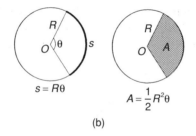

(b)

Example 2. Amanda is schooling up her horse George for a show. He is trotting in a circle at the end of a rope held by Amanda, who is at point A in figure c. The length of the rope is 10 meters.

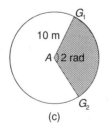

(c)

(a) As Amanda turns through an angle of 2 radians, find the distance trotted by George along the minor arc of the circle from G_1 to G_2.

(b) Find the area of the minor sector swept out by the rotation of the rope.

Solution. (a) The arc length is the distance G_1 to G_2 measured around the circumference of the circle. Write

$G_1G_2 = 10 \times 2$ Substitute $R = 10, \theta = 2$ into the formula $s = R\theta$

$G_1G_2 = 20$ meters

(b) For the area, write

$A = \frac{1}{2} \times 10^2 \times 2$ Substituting $R = 10$ and $\theta = 2$ into the formula $A = \frac{1}{2}R^2\theta$

$A = 100 \text{ cm}^2$

If the angle is in degrees, it must be converted into radians in order to use these formulas for arc length and area of a sector.

References: Arc, Arc Length, Area, Circumference.

RADIOACTIVE DECAY

Reference: Exponential Decay.

RADIUS

References: Chord, Circle.

RANDOM NUMBERS

Random numbers are numbers that occur without any design or pattern. They are numbers that occur completely by chance. For example, when you roll a die all of the numbers 1, 2, 3, 4, 5, and 6 have an equal chance of turning up, provided the die is not biased. A die will only generate random numbers from 1 to 6. If you wanted random numbers from 1 to 9 there are various options open to you. One option is to place the numbers 1 to 9 in a hat and draw one number at a time without looking. Of course after each number is drawn it would have to be replaced, because at each draw every one of the numbers must have an equal chance of being drawn from the hat. Another option is to design a "spinner" with the numbers 1 to 9 equally spaced, as in the figure. A

scientific calculator can generate random numbers of any desired size, and the reader is referred to the handbook for instructions for the particular calculator they use.

Random numbers are used in statistics to select a random sample from a population, as explained in the following example. One day the Principal strides into Luke's classroom and decides to inspect the students' books to check them for neatness. He does not want to read through all the books, because he is very busy, so he decides to select just five books at random. There are various options. The first option is to choose the first five students he meets on entering at the back of the room. But he knows that lazy students may sit at the back and these books might be the worst in the class. The second option is to choose the 1st, the 6th, the 12th, the 18th, the 24th, and the 30th names on the roll. This option will give a random sample. The third option, which is also a random selection, is explained below.

The teacher decides to select 5 students from the class of 30, using random numbers. She writes down the names of all 30 students and allocates to each name a two-digit number, such as 01, 02, 03, 04,..., 10, 12,..., 28, 29, 30. Then she generates 5 random numbers, using her calculator. The calculator can be preset to only include numbers from 01 to 30. If the calculator is not preset, it means that any numbers that turn up that are outside the range 01 to 30 are discarded. Suppose the random numbers are 07, 18, 02, 27, and 11. This means the students that have been allocated those numbers are the ones whose books are sent to the Principal.

References: Bias, Digit, Population, Random sample.

RANDOM SAMPLE

This is a term used in statistics. A random sample is part of a population and is chosen without bias from the whole population, and is hoped to have the same characteristics as the whole population. For the sample to be random, every member of the population must have the same chance of being selected. Since the random sample is expected to have the same characteristics as the population, it is used to make predictions about, the whole population. Jack is the manager of a jam factory of 1500 workers and wanted to know how they felt about overtime. He could ask all of them, which would take a lot of time and expense, or he could select a random sample of workers and question the sample. The information obtained from the sample would not be as reliable as from the whole population, but provided the sample included a good cross section of the work force, it would have, it is hoped, pretty much the same characteristics. To ensure a good cross section of ages Jack may wish randomly to select, say, 20 workers from five different age groups.

References: Bias, Population, Random Numbers, Sample.

RANGE

References: Cartesian Coordinates, Correspondence.

RANGE (STATISTICS)

Reference: Standard Deviation.

RATE

A rate compares one quantity with another quantity, which has different units, by dividing the quantities. Rate is not to be confused with ratio, which is the division of one quantity by another quantity which has the *same* units. When we divide the quantity of distance in kilometers by the quantity of time in hours, we obtain a rate of kilometers per hour, which is speed. We can say that speed is the rate at which distance is changing with respect to time. There are many examples of rates in everyday life, such as the running speed measured in meters per second of an athlete in a race, an individual's heart beat measured in beats per second, and fuel consumption in kilometers per liter or miles per hour on a car trip.

Example 1. In a recent athletics meet David ran the 10,000 meters race in a time of 28 minutes, 55 seconds. Calculate his average speed for the race in meters per second.

Solution. The time must be expressed in seconds: 28 minutes, 55 seconds $= 1735$ seconds. Write

$$\text{Average speed} = \frac{10,000}{1,735} \qquad\qquad \text{Speed} = \frac{\text{distance}}{\text{time}}$$

$$= 5.76 \quad \text{(to 2 dp)}$$

David's running rate is 5.76 meters per second.

Example 2. Which is the better buy, a 3-kg bag of potatoes for \$1.55 or a 10-kg bag for \$5.30?

Solution. Write

Rate for the 3-kg bag $= \$0.52$ per kg.   Dividing \$1.55 by 3 kg to 2 dp

Rate for the 10-kg bag $= \$0.53$ per kg.   Dividing \$5.30 by 10 kg

The 3-kg bag is the better buy, because it has a lower cost per kilogram.

Example 3. Darren uses 14 liters of gasoline for a trip of 189 km in his car. How far would he travel on 37 liters at the same rate?

Solution. Write

$$\text{Rate} = \tfrac{189}{14} \qquad\qquad \text{Dividing kilometers by liters}$$

$$= 13.5 \text{ km per liter}$$

For a trip of 37 liters he would travel

$$13.5 \times 37 = 499.5 \text{ km} \qquad \text{Distance} = \text{rate} \times \text{number of liters}$$

Darren would travel 499.5 km on 37 liters of gasoline.

Example 4. Pat and her friends are doing a vacation job in an orchard picking apricots. Three girls can pick six baskets of fruit in 5 hours. How many baskets of fruit could four girls pick in 4 hours, working at the same average rate?

Solution. Using a table is a good way to keep track of the information in this difficult problem. Reducing the number in the Number of girls column to 1 while keeping the number in the Number of hours column fixed is a suitable way to begin solving the problem. Restrict calculations to two columns at a time.

Number of Girls	Number of Baskets	Number of Hours
3	6	5
$3 \div 3 = 1$	$6 \div 3 = 2$	5
$1 \times 4 = 4$	$2 \times 4 = 8$	5
4	$8 \div 5 = 1.6$	$5 \div 5 = 1$
4	$1.6 \times 4 = 6.4$	$1 \times 4 = 4$

Result: Four girls pick 6.4 baskets of apricots in 4 hours.

If a problem on rates involve a graph it should be remembered that the gradient of the graph is a rate.

Example 5. The graph in the figure is about Nathan's bath time. It shows the volume of water in the bath after a certain time. V is the volume of water in the bath in liters and T is the time in minutes. The graph is made up of three straight lines. The first line shows the bath being filled. The second line represents Nathan being in the bath. The third line shows the bath emptying. Find the rate at which the bath is filling and the rate at which it is emptying.

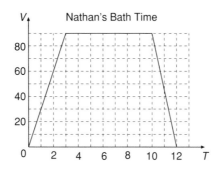

Solution. The rate at which the bath is filling is the gradient of the first straight line:

$$\text{Gradient} = \frac{\text{rise}}{\text{run}}$$

$$= \frac{90 \text{ liters}}{3 \text{ minutes}}$$

$$= 30 \text{ liters/minute}$$

The bath is filling at a rate of 30 liters per minute.

The rate at which the bath is emptying is the gradient of the third straight line, which is negative. The negative sign indicates that the bath is emptying rather than filling:

$$\text{Gradient} = \frac{\text{rise}}{\text{run}}$$

$$= \frac{90 \text{ liters}}{2 \text{ minutes}}$$

$$= -45 \text{ liters/minute}$$

The bath is emptying at a rate of 45 liters per minute. (the negative sign is not written in the final answer, because it is stated that the bath is emptying).

For straight-line graphs the rates are constant throughout the length of the straight line. Rates can also be calculated for curved graphs. For curved graphs the rate is changing and has a different value at each point on the graph. Therefore, the rate can only be found at an instant. For this reason rates for curved graphs are called instantaneous rates. An example of instantaneous rates is included in the entries Gradient of a Curve and Rate of Change.

References: Average, Gradient of a Curve, Proportion, Ratio.

RATE OF CHANGE

Suppose you are painting a wall of your house and observe how the area painted and the time spent on the job are related to each other. At different times of the day you are probably painting at a faster rate than at other times. A rate of change is the rate at which one quantity changes (say the area of the wall you have painted) with respect to the change in another quantity (say the time spent on painting that area of the wall). If a graph is drawn of the two quantities with area on the vertical axis and time on the horizontal axis, the rate of change of area with respect of time is the gradient of the graph. If the graph is a straight line, the rate of change is the gradient of the straight line, which is constant throughout its length. For a curved graph the gradient is changing throughout its length, so the gradient can only be calculated at a specific point on the graph and is called an instantaneous rate of change at that point.

Suppose a tap is turned on so that the water flows at a steady rate, and the water is caught in a container that is in the shape of a cylinder (see figure a). Let the depth of water in the container be d centimeters when the water has been flowing for t seconds. A graph is then drawn with time t seconds on the horizontal axis and depth d centimeters on the vertical axis. The resulting graph for this kind of container is a sloping straight line as shown in figure a.

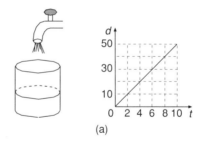

(a)

The rate of change of the depth of water in the cylinder with respect to time is the gradient of the graph. Write

$$\text{Rate of change} = \tfrac{50}{10} \qquad\qquad \text{Gradient} = \frac{\text{rise}}{\text{run}}$$

$$= 5$$

The rate of change of depth with respect to time is 5 centimeters/second.

For containers of different shapes the graphs are different, and a graph has a negative gradient if the container is emptying. The containers and their graphs are shown side by side in figure b. In each case the water is flowing into the containers at a constant rate and the graphs show how the depth of water in the containers changes with time. When the flow of water stops the graph will be horizontal, which has a zero gradient.

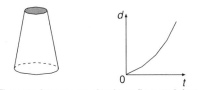

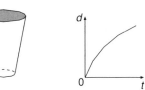

The rate increases slowly at first and then more rapidly as the container gets narrower.

The rate increases rapidly at first and then more slowly as the container gets wider

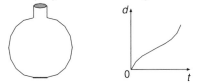

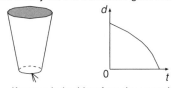

The rate increasesly quite rapidly at first, then slows as the container widens, then increases again. Finally the rate becomes steady and the graph straight at the neck of the container.

If water is leaking fom the container at a steady rate the gradient of the graph will be negative.

(b)

We now look at the method of finding the rate, which is the gradient, when a graph is curved. Also see the entry Acceleration.

Example. Farmer John has a pig that he is fattening up for Christmas. He weighs the pig every week and enters the weight in a table:

Week	0	1	2	3	4	5
Weight in kg	30	32	38	48	62	80

Using the data in the table, draw a graph with the number of weeks (N) on the horizontal axis and weight (W) on the vertical axis (see figure c). From the graph estimate the rate of increase in the weight of the pig at 3 weeks.

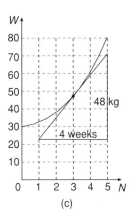

(c)

Solution. The graph is curved, so the rate is changing with time. To find the rate at 3 weeks we need to draw a tangent to the curve at the point on the curve when $t = 3$ weeks. A right-angled triangle is completed with the tangent as its hypotenuse. Try to arrange the size of the triangle so that its horizontal side is a whole number. This will make the next process easier. The sides of the triangle are 4 weeks and 48 kg. Write

$$\text{Gradient of tangent} = \tfrac{48}{4} \qquad \text{Gradient} = \frac{\text{rise}}{\text{run}}$$

After 3 weeks, the pig is growing at a rate of 12 kg per week.

References: Acceleration, Gradient of a Curve, Rate.

RATE OF INTEREST

Reference: Interest.

RATIO

A ratio is the comparison of one quantity with another quantity which has the same units, obtained by dividing the quantities. Suppose the two quantities are 10 cm and 5 cm, which have the same units. The ratio of the two quantities is 10/5 or more simply 2/1. The same ratio is also written as 10:5, which is simplified to 2:1. This ratio 2:1 tells us that one quantity is twice the size of the other. This fact is only true because the two quantities in the ratio are expressed in the same units. A ratio does not have units; it is incorrect to state a ratio as 2:1 cm. The numbers in a ratio must be counting numbers, and not fractions. For example, a ratio is correctly expressed as 5:2, but $2\tfrac{1}{2}$:1 is not correctly expressed.

Example 1. The scale for a map is 5 cm to 1 km. Express this scale as a ratio.

Solution. It is best to express both quantities in the smaller of the two units, in this case centimeters. The two quantities are 5 cm and 100,000 cm, since 1 km = 100,000 cm. Write

$$\text{Ratio} = \tfrac{5}{100,000} \qquad \text{A ratio is the division of one quantity by another}$$

$$\text{Ratio} = \tfrac{1}{20,000} \qquad \text{Canceling the fraction}$$

The scale of the map is 1:20,000.

Ratios are frequently used in everyday life, as illustrated in the following examples.

Example 2. Jacob and Nathan buy a $10 Lotto ticket. Jacob contributes $6.50 and Nathan $3.50.
(a) Write their contributions as a ratio.
(b) If they win $600, how much of the winnings should Jacob receive?

Solution. For (a), write

$$\text{Ratio} = \frac{6.5}{3.5} \qquad \text{This ratio is not written with counting numbers}$$

$$= \frac{6.5}{3.5} \times \frac{2}{2} \qquad \text{Rearranging the ratio to obtain counting numbers}$$

$$= 13{:}7$$

For (b), write

$$13 + 7 = 20. \qquad \text{Adding the numbers in the ratio}$$

$$\$600 \div 20 = \$30 \qquad \text{The winnings are divided into 20 equal parts of \$30}$$

Therefore

$$\text{Jacob gets 13 lots of } \$30 = \$390$$

$$\text{Nathan gets 7 lots of } \$30 = \$210$$

$$\text{Total} = \$600$$

Jacob should receive $390.

Example 3. William and his classmates are going on a camping holiday. The school regulations state that the teacher-to-student ratio must be 1:5. If there are 30 students in William's class, how many teachers must go with them?

Solution. Write

$$\frac{\text{Number of teachers}}{\text{Number of students}} = \frac{1}{5}$$

$$= \frac{1}{5} \times \frac{6}{6} \qquad \text{Arranging the fraction with number of students} = 30$$

$$= \frac{6}{30}$$

Six teachers are needed.

References: Equivalent Fractions, Gears, Proportion, Rate.

RATIONAL EXPRESSION

References: Algebraic Fractions, Canceling.

RATIONAL NUMBERS

Reference: Integers.

RAY

Reference: Line.

REAL NUMBERS

Reference: Integers.

RECIPROCAL

One number is the reciprocal of another number if their product is equal to 1. The word product here means that we multiply. For example, the reciprocal of the number $\frac{3}{2}$ is $\frac{2}{3}$, because $\frac{3}{2} \times \frac{2}{3} = 1$. The easiest way to obtain the reciprocal of a fraction is to turn it upside down. The reciprocal of $2\frac{3}{4}$ is $\frac{4}{11}$, because $2\frac{3}{4} = \frac{11}{4}$. The reciprocal of 5 is $\frac{1}{5}$.

Reciprocals can be used to solve equations, as set out in the following example.

Example. Solve the equation

$$\frac{1}{x} = \frac{2}{3}$$

Solution. Write

$$\frac{1}{x} = \frac{2}{3}$$

$$\frac{x}{1} = \frac{3}{2} \qquad \text{Taking the reciprocal of both sides of the equation}$$

$$x = 1.5$$

References: Fractions, Recurring Decimal, Solving an Equation.

RECTANGLE

A rectangle is a four-sided polygon, called a quadrilateral, which has its opposite sides parallel and all its four angles are right angles. An everyday word for rectangle is oblong. A rectangle has opposite sides equal in length. It also has two axes of symmetry and rotational symmetry of order two.

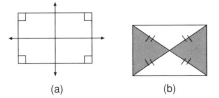

(a)   (b)

The two diagonals of the rectangle are equal in length and bisect each other (see figure a). Both pairs of "opposite" triangles formed by the two diagonals are congruent isosceles triangles, as shown in figure b. A rectangle that has all four sides equal in length is called a square.

If the lengths of two adjacent sides of a rectangle are known, we can find the perimeter of the rectangle, and we can also find the area of the rectangle.

Example. Find the perimeter and the area of the rectangle drawn in figure c.

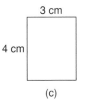

(c)

Solution. The opposite sides of a rectangle are equal in length, so the other two sides are of lengths 3 cm and 4 cm, respectively. Write

$$\text{Perimeter} = 3 + 4 + 3 + 4$$
$$= 14\,\text{cm}$$

The perimeter is 14 cm.

Alternatively, we could find the sum of the two adjacent sides, which is 7 cm, and double the answer.

For the area, write

$$\text{Area} = 4 \times 3 \qquad \text{Area of rectangle} = \text{length} \times \text{width}$$
$$= 12 \text{ cm}^2$$

The area is 12 cm^2.

The length of the diagonal of the rectangle can be calculated using the theorem of Pythagoras. If the length of the diagonal is denoted by d, then

$$
\begin{aligned}
d^2 &= 3^2 + 4^2 \qquad \text{Pythagoras' theorem} \\
&= 9 + 16 \\
&= 25 \\
&= \sqrt{25} \qquad \text{Using square roots} \\
&= 5
\end{aligned}
$$

The length of the diagonal of the rectangle is 5 cm.

References: Algebra, Congruent Figures, Constructions, Diagonal, Isosceles Triangle, Linear Equation, Polygon, Pythagoras' Theorem, Quadrilateral, Square, Square Root.

RECTANGULAR BLOCK

This is another name for a cuboid.

Reference: Cuboid.

RECTANGULAR HYPERBOLA

Reference: Hyperbola.

RECURRING DECIMAL

Reference: Decimals.

RE-ENTRANT QUADRILATERAL

For re-entrant quadrilaterals see the entry Acute Angle.

References: Acute Angle, Concave.

REFLECTION

Reflection is a geometrical transformation that occurs when a shape, called the object, is reflected in a mirror to obtain the image. Suppose we reflect the triangle labeled

ABC in figure a in the mirror m to obtain the image $A'B'C'$. The object on the left of the mirror is labeled with the letters A, B, and C in a clockwise direction, but the image is labeled A', B', and C' in a counterclockwise direction. The orientation, or sense, has changed. Reflection is the only transformation that changes sense. We sometimes say that reflection is an indirect transformation, because it changes sense.

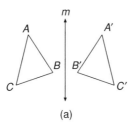

(a)

In figure b the point A' is the image of point A under reflection in the mirror m. It is important to realize that the perpendicular distance of point A from the mirror is equal to the perpendicular distance of point A' from the mirror, and the straight line joining A to A' is at right angles to the mirror.

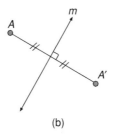

(b)

Notation Under a reflection we write that A' is the image of A. If the reflection is called M, then we write $M(A) = A'$. Alternatively we write: A maps to A' under the reflection M, or in symbols, $A \to A'$ under M.

Sometimes we reflect a shape that lies across the line of the mirror. In figure c, the shaded rectangle has been reflected in the mirror line m.

(c)

We can reflect shapes in sloping mirror lines. In figure d, a yacht is reflected in a sloping mirror line *m*. If we consider the object shape, the image shape, and the mirror line as one composite figure, then the mirror line is an axis of symmetry for this figure.

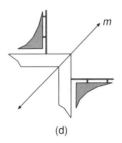

(d)

Example 1. Find the image of the triangle *T* in figure e after it has been reflected in the two mirrors that are drawn with arrows.

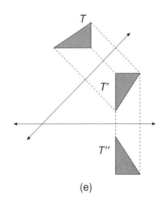

(e)

Solution. Draw lines from each vertex of the triangle at right angles to the mirror and produce the lines the same distance on the other side of the mirror. The final image is *T″*. A clockwise rotation about the point where the two mirrors intersect through an angle that is twice the angle between the mirrors will map *T* onto *T″*.

See the entry; Composite Transformations for further reflections in two mirrors.

Example 2. Harry lives in a house that is 20 meters from a river, and his uncle's house is 45 meters from the river (see figure f). These are the shortest distances from the river. The distance apart of the houses, measured along the river bank is 60 meters. Once each week Harry goes down to the river, fills a bucket with river water, and takes it to his uncle's house to replenish the water in his fish tank. Find the shortest distance that Harry has to walk to accomplish this task.

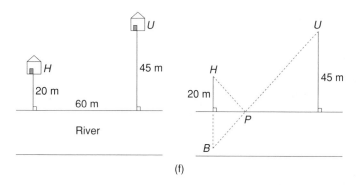

(f)

Solution. Using the bank of the river as a mirror, draw the image of Harry's house, which is the point B. Join point B to point U with a straight line. The shortest route is from H to P, and then from P to U. This is the same distance as a direct line from B to U. The solution to the problem will be the length of the straight line BU in figure f.

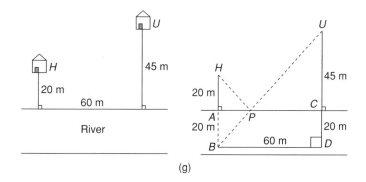

(g)

We now extend the line UC (see figure g), which is from uncle's house to the river, to meet the line through point B and parallel to the bank of the river. In this way the triangle UDB is created, with the length of $UD = 65$ meters and $BD = 60$ meters. The route for Harry to follow is from H to P, then from P to U. The length of this route is the same as the length of the straight line BPU, since $HP = BP$.

We apply the theorem of Pythagoras to triangle UBD:

$UB^2 = 65^2 + 60^2$

$UB^2 = 4225 + 3600$ Squaring 65 and 60

$UB^2 = 7825$

$UB = 88.5$ (to 1 dp)   Using the square root key on the calculator

The shortest distance from Harry's house to his uncle's house, via the river, is 88.5 meters.

Properties of reflection

- The mirror line is an invariant line.
- Any point on the mirror line is an invariant point.
- The perpendicular distance of a point from the mirror line is equal to the perpendicular distance of its image from the mirror line.
- The straight line that joins a point to its image is at right angles to the mirror line.
- Angle sizes are invariant.
- Area is invariant.
- The object and image are congruent shapes.
- Reflection is an indirect transformation.

References: Cross-multiply, Composite Transformations, Enlargement, Escher, Image, Indirect Transformation, Perpendicular Lines, Pythagoras' Theorem, Similar Figures, Symmetry, Transformation Geometry.

REFLEX ANGLE

Reference: Acute Angle.

REGIONS

References: Euler's Formula, Networks.

REGULAR POLYGON

References: Equiangular, Polygon.

REGULAR POLYHEDRON

Reference: Platonic Solids.

RELATION

Reference: Correspondence.

REMAINDER

Reference: Dividend.

REPEATED ROOTS

References: Cubic Equations, Quadratic Graphs.

REPEATING DECIMAL

Reference: Recurring Decimal.

REPRESENTATIVE VALUES

Reference: Central Tendency.

RESULTANT OF A VECTOR

Reference: Components of a Vector.

RETARDATION

Reference: Acceleration.

REVOLUTION

A revolution is a complete turn, which is an angle of 360° or four right angles.

Example 1. If the wheel of a cart has a diameter of 20 cm, find how far the cart travels if the wheel makes one revolution.

Solution. As the wheel makes one revolution, the cart will move forward a distance equal to the circumference of the wheel (see figure). Write

$C = \pi \times d$	Formula for circumference of a circle
$C = \pi \times 20$	Substituting $d = 20$
$C = 62.8 \quad \text{(to 1 dp)}$	Using a scientific calculator

The cart travels a distance of 62.8 cm.

Example 2. If the same cart wheel makes 400 revolutions per minute, find the speed of the cart in meters per minute.

Solution. Using the answer from the previous example, we know that the cart travels $\pi \times 20$ cm in one revolution. if we use 62.8 cm instead, the result will not be as accurate. Write

$$400 \times \pi \times 20 = 25,133 \text{ cm}$$ Distance cart travels in 1 minute, to nearest whole number

This means that the cart travels 25,133 cm in 1 minute. Now write

$$25,133 \div 100 = 251.33 \text{ meters}$$ 100 cm = 1 meter

The speed of the cart is 251.33 meters/minute, or 15.1 kilometers/hour, to 1 dp. Multiplying meters/minute by 0.06 changes it to kilometers/hour.

References: Circumference of Circle, Right Angle.

RHOMBUS

A rhombus is a parallelogram with the extra property that all of its four sides are the same length. In everyday life it is sometimes called a diamond. Another property of the rhombus is that its two diagonals bisect each other at right angles, but the two diagonals are not the same length (see figure a).

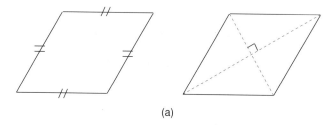

(a)

The rhombus has a rotational symmetry order of two, like the parallelogram, but unlike the parallelogram, the rhombus has two axes of symmetry, which are its two diagonals. These axes of symmetry can be clearly seen if the rhombus is drawn on a square grid, as shown in figure b.

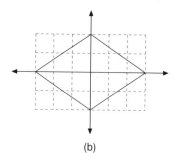

(b)

If the lengths of the diagonals of a rhombus are a and b, then the area of the rhombus is given by

$$\text{Area} = \tfrac{1}{2}\, ab$$

The lengths of the diagonals of the rhombus that are drawn on the grid in figure b are $a = 6$ and $b = 4$. Therefore

Area of this rhombus $= \tfrac{1}{2} \times 6 \times 4$ Using the formula for the area of a
                                          rhombus

$$= 12$$

The area of this rhombus is 12 square units.

The area of a rhombus is also equal to base × height, like the parallelogram.

Since the rhombus is made up of four congruent right-angled triangles, it is possible to use the Pythagoras Theorem in one of these triangles, as in the following example.

Example. Ron is having a diamond-shaped stained glass window put into a wall of the church to give light to a particularly dark corner. He wants the width of the window to be 3 meters and the height to be 4 meters, as shown in his preliminary sketch in the figure. How long will the total length of the window frame be?

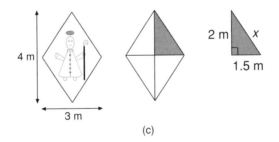

(c)

Solution. A diamond is taken to mean a rhombus, and the lengths of the diagonals are 4 and 3 meters, respectively. The diagonals divide the rhombus into four right-angled triangles that are all congruent. One of these triangles is drawn shaded in the figure. The lengths of the sides are 2 and 1.5 meters, respectively. Suppose we let the length of the hypotenuse of this triangle be x meters. Write

$$x^2 = 2^2 + 1.5^2 \qquad \text{Pythagoras' theorem}$$

$$x^2 = 4 + 2.25 \qquad \text{Squaring the numbers}$$

$$x^2 = 6.25$$

$$x = \sqrt{6.25} \qquad \text{Taking square roots}$$

$$x = 2.5$$

Then

Total length $= 4 \times 2.5$ The total length of four sides $= 4 \times$ length of one side

$\qquad = 10$

The total length of the window frame is 10 meters.

References: Congruent Figures, Diagonal, Hexagon, Hypotenuse, Parallelogram, Regular Polygon, Right Angle, Rotational Symmetry, Symmetry.

RIGHT ANGLE

Reference: Acute Angle.

RIGHT PRISM

This is a prism in which the sides are at right angles to the uniform cross section. The figure shows a right triangular prism in which the uniform cross section (drawn shaded) is at right angles to each of the three sides. Another example of a right prism is a cylinder.

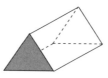

References: Cross Section, Cylinder, Prism.

RIGHT PYRAMID

This is a pyramid that has an axis of symmetry at right angles to its base. The figure shows a right square-based pyramid in which the axis of symmetry *VO* is perpendicular to the base. Another example of a right pyramid is a cone.

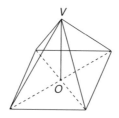

References: Cone, Pyramid, Right Prism.

RIGHT TRIANGLE

This is a triangle in which one of the angles is a right angle, and is used frequently in mathematics, especially in trigonometry and the Pythagoras Theorem. The longest side of a right triangle is called the hypotenuse. Another name for a right triangle is a right-angled triangle.

References: Pythagoras' Theorem, Trigonometry.

ROMAN NUMERALS

The number system that we use today enables us to express all numbers by means of the 10 symbols $\{0, 1, 2, 3, 4, 5, 6, 7, 8, 9\}$ with each symbol varying in value according to the position or place it occupies. For example, in the number 32 the three stands for 3×10, and in the number 324 the three stands for 3×100.

The mathematicians of ancient Rome did not use that principle of place value, and consequently their number system was clumsy and difficult to use. The Roman numerals and their equivalents are as follows:

- I is 1
- V is 5
- X is 10
- L is 50
- C is 100
- D is 500
- M is 1000

Other numbers are formed from these numerals. Roman numerals have a fixed value wherever they appear, unlike our present system. When two numerals appear side by side we add them. Some examples are

$$MM = 1000 + 1000$$
$$= 2000$$
$$CCC = 100 + 100 + 100$$
$$= 300$$
$$DC = 500 + 100$$
$$= 600$$

You will note that the above numerals are written such that the order of values is either descending or the same:

$$1943 = 1000 + 500 + 100 + 100 + 100 + 100 + 10 + 10 + 10 + 10 + 1 + 1 + 1$$
$$= M + D + C + C + C + C + X + X + X + X + I + I + I$$
$$= \text{MDCCCCXXXXIII}$$

Later, another principle was developed in order to reduce the number of letters involved in expressing large numbers such as 1943 as Roman numerals. When the letter for a smaller numeral is placed before the letter for a larger numeral, we subtract the two numerals. For example, IV means $5 - 1 = 4$, which is preferable to IIII. The number 1943 can now be expressed more briefly as M(CM)(XL)III, with the numerals that are enclosed in brackets being subtracted. Now remove the brackets to give $1943 =$ MCMXLIII.

Example. Write MCM as a number in present-day numerals.

Solution. Write

$$\text{MCM} = M + (M - C) \qquad \text{C written before M means we subtract, since C is smaller than M}$$

$$= 500 + (500 - 100)$$

$$= 900$$

The first 20 numbers in Roman numerals are

1 = I	6 = VI	11 = XI	16 = XVI
2 = II	7 = VII	12 = XII	17 = XVII
3 = III	8 = VIII	13 = XIII	18 = XVIII
4 = IV	9 = IX	14 = XIV	19 = XIX
5 = V	10 = X	15 = XV	20 = XX

Reference: Numerals.

ROOTS

When we solve an equation we obtain solutions, and these solutions are sometimes called the roots of the equation. For example, $x = 4$ is a root of the linear equation $2x - 3 = 5$, and $x = 3$ and $x = -4$ are the roots of the quadratic equation $x^2 + x - 12 = 0$.

References: Linear Equation, Quadratic Equations.

ROTATION

A rotation is a geometrical transformation that frequently occurs in everyday life. A rotation turns a shape around. An example is a child's carousel at a fair. The carousel spins about its center O, which is fixed (figure a). The carousel can turn clockwise or counterclockwise through a full turn or through an angle which may be part of a full turn. Amanda decides to ride on the pig P.

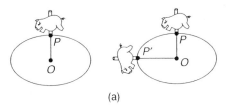

(a)

If the carousel turns counterclockwise about the center O through an angle of $90°$, the pig finishes up at the point P'. When the pig is in its starting position at P we say it is the "object." When the pig is in its finishing position at P' we say it is the "image." This is written as $P \rightarrow P'$; in words we say P maps onto P'. The whole transformation is described in the following way:

P' is the image of P after a rotation of $90°$ counterclockwise about the point O. This point O is called the "center of rotation" and $90°$ is the angle of rotation. The angle of rotation is regarded as a positive angle when it is measured counterclockwise and a negative angle when it is measured clockwise.

If the pig starts at P again and this time the rotation is $180°$ about O, the image of P is as shown in figure b as P'. A rotation of $180°$ can be regarded as clockwise or counterclockwise.

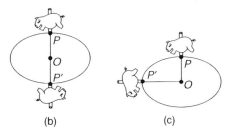

(b)   (c)

After a rotation about O through an angle of $270°$ clockwise the image of P is P', as shown in figure c. It can be seen that a rotation of $270°$ clockwise is the same as a rotation of $90°$ counterclockwise.

Example 1. Draw the image of triangle ABC after a rotation of $-90°$ about the origin. Write down the coordinates of the images of the points A, B, and C.

Solution. The rotation of the triangle ABC about the origin is in a clockwise direction, because the angle of rotation is negative, and is achieved by rotating about 0 each of the points A, B, and C in turn (see figure d). To rotate the point B, join the points 0 and B with a straight line $0B$ and draw a circle with center at 0 and radius $0B$. The points B and B' lie on this circle.

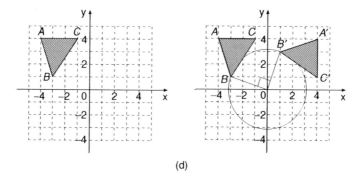

(d)

With a set square, or protractor, measure a clockwise turn of 90° from $0B$ and draw the line $0B'$. Repeat the process for each of the points A and C to obtain the image points A' and C'. Now join up the points A', B', and C' to make the image triangle. As you can see, we rotate the points of the shape one at a time and then join them up to obtain the complete image. An alternative method is to make a tracing of the figure and turn the tracing paper clockwise through 90° and trace the image triangle.

In some problems the center of rotation may be a point that is part of the shape that is being rotated, as in the following example:

A compound pendulum is rotated counterclockwise through an angle of 30° about a point that is the midpoint of one of its sides. The object and image are drawn in figure e.

(e)

In the next example you are given the object and the image shapes and asked to find the center of rotation and the angle of rotation.

Example 2. The shaded triangle ABC in figure f is rotated about a certain point and $A'B'C'$ is the image. Find the center of the rotation and also the angle of rotation.

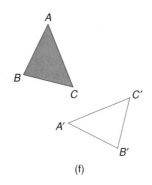

(f)

Solution. Join A to its image point A' (see figure g). Construct the perpendicular bisector of the line AA' and call it p. Join C to its image point C'. Construct the perpendicular bisector of the line CC' and call it q. The point O where the two lines p and q meet is the center of rotation.

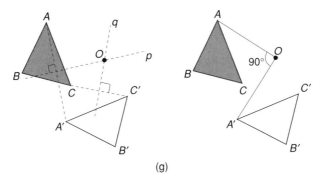

(g)

To find the angle of rotation, join O to A with a straight line and join O to A'. The angle AOA' between these two lines is the angle of rotation, which can be measured with a protractor. In this example, the angle of rotation is $90°$ counterclockwise.

Properties of Rotations

- The center of rotation is the only invariant point.
- A rotation of $180°$ about the center of rotation O is equivalent to an enlargement with a scale factor of $k = -1$ with its center of enlargement at O.
- The object shape and the image shape are congruent shapes. We call this an *isometry transformation.*
- If the lines are parallel in the object, they will also be parallel in the image. We call this an *affine transformation,* which preserves parallelism.
- A point in the object is the same distance from the center of rotation as its image point. This means that $OA = OA'$.
- Rotation is a direct transformation, which means that the object and image have the same sense i.e. it is a direct transformation.

References: Composite Transformations, Congruent Figures, Enlargement, Escher, Indirect Transformation, Invariant Points, Perpendicular Bisector, Rotational Symmetry.

ROTATIONAL SYMMETRY

Reference: Order of rotational symmetry.

ROUNDING

Reference: Accuracy.

ROW

Reference: Column.

RULE

References: Difference Tables, Formula, Patterns.

$$S$$

SAMPLE

A sample is a term used in statistics to refer to part of a population. If the sample is a random sample, it is chosen without bias from the whole population, and, it is hoped, has the same characteristics as the whole population. These terms are discussed more fully under the entries listed in the references.

References: Bias, Population, Random Sample.

SAMPLE SPACE

Reference: Event.

SCALE

In everyday life we use scales to measure such quantities as:

- Length, using a ruler or tape measure
- Temperature, using a thermometer
- Angles in degrees, using a protractor
- Weight, using weighing scales
- Revolutions per minute, using a revolution counter
- Volts, using a voltmeter
- Altitude, using an altimeter

A scale may be uniform or nonuniform, but in this entry we will study only uniform scales. For these scales equal distances on the scale represent equal quantities that are being measuring. The following examples illustrate how to read uniform scales.

On the voltmeter shown in figure a, 5 volts is divided up into four equal parts, so each part is $5 \div 4 = 1.25$ volts. The reading for point A is $5 + 3 \times 1.25 = 8.75$ volts.

The reading for point B is $20 + 2 \times 1.25 = 22.5$ volts. The reading for point C is $25 + 2.5 \times 1.25 = 28.125$ volts.

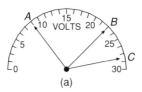

(a)

Figure b shows a measuring tape with 1 meter divided into five equal parts, so each part is $1 \div 5 = 0.2$ meters. The reading for point A is $3 + 2 \times 0.2 = 3.4$ meters. The reading for point B is $4 + 3 \times 0.2 = 4.6$ meters. The reading for point C is $2 + 3.5 \times 0.2 = 2.7$ meters.

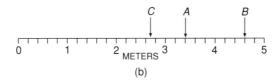

(b)

Figure c shows a measuring jug for liquids. Each 50 milliliters is divided into two equal parts, so each part is $50 \div 2 = 25$ milliliters. The reading for point A is $0 + 25 = 25$ ml. The reading for point B is $150 + 25 = 175$ ml. The reading for point C is $100 + 0.5 \times 25 = 112.5$ ml.

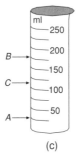

(c)

References: Meters, Milliliters.

SCALE DRAWING

Pat wanted a new house, so she visited Rob, an architect, who drew up on paper the plans for the house. The plans were obviously not the full size of the house and Rob used a scale of 1:150 for some of the drawings and a scale of 1:50 for those that required more detail. Rob's plans are called the scale drawings of the house. The scale must be included with every scale drawing. Scales can be written using units, for example a scale of 1:150 can be written as 1 cm to 150 cm, or 1 cm to 1.5 meters.

Example. The scale drawing in the figure is part of Pat's house. What is (a) the length of bedroom 3? (b) The length of the bathroom? (c) The area of bedroom 3?

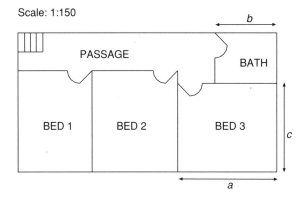

Scale: 1:150

Solution. A scale of 1:150 is equivalent to a scale of 1 cm to 1.5 meters.

(a) Using a ruler, we find that the length of bedroom 3 on the plan is 2.7 cm. Write

Actual length of bedroom 3 $= 2.7 \times 1.5$ Scale is 1 cm is equivalent to 1.5 m

$$= 4.05 \text{ meters}$$

(b) The measured length of the bathroom on the plan is 1.7 cm. Write

$$\text{Actual length of bathroom} = 1.7 \times 1.5$$
$$= 2.55 \text{ meters}$$

(c) The measured width of bedroom 3 on the plan is 2.4 cm. Write

$$\text{Actual width of bedroom 3} = 2.4 \times 1.5$$
$$= 3.6 \text{ meters}$$

Then

$$\text{Area of bedroom 3} = \text{length} \times \text{width}$$
$$= 4.05 \times 3.6$$
$$= 14.58 \text{ square meters}$$

The area of bedroom 3 is 14.58 square meters.

Reference: Ratio.

SCALE FACTOR

For information on the ratio of areas and volumes of similar shapes, see the entry Similar Figures.

References: Enlargement, Ratio, Similar Figures.

SCALE OF A MAP

Reference: Ratio.

SCALENE TRIANGLE

A scalene triangle is one that has three sides of different lengths. It does not have an axis of symmetry, and its order of rotational symmetry is one. The figure shows three examples of scalene triangles.

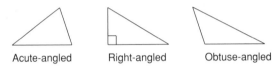

Acute-angled   Right-angled   Obtuse-angled

References: Axis of Symmetry, Equilateral Triangle, Isosceles Triangle, Rotational Symmetry.

SCATTER DIAGRAM

A scatter diagram is used in statistics to see if two variables are related to each other. The two variables are the two coordinates of a point, and the points are plotted on a Cartesian graph. A scatter diagram is used when both sets of data are continuous and when one set of data does not seem to be dependent on the other. This means that there is not a formula connecting the two sets of data. If the two sets of data were related by a formula, then a line graph would be drawn, instead of a scatter diagram. The processes involved in drawing a scatter diagram are outlined in the following example.

Example. Bill is a math teacher and decided to measure the heights (H) in meters and weights (W) in kilograms of 24 boys in his class to see if there was a correlation between the two variables. The results of his measurements were recorded in a table.

H	1.30	1.34	1.36	1.36	1.38	1.40	1.42	1.42	1.44	1.46	1.48	1.48	1.50
W	50	50	50	52	53	54	54	55	56	55	56	58	57

H	1.52	1.53	1.54	1.56	1.58	1.62	1.66	1.66	1.68	1.74	1.80
W	58	60	62	62	61	62	64	65	66	68	72

Using height on the horizontal axis and weight on the vertical axis, he plotted the 24 points. For example the first two points to be plotted on the graph are (1.30, 50) and (1.34, 50). See the figure. The broken axes indicate they do not start at (0, 0).

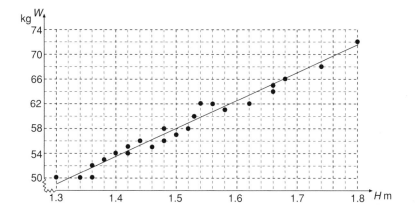

If there is a good correlation between H and W, the points should roughly lie on, or near, a straight line. This has proved to be the case with Bill's investigation. As a result of this investigation Bill can state that the heights of students in his class are related to their weights. In other words, there is a strong correlation between the heights and weights of students in his class. A straight line of "best fit" is drawn through, or near, as many points as possible, trying to get about the same number of points on either side of this line. If the points on the graph do not lie close to a line, but seem to be randomly scattered, then we assume there is not a correlation between the two variables H and W.

References: Cartesian Coordinates, Random Sample, Statistics.

SCIENTIFIC NOTATION

Reference: Standard Form.

SCORE

This is a set of 20. For example, a score of soldiers is a set of 20 soldiers.

SECANT

Reference: Chord.

SECOND

A unit of time.

References: CGS System of Units, Système International d'Unités.

SECOND

A second is a very small unit of angle measure. The angle of 1 second is $\frac{1}{60}$ of 1 minute, and the angle of 1 minute is $\frac{1}{60}$ of 1 degree. There are 360 degrees in a full turn.

Reference: Degree, Minute.

SECTION

Reference: Cross Section.

SECTOR OF A CIRCLE

A sector is that part of a circle that lies between two radii *OA* and *OB* and the arc *AB*, as shown in figure a. The shaded region *OAB* is a sector of the circle that has its center at the point *O*. The remainder of the circle that is not shaded is also a sector of the same circle. To distinguish between the two sectors, the smaller sector is called the *minor sector* and the larger sector is called the *major sector*. The correct way to cut up a pizza is to cut it up into sectors, one sector for each person.

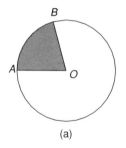

(a)

Example. Jo bought a family-size pizza. Her son Luke measures the diameter of the pizza to be 44 cm. Jo cuts the pizza into five equal pieces, one for each member of her family. What is the area of each piece?

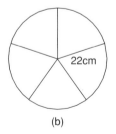

22cm

(b)

Solution. If the diameter is 44 cm, then the radius of the pizza is $R = 22$ cm (see figure b). There are five equal pieces of pizza, so one piece has an area of $\frac{1}{5}$ of the area of the whole pizza. Write

$$A = \pi R^2 \qquad \text{Formula for the area of a circle}$$

Therefore

$$\text{Area of one piece} = \tfrac{1}{5} \times \pi R^2 \qquad \text{There are five equal pieces in the pizza}$$

$$= \tfrac{1}{5} \times \pi \times 22^2 \qquad \text{Substitute } R = 22$$

$$= 304 \quad \text{to nearest whole number}$$

The area of each piece of pizza is 304 cm^2.

A word often confused with sector is *segment*. The chord of a circle divides the circle into a minor segment and a major segment (see figure c). The diameter divides the circle into two congruent half-circles called semicircles. The correct way to cut up a pizza is into sectors and not into segments!

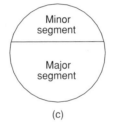

(c)

References: Arc Length, Area, Chord, Circle, Geometry theorems, Perimeter, Radian, Radius, Semicircle.

SEGMENT OF A CIRCLE

Reference: Sector of a Circle.

SELECTION

Reference: Combinations.

SELLING PRICE

Reference: Cost Price.

SEMICIRCLE

A diameter of a circle divides the circle into two equal sectors, and each sector is called a semicircle.

Example. Pat loves fans and has a particular favorite. One day she opens out her favorite fan into the shape of a semicircle (see the figure). Her son David measures the diameter of the fan to be 27 cm. Calculate the area and perimeter of the fan.

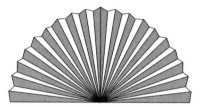

Solution. The area of the fan will be half the area of the full circle of diameter 27 cm. Write

$$\text{Area of semicircle} = \tfrac{1}{2}\pi R^2 \qquad\qquad \text{Area of a whole circle} = \pi R^2$$

$$= \tfrac{1}{2} \times \pi \times 13.5^2 \quad \text{The radius of a circle is half the diameter}$$

$$= 286 \quad \text{to nearest whole number}$$

The area of the fan is 286 cm^2.

The perimeter of the fan is half the circumference of the circle + the diameter of the circle. Write

$$\text{Perimeter of fan} = \tfrac{1}{2}\pi D + 27 \qquad\qquad \text{Circumference of circle} = \pi D$$

$$= \tfrac{1}{2} \times \pi \times 27 + 27$$

$$= 69 \quad \text{to nearest whole number}$$

The perimeter of the fan is 69 cm.

References: Area, Chord, Circle, Radius, Sector of a Circle.

SENSE

References: Indirect Transformation, Reflection.

SEPTAGON

A septagon is a polygon with seven sides. It is more often called a heptagon. See the entry Polygon for the angle sum of the septagon. Some septagons are drawn in figure a; the last one is a regular septagon.

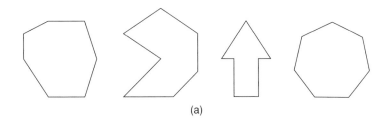

(a)

The properties of the regular septagon are as follows: Its seven sides are of equal length, and its seven angles are all equal to $128\frac{4}{7}$ degrees, or $128.6°$, to 1 dp. It has seven axes of symmetry, and the order of rotational symmetry is seven.

The regular septagon is made up of seven isosceles triangles. An enlarged view of one of these triangles is shown in figure b. The sizes of the angles have been rounded to 1 dp.

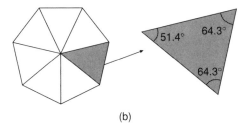

(b)

References: Axis of Symmetry, Heptagon, Isosceles Triangle, Polygon, Rotational Symmetry, Rounding.

SEQUENCE

References: Difference Tables, Patterns.

SI UNITS

References: Système International d'Unités, CGS System of Units.

SIDE

A side is one of the line segments that form a polygon. For example, in the figure, one side of the triangle *ABC* is the line segment *AB*. Alternatively, a side can refer to one of the faces of a polyhedron. For example, the shaded region of the cuboid in the figure is called a side of the cuboid. The cuboid has six sides.

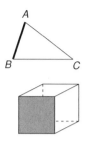

References: Cuboid, Line segment, Polygon, Polyhedron.

SIEVE OF ERATOSTHENES

References: Eratosthenes' Sieve, Prime Number.

SIGNIFICANT FIGURES

Abbreviated sf.

Reference: Accuracy.

SIMILAR FIGURES

When one plane figure is an enlargement of another plane figure we say the two figures are similar. If two figures are similar, they are the same shape; for example, both figures may be triangles, or both figures may be hexagons. Similar figures have the following properties:

- The angles in one figure are equal to the corresponding angles in the other figure.
- Corresponding sides in the two figures are in the same ratio.

Example 1. David and his son William have tee shirts of the same design, but different sizes. The patterns that William's mom, Jane, needs for making the shirts are shown in figure a. All the lengths on David's shirt are two times the lengths on

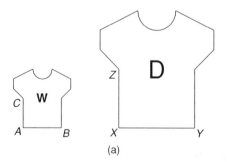

(a)

William's shirt. Therefore the shirt patterns are similar figures, except for the motifs on the fronts.

The lengths of corresponding sides of the patterns are in the same ratio. If the waist length AB on William's pattern is 30 cm and the waist length XY on David's pattern is 60 cm, we can write down the equal ratios:

$$\frac{\text{Length of } AB}{\text{Length of } XY} = \frac{30}{60}$$

$$= \frac{1}{2}$$

This ratio of $\frac{1}{2}$ can be called the scale factor of the enlargement of David's pattern to William's. This means that the lengths on William's pattern are one-half of the corresponding lengths on David's pattern.

On William's shirt pattern the length of $AC = 22.5$ cm, so the corresponding length on David's shirt pattern is XZ, which is $2 \times 22.5 = 45$ cm.

The ratio of the areas of the two shirts is the square of the ratio of the lengths:

$$\text{Ratio of areas} = \left(\tfrac{1}{2}\right)^2 \qquad \text{Squaring the scale factor for lengths}$$

$$\text{Ratio of areas} = \tfrac{1}{4}$$

This ratio is also true for areas of corresponding parts of the two patterns. For example, the ratio of the areas of the two semicircular necklines which have been cut out is also 1:4.

Suppose we are comparing two solid shapes where one is an enlargement of the other, like two bottles. When one solid shape is an enlargement of the other we say that the two solids are *similar shapes*.

Example 2. The two perfume bottles in figure b are similar shapes. The diameter of the base of the smaller bottle is one-third the diameter of the base of the larger bottle. If the smaller bottle holds 10 milliliters, how much does the larger bottle hold?

(b)

Solution. The ratio of the volumes (or capacities) of the two bottles is equal to the cube of the ratio of the lengths, which is the scale factor. Write

$$\text{Ratio of lengths} = \tfrac{1}{3}$$

Then

$$\text{Ratio of volumes} = \left(\frac{1}{3}\right)^3 \qquad \text{Cubing the scale factor}$$

$$= \frac{1}{27}$$

The larger bottle holds 27 times the volume (or capacity) of the smaller bottle, which holds 10 ml. Write

$$\text{Capacity of the larger bottle} = 27 \times 10$$

$$= 270 \text{ ml}$$

The capacity of the larger bottle is 270 milliliters, or 0.27 liters.

References: Area, Capacity, Corresponding Angles, Corresponding Sides, Cross Multiply, Cube, Enlargement, Liters, Milliliters, Ratio, Reflection, Scale Factor, Square, Volume.

SIMPLE INTEREST

References: Compound Interest, Percentage.

SIMULTANEOUS EQUATIONS

Simultaneous equations are two or more equations that are to be solved at the same time. The simultaneous equations we deal with here are sets of two equations in two variables, which represent two unknown quantities. A solution to the simultaneous equations is a value for each of the two variables that make both of the equations true at the same time hence the word simultaneous. An example best illustrates the concept.

Example 1. Nathan and Jacob go into the Cafe Roma and have lunch. Nathan has two sausage rolls and three small muffins at a cost of $3.80. Jacob buys two sausage rolls and two small muffins and his bill is $3.10. From this information find the price of one sausage roll and the price of one small muffin.

Solution. Jacob and Nathan buy the same number of sausage rolls, two, but Nathan has one extra muffin. Therefore the difference in what they pay must be the price of one muffin:

$$\text{Price of one muffin} = \$3.80 - \$3.10$$
$$= \$0.70$$

The price of one muffin is 70 cents.

To find the price of a sausage roll we substitute the price of a muffin into either Nathan's bill or Jacob's bill:

Cost of 2 sausage rolls + 3 muffins = $3.80	This is Nathan's bill
Cost of 2 sausage rolls + 3 × $0.70 = $3.80	Substituting $0.70 for the price of one muffin

$$\text{Cost of 2 sausage rolls} = \$3.80 - \$2.10$$
$$= \$1.70$$
$$\text{Cost of 1 sausage roll} = \$0.85$$

The price of one sausage roll is 85 cents.

Summary: A muffin costs 70 cents and a sausage roll costs 85 cents.

Instead of solving the problem in a descriptive way, we can model the problem with two equations and use the following alternative way of setting out.

Solution. Suppose the price of one sausage roll is r cents, so two will cost Nathan $2r$ cents, and the price of one muffin is m cents, so three muffins will cost him $3m$ cents. For Nathan's order, write

$$2r + 3m = 380 \qquad \text{A}$$

Similarly, for Jacob's order, Write

$$2r + 2m = 310 \qquad \text{B}$$

We refer to the two equations as A and B.

Subtract the two equations A − B:

$2r - 2r + 3m - 2m = 380 - 310$	The two equations are subtracted to eliminate r.
$m = 70 \quad 2r - 2r = 0$, and $3m - 2m = m$	

Substituting $m = 70$ into one of the equations, say A, enables us to find r:

$$2r + 3 \times 70 = 380$$
$$2r + 210 = 380$$
$$2r = 170$$
$$r = 85 \qquad \text{Dividing both sides of the equation by 2}$$

The two solutions $m = 70$ and $r = 85$ are the only solutions that solve this problem. We say these solutions are *unique*.

The following examples explain the various strategies in solving other simultaneous equations.

Example 2. Solve the simultaneous equations

$$2x + 3y = 3 \qquad \text{A}$$
$$4x - y = 20 \qquad \text{B}$$

Solution. The strategy is to eliminate the variable y by multiplying equation B by 3, so that the two coefficients of y are 3 and -3, which are equal except for being opposite in sign:

$$2x + 3y = 3 \qquad \text{A}$$
$$12x - 3y = 60 \qquad \text{C} \qquad \text{This equation is now called C}$$

Since the coefficients of y in each equation are now both equal, but have opposite signs, we add the two equations to eliminate y:

$$2x + 12x + 3y - 3y = 3 + 60 \qquad \text{A} + \text{C}$$
$$14x = 63 \qquad\qquad 2x + 12x = 14x, \ 3y - 3y = 0$$
$$x = 4.5 \qquad\qquad \text{Dividing both sides of the equation by 14}$$

In order to find y, we substitute $x = 4.5$ in one of the equations, say equation A:

$$2 \times 4.5 + 3y = 3$$
$$9 + 3y = 3$$
$$3y = -6 \qquad \text{Subtracting 9 from both sides of the equation}$$
$$y = -2$$

The solutions are $x = 4.5$, $y = -2$.

Example 3. Solve the simultaneous equations

$$2x + 3y = 4 \qquad \text{A}$$
$$3x + 4y = 6 \qquad \text{B}$$

Solution. The strategy is to eliminate the variable y by multiplying equation A by 4 and equation B by 3, so that the coefficients of y in both equations are the same, which is equal to 12:

$$8x + 12y = 16 \qquad \text{C}$$

$$9x + 12y = 18 \qquad \text{D} \qquad \text{The equations are renamed C and D}$$

Since the coefficients of y in each equation are now both the same and have the same signs, we subtract the two equations to eliminate y:

$$8x - 9x + 12y - 12y = 16 - 18 \qquad \text{C} - \text{D}$$

$$-x = -2$$

$$x = 2 \qquad \text{Multiplying both sides of the equation by } -1$$

In order to find y we substitute $x = 2$ in one of the equations, say equation A:

$$2 \times 2 + 3y = 4$$

$$4 + 3y = 4$$

$$3y = 0 \qquad \text{Subtracting 4 from both sides of the equation}$$

$$y = 0 \qquad \text{Dividing both sides of the equation by 3.}$$

The solutions are $x = 2$, $y = 0$.

Some questions are more easily solved by substitution instead of the "elimination" method described above. The substitution method is explained here.

Example 4. Solve the simultaneous equations

$$y = -2x + 5 \qquad \text{A}$$

$$3y - 2x = 47 \qquad \text{B}$$

Solution. One of the equations is written as $y = \ldots$, so the strategy is to substitute $(-2x + 5)$ for y in equation B; it is advised to use brackets around $-2x + 5$. Write

$$3(-2x + 5) - 2x = 47 \qquad \text{Substituting for } y \text{ in equation B}$$

$$-6x + 15 - 2x = 47 \qquad \text{Expanding the brackets}$$

$$-8x + 15 = 47 \qquad -6x - 2x = -8x$$

$$-8x = 32 \qquad \text{Subtracting 15 from both sides of the equation}$$

$$x = -4 \qquad \text{Dividing both sides of the equation by } -8$$

To find y we substitute $x = -4$ into equation A:

$$y = -2 \times -4 + 5$$
$$y = 8 + 5$$
$$y = 13$$

The solutions are $x = -4$, $y = 13$.

So far we have used the "elimination" method and the "substitution" method for solving simultaneous equations. Another method is to draw straight-line graphs of the two equations and read off from the graphs the coordinates of the point where the two straight lines intersect. Naturally if the answers involve decimals, this method may not give perfectly accurate results. Before continuing with this explanation, make sure you know how to draw linear graphs, which treated is under the entry Gradient-Intercept Form.

Example 5. Using a graphical method, solve the simultaneous equations $y = 2x + 3$ and $x + y = 9$.

Solution. In order to graph these two equations, we need to know the gradient and the y-intercept of each one. Compare $y = 2x + 3$ with $y = mx + c$; the gradient is $m = 2/1$ and the y-intercept is $c = 3$.
For $x + y = 9$ we need to express it in the form $y = mx + c$ first:

$y = -x + 9$	Subtracting x from both sides of the equation
$m = -1/1$ and $c = 9$	Matching this equation with $y = mx + c$

This process is explained more fully in the entry Gradient-Intercept Form. The graphs are drawn on the same axes, as shown in the figure. It can be seen from the figure that the two straight lines intersect at the point (2, 7). The solutions to the simultaneous equations are $x = 2$ and $y = 7$.

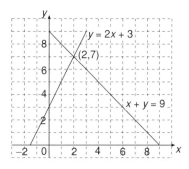

References: Gradient-Intercept Form, Like Terms, Linear Equation, Substitution, Variable.

SINE

Reference: Trigonometry.

SINE CURVE

References: Circular Functions, Trigonometric Graphs.

SINE RULE

The sine rule is a set of trigonometric formulas connecting the lengths of the three sides of a triangle with the sines of its angles. It is mainly used in triangles that are not right-angled, because simpler methods are used for right-angled triangles, as explained under the entry Trigonometry. The notation used for the sine rule, and also for the cosine rule, is that the sizes of the three angles of the triangle are referred to using capital letters, say A, B, and C, and the lengths of the sides are small letters, a, b, and c (see figure a). You need to be familiar with this notation in order to use the sine rule and the cosine rule.

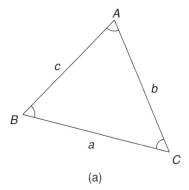

(a)

Notation. The length of the side opposite to angle A is called a, and similarly the side of length b is opposite to angle B and the side of length c is opposite to angle C.

The sine rule states that in any triangle

$$\frac{a}{\sin A} = \frac{b}{\sin B} = \frac{c}{\sin C}$$

This rule is unusual in that it contains two equal signs. In fact the sine rule represents three rules, which are

$$\frac{a}{\sin A} = \frac{b}{\sin B}, \quad \frac{a}{\sin A} = \frac{c}{\sin C}, \quad \frac{b}{\sin B} = \frac{c}{\sin C}$$

When we use the sine rule only one of these three rules is used at a time. The sine rule is used in a triangle for calculating the length of a side if two angles and a side are known, or for calculating the size of an angle if two sides and an angle are known. This may sound complicated, but in practice the procedure is straightforward, as explained in the first example, in which the length of a side is calculated.

Example 1. The line QR in figure b represents a grassy bank; a ladder QP which is 4.1 meters long has one end on the bank and its other end resting against a brick wall. The ladder makes an angle of 44° with the wall and an angle of 36° with the grassy bank. Find the distance the ladder reaches up the wall, which is the length PR. There is a peg at the point Q to stop the ladder slipping down the grassy bank.

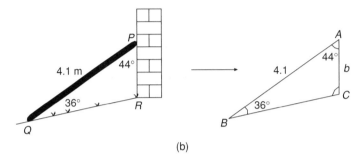

(b)

Solution. Draw the triangle PQR, but rename it ABC, because the sine rule refers to a triangle ABC. Write down what we know about triangle ABC: $A = 44°$, $B = 36°$, and $c = 4.1$. In order to use the sine rule we need to pair off sides with the angles opposite to them, so we need to find the size of angle C:

$$C = 100° \qquad \text{Sum of angles of triangle} = 180°$$

The sine rule can now be used. Write

$$\frac{b}{\sin 36°} = \frac{4.1}{\sin 100°}$$

Substituting $c = 4.1$, $C = 100°$, $B = 36°$

into $\dfrac{b}{\sin B} = \dfrac{c}{\sin C}$

$$b = \sin 36° \times \frac{4.1}{\sin 100°}$$

Multiplying both sides of the equation by $\sin 36°$

$$b = 2.4 \quad \text{(to 1 dp)}$$

Using a calculator

The ladder reaches 2.4 meters up the wall.

The sine rule can be used to find an angle in a triangle, as explained in the following example.

Example 2. John and Bill are using two ropes tied to the top of a heavy pole to hold it upright. John's rope is 12 meters long and makes an angle of 35° with the level ground (see figure c). John and Bill are on eactly opposite sides of the pole. If Bill's rope is 10 meters long, calculate the angle it makes with the ground.

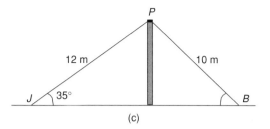

(c)

In the triangle *PJB*, $b = 12$, $j = 10$, and $J = 35°$. Write

$$\frac{12}{\sin B} = \frac{10}{\sin 35°}$$

Substituting $b = 12$, $j = 10$, and $J = 35°$

in $\dfrac{b}{\sin B} = \dfrac{j}{\sin J}$

$$\frac{\sin B}{12} = \frac{\sin 35°}{10}$$

Inverting both sides of the equation makes the next step easier

$$\sin B = 12 \times \frac{\sin 35°}{10}$$

Multiplying both sides of the equation by 12

$$\sin B = 0.68829\ldots$$

$$B = \sin^{-1}(0.68829\ldots)$$

$$B = 43.5° \quad \text{(to 1 dp)}$$

Using inv sin on the calculator

Bill's rope makes an angle of 43.5° with the level ground.

References: Cosine Rule, Trigonometry.

SIZE TRANSFORMATION

Reference: Enlargement.

SKETCH GRAPHS

Reference: Graphs.

SKEW LINES

Reference: Collinear.

SLANT HEIGHT OF CONE

Reference: Height.

SLOPE

Reference: Gradient.

SOLIDS

A solid is a three-dimensional object. Some solids are made up of flat surfaces called planes, such as the cube, the cuboid, the pyramid, and the tetrahedron. These are called polyhedrons, and have faces, vertices, and edges that are related by Euler's formula. Some solids are made up of a mixture of flat surfaces and curved surfaces, such as the hemisphere, the cone, and the cylinder. Other solids are made up of only curved surfaces, such as the sphere, the torus, and the ellipsoid. For more details about some of these solids see under their respective entries.

References: Circle, Cone, Cross Section, Cube, Cuboid, Cylinder, Edge, Ellipse, Euler's Formula, Face, Hemisphere, Plane, Polyhedron, Pyramid, Tetrahedron, Sphere, Vertex.

SOLVING AN EQUATION

References: Balancing an Equation, Equations, Linear Equation, Quadratic Equations.

SOUTH

Reference: East.

SPEED

Reference: Velocity.

SPEED–TIME GRAPH

Reference: Acceleration.

SPHERE

This is a solid shape in which every point on its surface is the same distance from its center. The "same distance" is the radius of the sphere. A well-known example of a sphere is a ball. The earth is approximately a sphere. Any cross section of a sphere is a circle. In the figure, the large dot is at the center, and the length of the arrow is the radius (R) of the sphere.

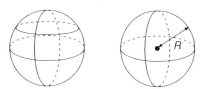

The formula for the volume of a sphere of radius R is

$$\text{Volume} = \tfrac{4}{3}\pi R^3$$

The formula for the surface area of a sphere of radius R is

$$\text{Area} = 4\pi R^2$$

It is not possible to draw a net for a sphere.

Example 1. Find the volume of a soccer ball if its radius is 11.3 cm (to 1 dp).

Solution. Write

$V = \tfrac{4}{3}\pi R^3$ Formula for the volume of a sphere

$V = \tfrac{4}{3} \times \pi \times 11.3^3$ Substituting $R = 11.3$ cm

$V = 6044$ to nearest whole number   Using the calculator value for π

The volume of the soccer ball is 6044 cm^3.

Example 2. What is the surface area of the earth if it has a diameter of 12,700 km?

Solution. The radius of the earth is half its diameter, so $R = 6350$ km. Write

$A = 4\pi R^2$ Formula for surface area of a sphere

$A = 4 \times \pi \times 6350^2$ Substituting $R = 6350$ km

$A = 507,000,000$ (to 3 sf)   Using the calculator value for π

The surface area of the earth is approximately 507 million km^2, which is 5.07×10^8 in standard form.

SPIRAL

A spiral is a curve traced out by a point that moves around a fixed point and at an ever-increasing distance from the fixed point (see the figure). It is the effect produced when a thick piece of rubber or carpet is rolled up. Spirals occur in nature; for example, some seashells are in the shape of spirals.

SPREAD

References: Box and Whisker Graph, Quartiles, Range, Standard Deviation.

SQUARE

A square is a quadrilateral with each of its four angles equal to 90° and all four sides equal in length. The square has four axes of symmetry and has a rotational symmetry of order four (see figure a). The diagonals of the square bisect each other at right angles, forming four congruent right-angled isosceles triangles. The diagonals make angles of 45° with the sides of the square.

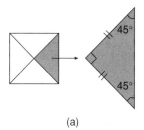

(a)

The length of a side of a square is one-fourth the length of its perimeter. The length of the side of the square is the square root of its area. For example, if the area of a square is 36 cm², then the length of the side of the square is $\sqrt{36} = 6$ cm.

Example. If the length of the diagonal of a square is 8 cm, find the length of a side of the square.

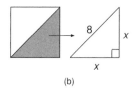

(b)

Solution. Let the length of each side of the square be x cm. The diagonal of the square divides it up into two right-angled isosceles triangles (see figure b). We can use the theorem of Pythagoras in one of these triangles:

$$x^2 + x^2 = 8^2 \qquad \text{Pythagoras' theorem}$$

$$2x^2 = 64 \qquad x^2 + x^2 = 2x^2 \text{ and } 8^2 = 64$$

$$x^2 = 32 \qquad \text{Dividing both sides of the equation by 2}$$

$$x = \sqrt{32}$$ Taking the square root of both sides of the equation

$$x = 5.66 \quad \text{(to 2 dp)}$$

The length of one side of the square is 5.66 cm.

The construction of a square, which is similar to the construction of a rectangle, is explained under the entry Constructions.

References: Axis of Symmetry, Bisect, Constructions, Equiangular Triangle, Isosceles Triangle, Right Angle, Rotational Symmetry, Squaring, Pythagoras' Theorem.

SQUARE NUMBERS

This is the set of numbers that are the squares of the natural numbers. To square a number, we multiply the number by itself. For example, $5^2 = 5 \times 5 = 25$. The squares of the first eight natural numbers are as follows:

Natural numbers	1	2	3	4	5	6	7	8
Square numbers	$1^2 = 1$	$2^2 = 4$	$3^2 = 9$	$4^2 = 16$	$5^2 = 25$	$6^2 = 36$	$7^2 = 49$	$8^2 = 64$

None of the square numbers is a prime number.

It is interesting to note that a square number n^2 is equal to the sum of the first n odd numbers. For example, $6^2 = 1 + 3 + 5 + 7 + 9 + 11$.

References: Natural Numbers, Prime Number.

SQUARE ROOT

This is an operation that occurs frequently in mathematics, especially in the use of the theorem of Pythagoras. When we take the square root of a number (or a term) we undo the "squaring" operation. We say that taking the square root is the inverse operation of squaring. For example, $5^2 = 25$, and the square root of 25 is 5. The square root of 25 is written as $\sqrt{25}$. The square root of a number can also be written in index form: for example, $\sqrt{25}$ can be written as $25^{1/2}$. Square roots of numbers can be found using a scientific calculator.

When we take the square root of terms in algebra that are expressed as indices we halve the index; for example: $\sqrt{x^8} = x^4$. The reason why this rule works is explained as follows:

$$\sqrt{x^8} = \left(x^8\right)^{1/2} \qquad \sqrt{a} = a^{1/2}$$
$$= x^4 \qquad \text{Multiplying the powers using laws of indices}$$

Note. When we square negative numbers, as well as positive numbers, we obtain positive answers. For example, $(-4)^2 = 16$ and $(+4)^2 = 16$. So if we are solving equations by taking the square root, we must write down both the positive and the negative solutions. The solutions of the equation $x^2 = 16$ are $x = +4$ and $x = -4$. These two answers are usually expressed as $x = \pm 4$.

References: Indices, Inverse Operations, Pythagoras' Theorem, Squaring.

SQUARING

This is an operation that multiplies any number, or term, by itself. For example, when we square 7, the result is $7 \times 7 = 49$. This can also be written as 7^2, which we say in words as seven squared.

Example 1. Square $2x + 3$.

Solution. The square of $2x + 3$ is $(2x + 3)^2$. Write

$$(2x + 3)^2 = (2x + 3) \times (2x + 3)$$
$$= 4x^2 + 12x + 9. \qquad \text{Expanding the brackets}$$

Example 2. Find the square $5xy^3$.

Solution. The square of $5xy^3$ is $(5xy^3)^2$. Write

$$(5xy^3)^2 = 5xy^3 \times 5xy^3 \qquad a^2 = a \times a$$
$$= 25x^2y^6 \qquad \text{Using laws of algebra}$$

References: Algebra, Expanding Brackets, Square Numbers.

STANDARD DEVIATION

Under this entry we study spread, range, and standard deviation, which are three statistical terms. Range and standard deviation both measure the spread of a set of data. Another measure of spread uses quartiles. The terms spread, range, and standard deviation are illustrated in the example that follows.

Luke plays junior rugby for Plato High School, and Will plays for Socrates High School. The weights of some players are very important in the game of rugby, and the weights of each player in the two teams are shown in the table in kilograms. There are 15 players in each team. The total weight of each team and the mean weight of each team are calculated as shown.

Plato rugby team
   Weights, in kg, of individual players: 56, 59, 62, 63, 66, 69, 70, 75, 78, 80, 83, 84, 85, 87, 93
   Total weight = 1110 kg, mean weight = 1110 ÷ 15 = 74 kg
Socrates rugby team
   Weights, in kg, of individual players: 62, 62, 63, 64, 69, 70, 72, 76, 77, 77, 79, 81, 85, 86, 87
   Total weight = 1110 kg, mean weight = 74 kg

What a coincidence! The mean weight of each team is the same.

We now compare the "spread" of the weights of each team, but first give an introduction to the term *spread*. When you are spreading jam on your toast you can spread it evenly over the whole slice or you can "load it up" in the middle and save the most jam for the final mouthful. Similarly, the weights of the players in a rugby team can be well spread out or they can be "clustered" closely together. We could say that the "center" of the range of weights is the mean. The weights of the players from each team are plotted on a number line in order to compare how spread out the data are (see the figure).

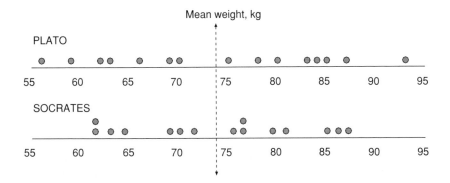

It can easily be seen that the weights of the Plato players are more spread about their mean than the weights of the Socrates players. One of the ways of measuring spread is called the *range*. The range is the difference between the largest value and the smallest value:

$$\text{Plato:} \qquad \text{Range} = 93 - 56$$
$$= 37$$
$$\text{Socrates:} \qquad \text{Range} = 87 - 62$$
$$= 25$$

We therefore conclude that the distribution of weights of the Plato players is more spread out than that for the Socrates players, since they have a greater range. Since range only involves 2 players and ignores the other 13 players, it does not give us an accurate picture of how spread out the whole data are. A more reliable figure to indicate spread is called *standard deviation*. This figure tells us how spread out the weights are from the mean by including every person's weight in the calculation. The steps involved in calculating standard deviation are as follows:

1. Calculate the mean weight of the players, which is $\bar{x} = 74$ in this example.
2. Calculate the deviation (difference) of each weight (x) from the mean $\bar{x}$, which is $(x - \bar{x})$.
3. Square the deviation of each weight from the mean, which is $(x - \bar{x})^2$ for each player.

4. Add together all these terms $(x - \bar{x})^2$, which is written as $\sum(x - \bar{x})^2$. Here $\sum$ is the capital letter of the Greek alphabet, and means "add together."

5. Divide this total by the number of players in the team, which is

$$\frac{\sum(x - \bar{x})^2}{15}$$

6. Take the square root of this answer to give the standard deviation of the weights of the players on the team. The standard deviation can be described as the square root of the mean of the squares of the deviations. The symbol for standard deviation is σ, which is lower case Greek sigma:

$$\sigma = \sqrt{\frac{\sum(x - \bar{x})^2}{15}}$$

These steps for calculating the standard deviation can be clearly set out in a table as shown, for each team of rugby players. Alternatively, the standard deviation can be found using a scientific calculator, with the processes explained in the calculator's handbook. In the table, the first line in each column sets out the steps of working for that column.

	Plato			Socrates	
x	$x - \bar{x}$	$(x - \bar{x})^2$	x	$x - \bar{x}$	$(x - \bar{x})^2$
56	$56 - 74 = -18$	$(-18)^2 = 324$	62	$62 - 74 = -12$	$(-12)^2 = 144$
59	-15	225	62	-12	144
62	-12	144	63	-11	121
63	-11	121	64	-10	100
66	-8	64	69	-5	25
69	-5	25	70	-4	16
70	-4	16	72	-2	4
75	1	1	76	2	4
78	4	16	77	3	9
80	6	36	77	3	9
83	9	81	79	5	25
84	10	100	81	7	49
85	11	121	85	11	121
87	13	169	86	12	144
93	19	361	87	13	169
		Total = 1804			Total = 1084

$$\sigma = \sqrt{\frac{\sum(x - \bar{x})^2}{15}}$$

$$\sigma = \sqrt{\frac{1804}{15}}$$

$\sigma = 11.0$ to 1 dp

$$\sigma = \sqrt{\frac{\sum(x - \bar{x})^2}{15}}$$

$$\sigma = \sqrt{\frac{1084}{15}}$$

$\sigma = 8.5$ to 1 dp

These figures confirm that the weights of the players from Plato High School are more spread about the mean than the weights of the players from Socrates High School, because the greater the value of the standard deviation, the greater is the spread.

In the example, the number of items of data for each team was 15. In the following general formula for standard deviation there are n items of data:

$$\sigma = \sqrt{\frac{\sum(x - \bar{x})^2}{n}}$$

References: Mean, Normal Distribution, Quartiles, Variance.

STANDARD FORM

Standard form is sometimes called scientific notation. It is a convenient way of writing very large or very small numbers in shortened form. This is especially useful in science, which often deals with very large and small numbers. In astronomy there is a unit of distance called the "light year," which is approximately equal to 6,000,000,000,000 miles. When it is written in that form it takes up a lot of room and cannot easily be compared with other numbers of similar size. In standard form a number is written as $a \times 10^n$, where n is an integer and a is a number between 1 and 10, or, more precisely $1 \le a < 10$. This makes the size of a clear: it can equal 1, but cannot equal 10.

Example 1. Write the number 6,000,000,000,000 in standard form.

Solution. Write

$$6,000,000,000,000 = 6 \times 1,000,000,000,000 \qquad \text{In this case } a \text{ is } 6$$
$$= 6 \times 10^{12} \qquad \text{The index } n \text{ is } 12$$

The number in standard form is 6×10^{12}.

Example 2. The greatest distance of the Moon from the Earth is 405,470 km. Write this distance in standard form correct to three significant figures.

Solution. Write

$$405,470 = 4.05470 \times 100,000$$
$$= 4.05 \times 100,000 \qquad \text{Writing the number to 3 sf}$$
$$= 4.05 \times 10^5$$

The number in standard form is 4.05×10^5.

Example 3. The length of a room is 4 meters to 1 sf. Write this length in standard form.

Solution. Write

$$4 = 4 \times 1$$
$$= 4 \times 10^0 \qquad \text{Since } 10^0 = 1$$

The length of the room, in standard form, is 4×10^0 meters.

Very small numbers can be written in standard form using negative indices, as illustrated with the following example.

Example 4. The length of the wavelength of ultraviolet light is 0.0000254 cm. Write this in standard form.

Solution. Write

$$0.0000254 = 2.54 \div 100,000 \qquad \div 100,000 \text{ is equivalent to } \times 10^{-6}$$
$$= 2.54 \times 10^{-5} \qquad \text{The index is negative because we are dividing}$$

The number in standard form is 2.54×10^{-5}.

Example 5. Write the number 3.4×10^{-4} in decimal form.

Solution. Write

$$3.4 \times 10^{-4} = 3.4 \div 10,000 \qquad \times 10^{-4} \text{ is equivalent to } \div 10,000$$
$$= 0.00034 \qquad \text{Moving the decimal point 4 places to the left.}$$

The number written in decimal form is 0.00034

We can easily compare the sizes of numbers that are in standard form, as illustrated by the following examples:

- 3.4×10^5 is greater than 8.5×10^4, because it has a greater index ($5 > 4$).
- 2.8×10^8 is less than 3.0×10^8, because 2.8 is less than 3.0 and the indices are the same.
- 5.9×10^{-6} is less than 2.8×10^{-5}, because it has a smaller index ($-6 < -5$).

References: Decimal, Indices.

STAR POLYGONS

These are star shapes formed by joining the vertices of regular polygons, as described below. Regular star polygons were well known to Pythagoras and his followers, who lived about 500 years BC. In fact, the pentagram was adopted as the badge of the ancient Pythagorean school.

Pentagon The first vertex is joined with a straight line to the third vertex, the second vertex is joined to the fourth vertex, the third vertex to the fifth vertex, and so on (figure a). Then the polygon is removed, and the result is a star pentagon, which is also known as a pentagram.

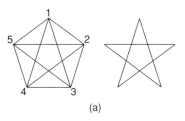

(a)

Hexagon The process is repeated for the hexagon.

Octagon There are two star octagons (figure b). In the first octagon the first vertex is drawn to the third vertex, the second to the fourth, the third to the fifth, and so on. In the second octagon the first vertex is joined to the fourth vertex, the second to the fifth, the third to the sixth, and so on.

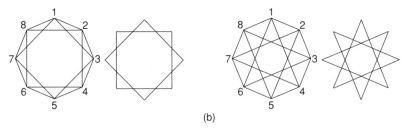

(b)

References: Hexagon, Pentagon, Pentagram, Regular Polygon, Vertex.

STATIONARY POINT

In this entry, a stationary point on a curve is defined in terms of the gradient of the curve at that point. See the entry Gradient of a Curve to ensure you understand the term. A stationary point on a curve is the point where the gradient of the tangent to the curve is zero. In the figure a the points A, B, and C are stationary points, because the tangents at those points, which are drawn dashed, have zero gradients.

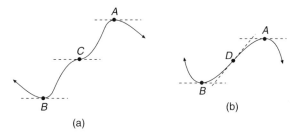

(a)

(b)

In the figure a there are three kinds of stationary points:

- Point A is a maximum point.
- Point B is a minimum point.
- Point C is a stationary point of inflection.

At a point of inflection the tangent changes from one side of the curve to the other. In the figure b point D is also a point of inflection, but since the gradient at that point is not zero, it is not a stationary point.

A stationary point can also be explained as a point where the function is neither increasing nor decreasing, hence stationary! A turning point on a curve is either a maximum point or a minimum point. It is a point where a function changes from increasing to decreasing, or vice versa. A critical point is also known as a stationary point.

References: Concave; Decreasing Function; Gradient; Gradient of a Curve; Inflection, Point of; Maximum Value; Slope; Tangent.

STATISTICS

This is a branch of mathematics that is about the collection, display, and analysis of numerical data in order to draw conclusions or make predictions. For information on the various terms in statistics, see the respective entries.

References: Arrangements, Average, Decile, Factorial, Frequency Distribution, Mean, Median of a Set of Data, Mode, Normal Distribution, Percentiles, Quartiles, Range, Scatter Diagram, Selection, Spread, Standard Deviation, Variance.

STEM AND LEAF GRAPH

This graph, which is used in statistics, is really a histogram in which each item of the data is shown in the graph. The stem and leaf graph has the appearance of a histogram, but none of the original data are lost. The examples illustrate the concept of a stem and leaf graph.

Example 1. Nathan collected data about the weights, in kilograms, of the students in his class of 25 students. The weights were recorded to the nearest kilogram. Draw a stem and leaf graph of the data, which are listed here:

39, 42, 45, 46, 46, 49, 51, 53, 53, 55, 57, 57, 58, 58, 59, 60, 61, 62, 63, 64, 64, 67, 68, 71, 75

Solution. Each item of data is made up of tens and units. For example, the first item of data is 39, which is three tens and nine units. The tens numbers form the "stem of the plant" and the units numbers form the "leaves of the plant." The result is shown in figure a and is called a stem and leaf graph. It is customary to write only one digit in the stem, so that 3 stands for 30, 4 stands for 40, and so on. The two graphs are drawn side by side. Also included is a histogram for comparison.

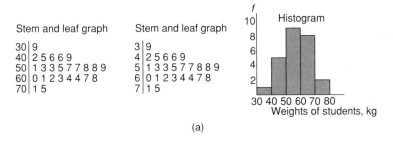

Stem and leaf graph

```
30 9
40 2 5 6 6 9
50 1 3 3 5 7 7 8 8 9
60 0 1 2 3 4 4 7 8
70 1 5
```

Stem and leaf graph

```
3 9
4 2 5 6 6 9
5 1 3 3 5 7 7 8 8 9
6 0 1 2 3 4 4 7 8
7 1 5
```

(a)

Back-to-back stem and leaf graphs are used to compare two sets of data, as shown in the next example.

Example 2. The back-to-back stem and leaf graph in figure b compares the results of the 1972 Olympic final in the 400 meters hurdles with the 1992 final 20 years later. There were eight finalists in each of the two events. The times are given in seconds, to 2 dp. The time for the first-place runner in 1992 (the fastest time in the graph) is 46.78 seconds; the time for the second-place runner in 1992 is 47.66 seconds; that for the first-place runner in 1972 is 47.82 seconds; etc.

Back to Back Stem and Leaf Graph

```
        1972        1992
                46 | 78
             82 47 | 66  82
       64 52 21 48 | 13  63  63  86
       66 66 65 49 | 26
             25 50 |
```

(b)

As a matter of interest, if all 16 athletes ran in the same race and finished the race in their Olympic times, then their places would be as shown in the table, where 92 and 72 indicate runners from those respective years. The best 1972 runner would have tied for a bronze medal in the 1992 final.

1st	2nd	3rd	5th	6th	7th	8th	10th	11th	12th	13th	14th	15th	16th
92	92	92	92	72	72	92	72	72	92	72	72	72	72
		72				92							

The back-to-back stem and leaf graph shows that athletic performance in the 400 meters hurdles did not improve by a great amount in those 20 years. The mean time for the 1972 race was 49.05 seconds and that for the 1992 race was 48.22 seconds. The percentage improvement is

$$\frac{0.83}{49.05} \times 100 = 1.7\% \quad \text{(to 1 dp)}$$

References: Histogram, Mean, Percentage, Statistics.

STRAIGHT ANGLE

Reference: Acute Angle.

STRAIGHT-LINE GRAPH

References: Cartesian Coordinates, Gradient-intercept Form, Graphs.

SUBJECT OF A FORMULA

Reference: Changing the Subject of a Formula.

SUBSTITUTION

Substitution usually means replacing a variable in a formula or equation by a number that represents the variable, or by another variable. Various examples of substitution are described in the entries given in the references. One type is explained here, which is replacement by numbers of the variables in algebraic terms.

Example. If $x = -2$, $y = 3$, and $z = 0$, find the values of each of the expressions (a) $2x - 3y$ and (b) $(x^2 + y^2)/z^2$.

Solution. For (a), write

$$2x - 3y = 2 \times (-2) - 3 \times 3 \qquad \text{Substituting } -2 \text{ for } x, \text{ in brackets, and 3 for } y$$

$$= -4 - 9$$

$$= -13$$

For (b), write

$$\frac{x^2 + y^2}{z^2} = \frac{(-2)^2 + 3^2}{0^2} \qquad \text{Substituting } -2 \text{ for } x, \text{ 3 for } y, \text{ and 0 for } z$$

$$= \frac{13}{0}$$

$$= \text{undefined} \qquad \text{The answer is undefined; we cannot divide a number by zero}$$

References: Algebra, Equations, Formula, Graphs, Table of Values, Variable.

SUBTEND

Subtend is a word used in circle geometry theorems. A subtended angle is an angle that "stands" on an arc of a circle. In the figure the angle x is subtended at a point P

by the arc *AB* of the circle. The arc *AB* referred to is highlighted in the figure. The angle *x* may be subtended at a point on the circumference of the circle or at its center.

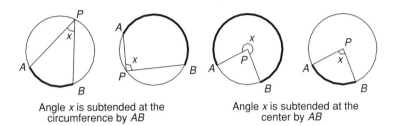

Angle *x* is subtended at the
circumference by *AB*

Angle *x* is subtended at the
center by *AB*

References: Arc, Circle, Circle Geometry Theorems, Circumference.

SUPPLEMENTARY ANGLES

Reference: Complementary Angles.

SURFACE

A surface is the complete boundary of a solid. A well-known solid is a cube, which has six square faces, and these six faces make up the surface of the cube.

References: Cube, Solids, Surface Area.

SURFACE AREA

The surface area of a solid is the total area of its surface. To find the surface area of a particular solid, see the entry under the name of the solid.

Reference: Area.

SURVEY

References: Census, Questionnaire.

SYMMETRY

References: Axis of Symmetry, Order of Rotational Symmetry.

SYSTÈME INTERNATIONAL D'UNITÉS

Abbreviated SI units. These SI units, along with some scientific units, were adopted by international agreement in the year 1960. The units in the SI system are as follows:

- Length: meter; abbreviation m
- Mass: kilogram; abbreviation kg (in everyday use, mass is sometimes called weight)
- Time: second; abbreviation s.

A full explanation of these metric units is given under the entry Metric Units.

References: CGS System of Units, Metric Units.

T

TABLE OF VALUES

References: Asymptote, Graphs.

TALLY CHART

Reference: Frequency.

TANGENT

Reference: Normal.

TANGENT AND RADIUS THEOREM

Suppose that T is a point on a circle with center O, and a tangent is drawn to the circle at that point T (see figure a). If a radius of the circle is also drawn through the same point T, then the angle between the tangent and the radius is a right angle. We say that the radius is perpendicular to the tangent.

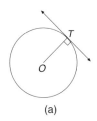

(a)

The next example involves the use of this theorem and also includes the use of another circle geometry theorem.

433

Example. In figure b, *PT* and *PQ* are tangents to the circle, and *O* is the center of the circle. Find *x* and *y*.

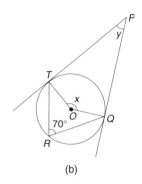

(b)

Solution. Write

$$x = 140°$$

     Angle at the center is twice the angle at the circumference

Angle *PTO* = angle *PQO*

$$= 90°$$

     Radius is perpendicular to the tangent

The figure *PTOQ* is a quadrilateral and the sum of its four angles is 360°:

$$x + 90° + y + 90° = 360°$$
$$140° + 90° + y + 90° = 360° \qquad \text{Substituting } x = 140°$$
$$y = 40°$$

See the entry Isosceles Triangle for another example of the theorem that the radius is perpendicular to the tangent.

References: Isosceles Triangle, Tangents from a Common Point.

TANGENT OF AN ANGLE

References: Gradient, Trigonometry.

TANGENT TO A CIRCLE

References: Chord, Circle.

TANGENTS FROM A COMMON POINT

This is a geometry theorem about two tangents drawn from one point to a circle. In figure a, the point common to the two tangents PT and PQ is the point P. The theorem states that the two tangents are equal in length. In figure a, $PT = PQ$, and this theorem is stated as follows:

- Tangents from a common point are equal.

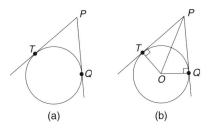

(a)   (b)

The properties of figure b are as follows: The line OP is an axis of symmetry, so the two triangles PTO and PQO are congruent. The figure $OTPQ$ is a cyclic quadrilateral because a pair of its opposite angles, T and P, add up to 180°. This means that angle TPQ + angle $TOQ = 180°$.

Example. A straight plank of wood AT rests across a circular garden roller, which stands on horizontal ground (see figure c). If the distance $AB = 2$ meters, what is the distance AT?

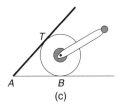

(c)

Solution. AT and AB are tangents to the circle which is the cross section of the roller. Write

$$AT = AB$$

$$= 2 \text{ meters} \qquad \text{Tangents from a common point are equal}$$

The distance AT is 2 meters.

References: Axis of Symmetry, Cyclic Quadrilateral, Tangent and Radius Theorem, Trigonometry.

TANGRAM

A tangram is a Chinese puzzle. A square is cut up into seven shapes, as shown in figure a. The object of the puzzle is to cut out the seven shapes and rearrange all of them to form interesting figures of animals, humans, etc. Figure b shows a pelican, made up of all the seven pieces.

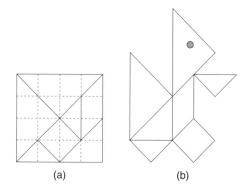

(a)   (b)

References: Area, Isosceles Triangle, Parallelogram, Right Angle, Square.

TEMPERATURE

Temperature is a measure of how hot or cold something is. The two scales of temperature in common use are Fahrenheit (F) and Celsius (C). Celsius used to be called centigrade. The units for measuring temperature are degrees, but degrees Celsius (°C) are not the same size as degrees Fahrenheit (°F). Conversions between the two temperatures can be done using formulas, as follows.

To change °C to °F we use the formula $F = \frac{9}{5}C + 32$.

Example 1. Convert 37°C to the Fahrenheit scale.

Solution. Write

$$F = \frac{9}{5} \times 37 + 32 \qquad \text{Substituting 37 for } C$$

$$F = 98.6$$

37°C converts to 98.6°F.

To convert °F to °C we use the formula $C = \frac{5}{9}(F - 32)$.

Example 2. Convert 212°F to the Celsius scale.

Solution. Write

$$C = \tfrac{5}{9} \times (212 - 32) \qquad \text{Substituting 212 for } F$$

$$C = 100$$

212°F converts to 100°C.

References: Conversion, Formula.

TERMINATING DECIMALS

Reference: Decimal.

TERMS

Reference: Algebra.

TESSELLATIONS

Tessellations are also known as mosaics. A tessellation is a tiling pattern that covers a flat surface. The tiles we often use are regular polygons such as equilateral triangles, squares, or regular hexagons, but a variety of shapes may be used. Some tessellations involve more than one shape, such as combining square tiles with regular octagons. A tessellation is only correct if the tiles fit together with no gaps and no overlapping and a definite repeated pattern can be easily recognized.

Tiling a kitchen floor with a square tile is a tessellation. Three examples are shown in figure a. If you know where to put the next tile, you probably have an easily recognizable pattern.

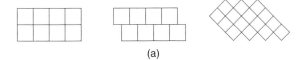

(a)

Tessellations occur in nature. A good example is the shape of the cells that form the honeycomb in a beehive. The tile is a regular hexagon (see figure b). Each interior angle of a regular hexagon is 120°. At the point where three hexagons meet they will completely "surround" the point with no gaps, because $3 \times 120° = 360°$, which is a full turn. To see how triangles, quadrilaterals, hexagons, and octagons tessellate, see

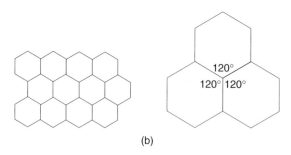

(b)

their respective entries. Some regular polygons do not tessellate, because when they meet at a point they do not surround the point in a way that leaves no gaps.

Example. Explain why regular pentagons do not tessellate.

Solution. Each interior angle of a regular pentagon is $108°$. If three regular pentagons meet at a point, the angle sum at the point is $3 \times 108° = 324°$, which is less than a full turn. If four of them meet at a point, the angle sum is $4 \times 108° = 432°$, which is more than a full turn. The regular pentagon will not tessellate, because each interior angle is $108°$ and there are no combinations of this angle that will make $360°$, so regular pentagons will not "surround" a point.

Two regular polygons that will combine to form a tessellation are the square and the octagon. Their tiling pattern is shown in figure d. If we examine how the shapes surround one point, say A, we can find out why these two shapes tessellate. Each interior angle of a regular octagon is $135°$ and each interior angle of a square is $90°$ (see figure d). At the point A, there are two octagons and one square. Therefore $135° + 135° + 90° = 360°$, which is a full turn. A square and two octagons surround the point A.

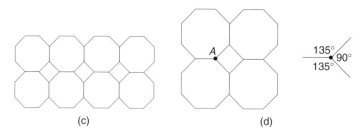

(c)   (d)

Combinations of other regular shapes, in addition to the octagons and squares described above that tessellate, are listed here. The number of the shapes that surround a point is stated and their angle sum.

- Hexagon, two squares, one equilateral triangle ($120° + 90° + 90° + 60° = 360°$)
- One dodecagon, one square, two equilateral triangles ($150° + 90° + 60° + 60° = 360°$)

- Two squares, three equilateral triangles $(90° + 90° + 60° + 60° + 60° = 360°)$
- Two hexagons, two equilateral triangles $(120° + 120° + 60° + 60° = 360°)$
- One dodecagon, one hexagon, one square $(150° + 120° + 90° = 360°)$
- One hexagon, four equilateral triangles $(120° + 60° + 60° + 60° + 60° = 360°)$

It is interesting to tessellate irregular shapes. In figure e, the curved shape that is shaded produces the attractive tessellation of a queue of bald-headed men. The partly shaded tile, when tessellated, produces a set of steps.

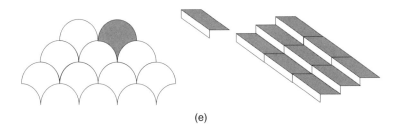

(e)

The Dutch artist M. Escher constructed some extraordinary tessellations. In particular, his picture of knights on horseback is very striking.

References: Dodecagon, Equilateral Triangle, Hexagon, Mosaics, Octagon, Pentagon, Polygon, Quadrilateral, Regular Polygon, Square, Triangle.

TETRAHEDRON

This is a pyramid with four plane faces, each of which is a triangle. The prefix "tetra" means four. One of the four faces is called the base, and the vertex that is opposite to the base is called the apex (see figure a). The volume of a tetrahedron $= \frac{1}{3}$ area of base $\times$ altitude.

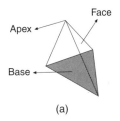

(a)

Example 1. Figure b shows a tetrahedron drawn inside a cube and occupying one of its corners. The apex of this tetrahedron is directly above the point which is one of the vertices of its base. If the length of each edge of the cube is 5 cm, find the volume of the tetrahedron.

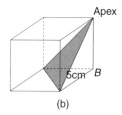

(b)

Solution. The area of the base of the cube is $5 \times 5 = 25$ cm^2. The base of the tetrahedron is obviously half the area of the base of the cube and is 12.5 cm^2. The altitude of the tetrahedron is 5 cm. Write

Volume of tetrahedron $= \frac{1}{3} \times 12.5 \times 5$ Using the formula for the volume of a tetrahedron

$= 20.83$ (to 2 dp)

The volume of the tetrahedron is 20.83 cm^2, which is $\frac{1}{6}$ of the volume of the cube.

The regular tetrahedron, which is one of the Platonic solids, has six edges all the same length. This means that each face is an equilateral triangle (see figure c). The apex V of a regular tetrahedron is directly above a point C in the base which is the centroid of the triangular base. The centroid of a triangle is the point of intersection of the medians of the triangle, and is shown in the drawing of the base on the right-hand side of figure c.

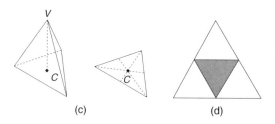

(c)   (d)

The net of a regular tetrahedron is shown in figure d with its base shaded. It is made up of four equilateral triangles, which themselves form a larger equilateral triangle. The area of the surface of a tetrahedron is the same as the area of its net.

Example 2. Find the surface area of a regular tetrahedron if the length of one edge is 6 cm.

Solution. The net of the tetrahedron is drawn in figure e. Each angle of the equilateral triangle is 60°. Write

Area of triangle ABC $= \frac{1}{2}bc \sin A$ See entry Cosine Rule

$$= \frac{1}{2} \times 12 \times 12 \times \sin 60° \quad \text{Substituting } b = c = 12, \text{ angle}$$
$$A = 60°$$

$$= 62.4 \quad \text{(to 1 dp)}$$

The surface area of the tetrahedron is 62.4 cm^2.

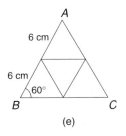

(e)

References: Apex, Cosine Rule, Equilateral Triangle, Median of a Triangle, Net, Platonic Solids, Plane, Polyhedron, Pyramid, Regular Polygon, Vertex.

TETROMINO

Reference: Polyominoes.

THEOREM

A theorem is a statement or formula that can be proved by steps of reasoning. The study of geometry as set out by Euclid begins with a few definitions and axioms (statements that we believe are obviously true and do not require proof) and uses them to prove theorems, which in turn are used to prove more theorems, and so on, building a chain. For an example on proving a geometry theorem refer to the entry Proof.

References: Circle Geometry Theorems, Euclid of Alexondria Geometry Theorems, Proof.

TILING PATTERNS

Reference: Tessellations.

TIME SERIES

In this entry a time series is taken to be a statistical graph that shows how a quantity varies as time changes. For example, a time series may be a line graph that shows the changes in temperature in New York City during the first 6 months of a given year. Other examples of quantities that may be measured are weight, height, cost, age, popularity, and so on. A time series is a record of what has happened in the past and is used so that a comparison can be made with another time series, or to make a prediction about future changes. A time series may be used in the following way. Suppose you own a supermarket and decide to employ an advertising agency to promote sales. It would be a good idea to first measure sales for a period of 1 year or more and have historical record of how well you are doing. Then bring in the agency and measure sales for the following year for a comparison. In this way you would be able to decide if the agency was effective. There are drawbacks to using historical data to make predictions about the future. You would have to be certain that there were no other influences that could affect your sales. For example, another supermarket may have closed down in your vicinity during the year of the advertising campaign, or there might have been a change in parking restrictions close by and customers suddenly found your supermarket more convenient than before. Whenever time series are used to measure changes and then used in this way for comparisons and predictions great care must be taken to ensure that there is a definite relation between cause and effect.

Example. Ken owns an ice cream company called Ken's Kones. His accountant provides him with his quarterly sales for spring, summer, fall, and winter over a 3-year period. Based on the following data, draw a time series graph of his sales over this 3-year period, and calculate the moving average on a yearly basis. The order of the moving average is four, because there are four quantities in each yearly cycle,

Sales for Ken's Kones in $100,000			
	1st year	2nd year	3rd year
Spring (Sp)	6.2	7.2	6.2
Summer (Su)	8.1	9.3	10.1
Fall (F)	4.9	6.0	6.5
Winter (W)	3.5	4.2	5.5

Solution. The time series line graph is shown in the figure. The points are joined up to form a series of continuous line segments, but there is no real validity in doing this, because there is no information provided by the accountant about sales from one season to the next. The points are joined up to provide a visual indication of any trends in sales. The average (mean) sales for each year are calculated as follows:

- 1st year: $(6.2 + 8.1 + 4.9 + 3.5) \div 4 = 5.7$ (to 1 dp)
- 2nd year: $(7.2 + 9.3 + 6.0 + 4.2) \div 4 = 6.7$
- 3rd year: $(6.2 + 10.1 + 6.5 + 5.5) \div 4 = 7.1$

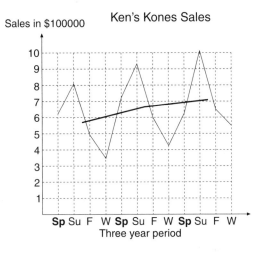

These values are tabulated as follows:

Average Sales per year in $100,000		
1st year	2nd year	3rd year
5.7	6.7	7.1

These three "moving averages" 5.7, 6.7, and 7.1 are plotted on the time series and joined with two line segments, as shown on the graph. In this example moving averages are used to represent the trend over the 3 years because there are seasonal variations in the time series that may distract from the overall pattern. We say that the moving averages "smooth" out the seasonal variations.

Ken needs to know whether or not his business is growing and why. He may want to employ more staff to keep up with sales, or lay off staff if the business is declining. The moving averages suggest that the business is growing, but at a slower rate in the third year. The spring sales have declined and the growth rates for summer and fall have slowed. But there has been a spurt in winter sales. This inconsistent pattern makes it difficult to predict the future with assurance. There is other information that may need to be examined and incorporated into the analysis, for example, the weather patterns, if it is true that people eat more ice cream in warm winters than cold winters. Has advertising had any impact, or has competition from other companies affected sales during the 3 years? Have there been any price changes? It is important not to rely solely on the trends in a time series when other influences may be at work.

References: Average, Graphs, Mean, Statistics.

TON

Reference: Imperial System of Units.

TONNE

A tonne is a unit of weight and is equal to 1000 kilograms. A builder having a truckload of sand delivered to a building site would expect several tonnes to arrive. One tonne of water has a volume of 1 cubic meter at 4°C.

References: CGS System of Units, Gram, Kilogram, Metric Units, SI Units.

TRANSFORMATION GEOMETRY

Transformation geometry in this book is the study of how sets of points and their positions in space are changed by enlargements, reflections, rotations, and translations. In fact the whole of space is transformed, but we focus on specific points or shapes. The original set of points or shape is called the object, and the set of points or shape that results from the transformation is called the image. Properties of the original shape that do not change after the transformation are called invariant properties. For example, the invariant properties may be area, shape, angle size, and so on. The transformations reflection, rotation, and translation leave size and shape invariant (this includes lengths of lines, the area of the shape, and angles in the shape). Transformations that do this are called isometries. It is clear that enlargement is not an isometry, because it changes the size of the shape. The following transformations are discussed in more detail under the appropriate entries: enlargement, reflection, rotation, and translation.

References: Congruent Figures, Image, Invariant Points, Object.

TRANSLATION

A translation is one of four transformations described in this book, and is defined in terms of a grid, such as the squares on a chessboard. Column vectors are used to represent translations. Figure a shows part of a chessboard, with the piece known as the queen on a white square. The queen has been chosen for this demonstration because it can move any number of squares in various directions, provided it lands

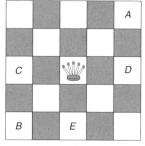

(a)

on a square. As the queen moves from one position to another its movement is a translation. The changes in the position of the queen are defined by two numbers in brackets, with one written above the other. The top number states how many squares the queen moves in a direction from side to side, and the bottom number describes how many squares it moves up or down. If the numbers are negative, the queen moves in the opposite direction.

Examples of sideways movement:

- 2 means two squares to the right
- −3 means three squares to the left

Examples of up and down movement:

- 5 means five squares up.
- −1 means one square down.

The translation that moves the queen to square A is 2 squares to the right and 2 squares up, which is written $\binom{2}{2}$. Queen to B is written $\binom{-2}{2}$. Queen to C is written $\binom{-2}{0}$. Queen to D is written $\binom{2}{0}$. Queen to E is written $\binom{0}{-2}$. If the Queen does not move at all, the translation is written $\binom{0}{0}$.

Translations of shapes can take place on a Cartesian plane, as illustrated by the following example. The initial position of the shape is called the object and the final position is called the image, which is the convention for all transformations. It is also the convention in transformation geometry that not only does the object change its position by the transformation, but so does the whole of the Cartesian plane. A translation has no invariant points.

Example. With reference to figure b, state the vector for each of the translations where F' is the image of F and F'' is the image of F.

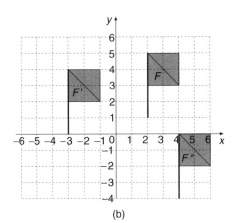

(b)

Solution. Every point on the flag undergoes the same translation. So we choose one point on F, say the top of the flagpole, and find its image on F'. The translation of the top of the flagpole is $\binom{-5}{-1}$, so this is the vector for the whole flag. Similarly, the vector that translates F to F'' is $\binom{2}{-5}$.

Notation We say that the flag F maps onto its image F' under the translation $\binom{-5}{-1}$, and F maps onto F'' under the translation $\binom{2}{-5}$. The symbol for "maps onto" is $\rightarrow$, so we write $F \rightarrow F'$.

In all transformations the image shape of F is denoted by F', using the appropriate letter.

Suppose the translation that maps a point A onto A' is defined by the vector $\binom{4}{-2}$ and $A' \rightarrow A''$ under a further translation $\binom{3}{2}$. The vector that maps A onto A'' is found by adding together the two vectors. Therefore, $A \rightarrow A''$ under the translation $\binom{7}{0}$. The translation that maps the image A' back to the object A is $\binom{-4}{2}$, and is found by changing the signs of the two numbers in the vector that maps A to A'.

Properties of Translations

- There are no invariant points.
- The object shape and the image shape are congruent shapes. We call this an *isometry transformation*.
- If the lines are parallel in the object, they will also be parallel in the image. We call this an *affine transformation,* which is one that preserves parallelism.
- Translation is a direct transformation.

References: Cartesian Coordinates, Composite Transformations, Congruent Figures, Escher, Indirect Transformation, Invariant Points, Transformation Geometry, Vector.

TRANSPOSE FORMULAS

Reference: Changing the Subject of a Formula.

TRANSVERSAL

Reference: Alternate Angle.

TRAPEZIUM

A trapezium is a quadrilateral, which means a four-sided polygon, with one pair of opposite sides parallel. In figure a, the angles a and b are called cointerior angles and their sum is $180°$, which means that $a + b = 180°$. The two triangles ACD and BCD are equal in area. The trapezium has rotational symmetry of order one. The trapezium on the right is an isosceles trapezium, which has two sides equal in length, two pairs of congruent angles, and one axis of symmetry.

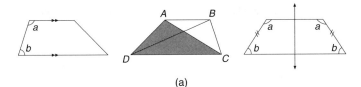

(a)

To find the area of a trapezium we need to know the lengths of each of the two parallel sides and their perpendicular distance apart. The formula for the area of the shaded trapezium in figure b is

$$\text{Area} = \frac{(a+b)}{2}h$$

In words, the area is expressed as the mean length of the two parallel sides multiplied by the perpendicular height.

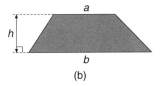

(b)

Example. The trapezium in figure c is drawn on a square grid, and the side of each square is 1 cm. Find the area and perimeter of the trapezium.

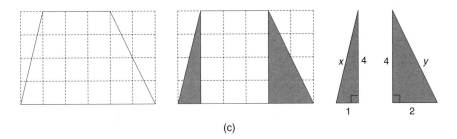

(c)

Solution. From figure c we can see that $a = 3$, $b = 6$, and $h = 4$, all in centimeters. Write

$$\text{Area} = \frac{(3+6)}{2} \times 4 \qquad \text{Substituting into the formula } A = \frac{(a+b)}{2}h$$

$$= 4.5 \times 4$$

$$= 18$$

The area of the trapezium is 18 cm^2.

The trapezium is divided into two right-angled triangles and a rectangle, as shown in figure c. Write

Perimeter $= 3 + y + 6 + x$ Adding together the four sides of the trapezium.

x is found in the first right-angled triangle and y in the second right-angled triangle:

$$x^2 = 1^2 + 4^2, \qquad y^2 = 2^2 + 4^2 \qquad \text{Pythagoras' theorem}$$

$$x^2 = 17, \qquad\qquad y^2 = 20$$

$$x = \sqrt{17}, \qquad\qquad y = \sqrt{20} \qquad \text{Taking square roots.}$$

Therefore

$$\text{Perimeter} = 3 + \sqrt{20} + 6 + \sqrt{17}$$

The perimeter of the trapezium is 17.60 cm, to 2 dp.

See the entry Prism for the volume of a swimming pool with a trapezium as the cross section.

References: Area, Base, Congruent Figures, Hexagon, Mean, Parallel, Perimeter, Perpendicular Lines, Polygon, Pythagoras' Theorem, Quadrilateral, Rectangle, Regular Polygon, Symmetry, Triangle.

TRAVEL GRAPHS

These graphs are also known as distance–time graphs, in which time is the horizontal axis and distance is the vertical axis. Travel graphs usually depict a journey. The first example explains how to set up a distance–time graph to solve a problem.

Example 1. David and his son William decide to have a race. David can run at a constant speed of 5 meters per second (m/s) and William can run at a constant speed of 2 m/s. David gives his son a 10 meters head start. After how many seconds will David catch up with his son?

Solution. We draw a distance–time graph for each runner using a suitable table of values. The distances are measured from the place where David starts the race; therefore we must add 10 meters to the distances run by William so the graph of each runner starts at $t = 0$.

Time in seconds (t)	0	1	2	3	4
David's distance in meters	0	5	10	15	20
William's distance in meters	10	12	14	16	18

The straight-line graphs of the two athletes are both drawn on the same axes, using the following sets of coordinates (see figure a):

- David: (0, 0), (1, 5), (2, 10), (3, 15), (4, 20)
- William: (0, 10), (1, 12), (2, 14), (3, 16), (4, 18)

David catches up with William at the point on the graph where the two lines intersect. This time is estimated to be about 3.3 seconds.

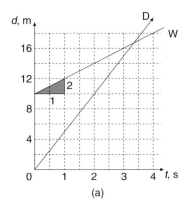

(a)

There are some important points to note from this problem.

- A distance–time graph of constant speed is a straight line.
- The greater the speed, the steeper is the line.
- The slope (gradient) of a distance–time graph gives the speed. For example, suppose we calculate the gradient of William's line graph:

$$\text{Gradient} = \frac{\text{rise}}{\text{run}} \qquad \text{The formula for gradient}$$

$$= \frac{2\ \text{meters}}{1\ \text{second}} \qquad \begin{array}{l}\text{In the right-angled triangle rise} = 2\ \text{meters,,}\\ \quad \text{run} = 1\ \text{second}\end{array}$$

$$= 2\ \text{m/s} \qquad \text{This is William's speed}$$

Example 2. Harry loves cycling and goes for a ride. For the first 3 hours he cycles at a speed of 20 km/h. He stops 1 h for lunch, and then continues at a more leisurely pace of 15 km/h for 2 h. Harry stops for $\frac{1}{2}$ h to take photographs of a waterfall. Realizing how late it is, he turns for home, and arrives there 3 h later. Draw a distance–time graph of his trip. At what constant speed did he return home? What was his average speed for the whole trip, assuming each of his speeds is constant?

Solution. The trip falls into five stages, and since the speeds are constant, each stage is represented by a straight-line graph (see figure b):

- Stage 1. Harry travels 60 km in 3 h.
- Stage 2. He stops for lunch, and the graph is horizontal for 1 h.
- Stage 3. He travels 30 km in 2 h.
- Stage 4. Sightseeing for $\frac{1}{2}$ h.
- Stage 5. From the point on the line at the end of stage 4, draw a straight line to the time axis to meet it 3 h later.

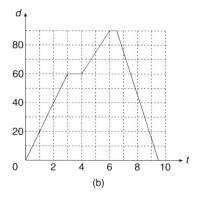

(b)

The travel graph can now be drawn, where d is the distance from home in kilometers and t is the time taken in hours.

For the return journey Harry travels 90 km in 3 h:

$$\text{Speed for return journey} = \frac{90\,\text{km}}{3\,\text{h}} \qquad \text{Speed} = \text{distance} \div \text{time}$$

$$= 30\,\text{km/h}$$

To find the average speed for the whole trip we need the total distance traveled and the total time taken.

Total distance $= 180\,\text{km}$ 90 km out and 90 km back

Total time $= 9.5\,\text{h}$ $3 + 1 + 2 + 0.5 + 3$

$$\text{Average speed} = \frac{180\,\text{km}}{9.5\,\text{h}} \qquad \text{Average speed} = \text{total distance} \div \text{total time}$$

$$= 18.9 \quad (\text{to 1 dp})$$

The average speed is 18.9 km/h.

References: Average Speed, Cartesian Coordinates, Gradient, Table of Values.

TRAVERSABLE NETWORKS

References: Konigsberg Bridge Problem, Networks.

TREE DIAGRAM

A tree diagram is a figure that resembles the branches of a tree, and is designed to show the events of an experiment and enable the probabilities of each event to be calculated. A tree diagram can be used in studying probability to list the sample space of an experiment, which is a list of all possible outcomes. This is explained in the entry

Event. It is also used to calculate probabilities. It is this latter use of tree diagrams that is described in this entry. The tree diagram is also known as a probability tree. The following examples explain how the tree diagram works in solving probability problems.

Example 1. Pat has three children, and none of them are twins. Each time a baby is born the chances of giving birth to a boy or a girl are equal. Calculate the following probabilities:

(a) They are all girls.
(b) There are two boys and one girl.
(c) There are two boys and one girl, born in that order.
(d) There are at least two girls.

Solution. The three births are represented by three sets of branches (see figure a; $G = $ girl, $B = $ boy) and at each birth the probability of having a boy or a girl is $\frac{1}{2}$. We have the following rules:

• To find the probability at the end of a branch we multiply the probabilities along the branch.
• If a solution is obtained from two or more sets of branches, we add together the probabilities at the end of each branch.

These two rules are put into practice as we solve the problem.
For (a),

$$\text{Prob(3 G)} = \tfrac{1}{8} \qquad\qquad \text{Multiplying along the branch } \tfrac{1}{2} \times \tfrac{1}{2} \times \tfrac{1}{2}$$

For (b),

$$\text{Prob(2 B \& 1 G)} = \tfrac{1}{8} + \tfrac{1}{8} + \tfrac{1}{8} \qquad \text{Adding ends of three branches}$$

$$= \tfrac{3}{8}$$

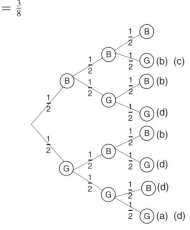

(a)

For (c),

    Prob(BBG in order) $= \frac{1}{8}$ Multiplying along the branch $\frac{1}{2} \times \frac{1}{2} \times \frac{1}{2}$

For (d),

    Prob(at least 2 G) $= \frac{4}{8}$ Adding the ends of the four branches
                                 with at least two girls in them

In the above problem the probability along each branch was the same value of $\frac{1}{2}$ because previous births do not affect the probability of the next birth. In the next example the probabilities change along the branches.

Example 2. Jim bought his grandson Luke a bag containing nine lollipops. There are four red, three green, and two yellow lollipops. He lets Luke choose one without looking, and then another. You may assume that Luke does not put the first one back before choosing the second! Work out the following probabilities:
  (a) They are both the same color.
  (b) They are different colors.
  (c) The first is green and the second is yellow.

Solution. At each choice the tree has three branches, which are R, G, Y (see figure b):

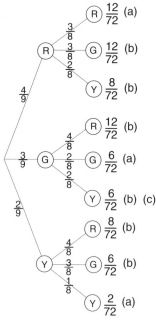

(b)

- First choice. There are four red, three green, and two yellow, making nine lollipops in the bag.
- Second choice. If one red has been taken, there are three red, three green, and two yellow left, making eight in all. On the other hand, if one green has been taken, there are four red, two green, and two yellow left, making eight in all, and so on.

The probabilities at the end of each of the nine branches are obtained by multiplying the probabilities along each of the branches.

For (a),

$$\text{Prob(same color)} = \tfrac{12}{72} + \tfrac{6}{72} + \tfrac{2}{72}$$

$$= \tfrac{20}{72} \quad \text{or} \quad \tfrac{5}{18}$$

(b) To obtain the probability of the two different colors it is convenient to subtract the probability of the same color from 1:

$$\text{Prob(different color)} = 1 - \tfrac{5}{18} \qquad \text{Using part (a)}$$

$$= \tfrac{13}{18}$$

For (c),

$$\text{Prob(G \& Y, in order)} = \tfrac{6}{72} \quad \text{or} \quad \tfrac{1}{12}$$

Note. The fractions representing probabilities may be simplified by canceling.

References: Event, Experiment, Sample Space.

TRIAL AND ERROR

This is a process that involves guessing or estimating the solution to a problem and then improving on the guess to obtain a more accurate answer. This is sometimes called trial and improvement, which seems a better description of the process.

Example. The length of a rectangular garden is 20 meters greater than its width. If the area of the garden is 900 square meters, find the width of the garden.

Solution. We begin with a guess. Suppose we guess the width of the garden to be 25 m. The length will be 45 m and the area $45 \times 25 = 1125 \, \text{m}^2$. The area is too big, so we reduce the width and try again. A convenient way of handling this problem is to put the numbers in a table so we have an orderly record of our progress.

Width, in m	Length (= width + 20 m), in m	Area, in m²	Comment
25	45	1125	Too big, reduce width
15	35	525	Too small, increase width
20	40	800	Too small, increase width
22	42	924	Too big, reduce width by a small amount
21.7	41.7	904.89	Too big, reduce width by a very small amount
21.6	41.6	898.56	Too small, increase width by an extremely small amount

You may consider the answer of 21.6 m to be accurate enough, or you may wish to try 21.65 m, say, and try further. The actual width, correct to four decimal places, is 21.6228 m.

References: Area, Rectangle.

TRIANGLE

A triangle is a polygon with three sides. See the entry Polygon for the angle sum of a triangle. The following triangles are studied under the respective entries: isosceles triangle, equilateral triangle, and scalene triangle.

Triangles will always tessellate. Suppose we draw any triangle. The one chosen in figure a is a scalene triangle and is shaded with an arrow drawn on it facing downward so the pattern of the tessellation is clear. To achieve the pattern the shaded tile is rotated through half a turn to obtain the white tile. In this way the shaded tile and the white tile placed together form a parallelogram, which is easy to tessellate. A tiling pattern is drawn in the figure.

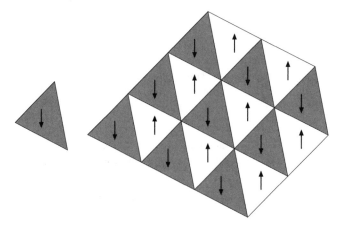

(a)

The area of a triangle is obtained using the length of its base and its height.

$$\text{Area} = \tfrac{1}{2} \text{ base} \times \text{height}$$

In the tessellation pattern in figure a, the area of the triangle is equal to half the area of the parallelogram. The area of a parallelogram is equal to base × height.

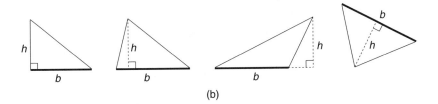

(b)

Figure b has different kinds of triangles that may help you to identify the base and the perpendicular height so that the area can be calculated. In fact any side of the triangle can be taken as the base, but choose one that matches a known perpendicular height.

Example. Find the area of the triangle drawn in figure c on the grid of squares of side 1 cm.

Solution. The triangle is scalene. Extensive calculations would have to be done to calculate the lengths of the base and the perpendicular height to find the area of the triangle. An alternative method is to draw a rectangle to enclose the triangle; in this example the rectangle happens to be a square.

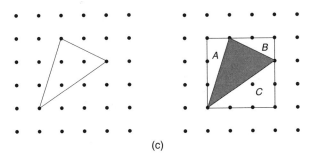

(c)

Write

Area of square = $9\,\text{cm}^2$	Area = length × width
Area of $A = 1.5\,\text{cm}^2$	Area = half the base (which is 3) × height (which is 1)
Area of $B = 1\,\text{cm}^2$	Similar to finding area of A
Area of $C = 3\,\text{cm}^2$	Similar to finding area of A

The areas in figure c are connected by the following equation:

$$\text{Area of square} = \text{Shaded triangle} + A + B + C$$
$$9 = \text{Shaded triangle} + 1.5 + 1 + 3$$
$$9 = \text{Shaded triangle} + 5.5$$

Area of shaded triangle $= 3.5 \, \text{cm}^2$ Solving the equation

The idea used to solve the above example is adapted to prove Pythagoras' theorem, as follows.

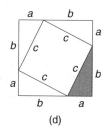

(d)

Figure d shows a larger square with sides of length $(a + b)$ units, a small one with sides of length c units, and four congruent right-angled triangles. First write

Area of larger square $= (a + b)(a + b)$	Area $=$ length $\times$ width
$= a^2 + 2ab + b^2$	Expanding the brackets
Area of smaller square $= c^2$	Area $=$ length $\times$ width
Area of each right-angled triangle $= \frac{1}{2}ab$	Area of triangle $=$ half-base $\times$ height
Sum of areas of 4 triangles $= 2ab$	

The areas in the figure are connected by the following equation:

Area of larger square $=$ Area of smaller square $+$ sum of areas of 4 triangles

$$a^2 + 2ab + b^2 = c^2 + 2ab$$
$$a^2 + b^2 = c^2 \qquad \text{Subtracting } 2ab \text{ from both sides of the equation}$$

This is the formula for Pythagoras' theorem in each of the four right-angled triangles.

For other methods of calculating the area of a triangle check see the entries Heron's Formula and Cosine Rule.

References: Angle Sum of a Triangle, Cosine Rule, Heron's Formula, Linear Equation, Polygon, Pythagoras' Theorem, Tessellations.

TRIANGLE NUMBERS

References: Difference Tables, Pascal's Triangle.

TRIGONOMETRIC GRAPHS

References: Amplitude, Circular Functions, Cycle, Frequency.

TRIGONOMETRY

Trigonometry, like Pythagoras' theorem, has a wide variety of uses and appears in many entries in this book. In this entry a brief introduction will be given; then trigonometry will be applied to solving right-angled triangles. It is essential to have a scientific calculator available, set in the degree mode. For other aspects of trigonometry see the following entries Angle between a Line and a Plane, Angle between Two Planes, Circular Functions, Cosine Rule, Sine Rule, Trigonometric Graphs.

A triangle has three angles and three sides, and the trigonometry we study in this entry uses two sides and one angle at a time. The sides of a right-angled triangle are given names in the following way (see figure a). Suppose the angle in the triangle we are focusing on is called theta, which is a letter of the Greek alphabet and is written as θ. The side of the triangle that is opposite to θ is called the opposite side, which is written as O. The side that is opposite to the right angle is called the hypotenuse, which is written as H. This is the longest side in the triangle. The remaining side, which is adjacent to angle θ, is called the adjacent side, and is written as A. In the two right-angled triangles in figure a the names of the sides are marked on the triangles. Depending on where angle θ is, the names of the sides O and A are interchanged.

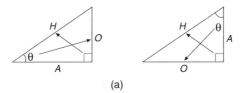

(a)

Now that we have established the names of the sides of the right-angled triangle in relation to a given angle, we can define the three trigonometric ratios sine, cosine, and tangent. Suppose we draw a right-angled triangle and let one of the acute angles be $\theta°$ (see figure b). The lengths of the sides are labeled O, A, and H in the usual way. This triangle is now enlarged to produce additional, similar right-angled triangles with sides of lengths O_1, A_1, and H_1, and so on. All the triangles are similar; they have a right angle and have an angle of $\theta°$. One of the properties of these triangles is that the ratios O/A are all equal to the same value, say k. This can be expressed as

$$\frac{O}{A} = \frac{O_1}{A_1} = \frac{O_2}{A_2} = \frac{O_3}{A_3} = k$$

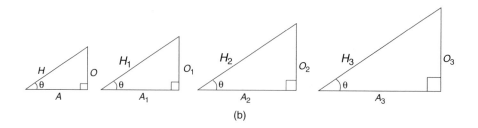

(b)

The value k is defined to be the *tangent* of the acute angle θ. Similarly, the ratios

$$\frac{O}{H} = \frac{O_1}{H_1} = \frac{O_2}{H_2} = \frac{O_3}{H_3}$$

are all equal to the *sine* of the angle θ, and the ratios

$$\frac{A}{H} = \frac{A_1}{H_1} = \frac{A_2}{H_2} = \frac{A_3}{H_3}$$

are all equal to the *cosine* of the angle θ. The abbreviation for sine is sin, that for cosine is cos, and that for tangent is tan.

These definitions are used to solve problems regarding right-angled triangles, and are summarized here (see figure c):

$$\sin\theta = \frac{O}{H}, \qquad \cos\theta = \frac{A}{H}, \qquad \tan\theta = \frac{O}{A}$$

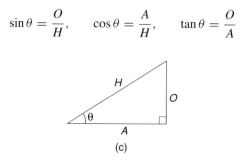

(c)

In the following examples ensure that your calculator is in degree mode, which will be indicated on the calculator screen as DEG.

Example 1. A ladder of length 3 meters is placed against a vertical wall and makes an angle of 70° with the horizontal ground (see figure d). Calculate how far up the wall the ladder reaches.

Solution. Figure d is a sketch of the problem, with the distance required marked as x. Write

$$O = x \qquad O \text{ is opposite to the angle of } \theta = 70°$$

$$H = 3 \qquad H \text{ is opposite the right angle}$$

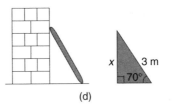

(d)

Select the trigonometric ratio that uses O and H, which is the sine:

$\sin 70° = \frac{x}{3}$ $\qquad\qquad$ $\sin \theta = \dfrac{O}{H}$

$3 \times \sin 70° = x$ $\qquad\qquad$ Multiplying both sides of the equation by 3

$\qquad x = 2.82$ (to 2 dp) $\qquad$ Using the calculator

The ladder reaches 2.82 meters up the wall.

Example 2. A road has a gradient of 1 in 7 (see figure e). Find the angle the road makes with the horizontal.

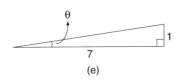

(e)

Solution. A gradient of 1 in 7 means that the road rises 1 meter vertically for every 7 meters in a horizontal direction. Suppose the road makes an angle of $\theta°$ with the horizontal; this is the angle we have to find. Write

$\qquad\qquad O = 1$ $\qquad$ O is opposite to the angle of θ

$\qquad\qquad A = 7$ $\qquad$ A is adjacent to the angle of θ

Select the trigonometric ratio that uses O and A, which is the tangent:

$\tan \theta = \frac{1}{7}$ $\qquad\qquad$ $\tan \theta = \dfrac{O}{A}$

$\theta = \tan^{-1}\left(\frac{1}{7}\right)$ $\qquad\qquad$ If $\tan \theta = a$, then $\tan^{-1} a = \theta$, provided θ is an acute angle

$\theta = 8.1°$ (to 1 dp) $\qquad$ Using the calculator

The road makes an angle of 8.1° with the horizontal.

Example 3. The length of a shadow cast by a tree is 24.5 meters. If the sun's rays are shining at an angle of 67° with the horizontal, calculate the height of the tree (see figure f).

Solution. The angle of 67° can be described as the angle of elevation of the sun. The right-angled triangle has its sides designated as O, A, H. Write

$$O = h \qquad O \text{ is opposite to the angle of } 47°$$

$$A = 24.5 \text{ m} \qquad A \text{ is adjacent to the angle of } 47°$$

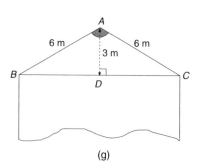

(f)

Select the trigonometric ratio that uses O and A. Write

$$\tan 47° = \frac{h}{24.5} \qquad\qquad \tan \theta = \frac{O}{A}$$

$$h = 24.5 \times \tan 47° \qquad \text{Multiplying both sides by 24.5}$$

$$h = 26.3 \quad \text{(to 1 dp)} \qquad \text{Using the calculator}$$

The height of the tree is 26.3 meters.

Example 4. Figure g shows the end view of a barn. The height of the roof is 3 meters and the length of the roofing is 6 meters. Find (a) the pitch of the roof, which is the angle shaded in the figure, and (b) the width of the barn, which is the distance BC.

(g)

Solution. Using the symmetry of the figure, we can see that angle *CAD* is half the shaded angle. The triangle *ACD* is right-angled, and we first find angle *CAD*.
   (a) Write (see figure h)

$$H = 6 \qquad H \text{ is opposite to the right angle}$$

$$A = 3 \qquad A \text{ is adjacent to angle } CAD$$

Select the trigonometric ratio that uses *H* and *A*, which is the cosine:

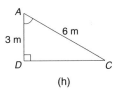

(h)

$$\cos (\text{angle } CAD) = \frac{3}{6} \qquad\qquad \cos \theta = \frac{A}{H}$$

$$\text{Angle } CAD = \cos^{-1}\left(\frac{3}{6}\right) \qquad \text{If } \cos \theta = a, \text{ then } \cos^{-1} a = \theta, \text{ provided } \theta \text{ is an acute angle}$$

$$\text{Angle } CAD = 60° \qquad \text{Using the calculator}$$

Since the shaded angle is twice angle *CAD*, the pitch of the roof is 120°.
   (b) The width of the barn is twice the length of *CD*. In triangle *ADC*

$$\text{Angle } CAD = 60° \qquad \text{Found in part (a)}$$

$$H = 6 \qquad H \text{ is opposite the right angle}$$

$$O = CD \qquad O \text{ is opposite the angle of } 60°$$

Select the trigonometric ratio that uses *H* and *O*, which is the sine:

$$\sin 60° = \frac{CD}{6} \qquad\qquad \sin \theta = \frac{O}{H}$$

$$6 \times \sin 60° = CD \qquad \text{Multiplying both sides of equation by 6}$$

$$CD = 5.196 \quad \text{(to 3 dp)} \qquad \text{Using the calculator}$$

Therefore

$$\text{Width of barn} = 10.392 \text{ meters} \qquad \text{Width} = 2 \times CD$$

References: Acute Angle, Angle of Elevation, Enlargement, Gradient, Horizontal, Pythagoras' Theorem, Ratio, Right Angle, Similar Figures, Vertical.

TRINOMIAL

A trinomial is an algebraic expression that contains three terms. Examples of trinomials are

$$x + 2a - b, \qquad 3ay - y^4 + 2, \qquad 2x^2 - 3x + 5$$

References: Binomial, Degree, Quadratic Equations.

TRIOMINO

Reference: Polyominoes.

TURNING POINTS

References: Concave, Maximum Value, Stationary Point.

$$U$$

UNITS

Reference: Metric Units.

UNITY

One.

UPPER QUARTILE

References: Cumulative Frequency, Interquartile Range.

V

VARIABLE

Reference: Constants.

VARIANCE

In statistics the variance of a set of data is the square of the standard deviation of the data:

$$\text{Variance} = \sigma^2$$

Reference: Standard Deviation.

VARIATION

There are two kinds of variation discussed in this text. One is direct variation, which is also called direct proportion, and the other is inverse variation, which is also called inverse proportion. Both of these terms are explained in the entry Proportion.

Reference: Proportion.

VECTOR

In this entry we discuss only two-dimensional vectors. Vectors are quantities that have both magnitude and direction. Examples of vectors are force, velocity, and acceleration. Speed is not a vector, because it has magnitude, but not direction. Suppose Ken is using a garden roller on the lawn and he is pulling with a force of 200 newtons. We have stated the magnitude of the force, but not the direction. We may add that Ken is pulling at an angle of 30° to the horizontal. Both the magnitude and the direction are needed to describe a vector quantity.

In this entry, a line segment represents a vector, and the length of the line segment represents the magnitude or size of the vector and the direction of the line segment is the direction of the vector. Ken is pulling the garden roller with a force of 200 newtons at an angle of 30° with the lawn. This vector can be represented by a line segment which is drawn at an angle of 30° with the horizontal (see figure a). An arrow on the line indicates the direction in which the force acts. Figure a is a scale drawing of the vector where 200 newtons is represented by the length of the line segment.

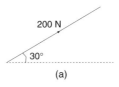

200 N

30°

(a)

There are different ways of writing vectors. Suppose a vector is represented on the grid in figure b by a line segment AB. In writing by hand, it is difficult to express vectors by thick letters as when using heavy type, and this notation will not be used in this text. Writing the vector $\underline{AB}$ means that its direction is from A to B. The vector $\underline{BA}$ is the negative of $\underline{AB}$, and its direction is from B to A.

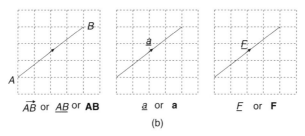

$\overrightarrow{AB}$ or $\underline{AB}$ or **AB** $\underline{a}$ or **a** $\underline{F}$ or **F**

(b)

Vectors are also expressed as 2 by 1 matrices, in a similar way to translations. The vector $\underline{AB}$ can be expressed as $\underline{AB} = \binom{4}{3}$, and the vector $\underline{BA} = \binom{-4}{-3}$. (For this positive and negative convention see the entry Translations.) When vectors are expressed as 2 by 1 matrices they are called column vectors, and can be added and subtracted in the following way.

Example 1. If $\underline{a} = \binom{-2}{1}$ and $\underline{b} = \binom{4}{-3}$, work out (a) $\underline{a} + \underline{b}$, (b) $\underline{a} - \underline{b}$, and (c) $2\underline{a} + 3\underline{b}$

Solution. For (a), write

$$\underline{a} + \underline{b} = \binom{-2}{1} + \binom{4}{-3}$$

$$= \binom{-2+4}{1+(-3)} \qquad \text{Adding the numbers in the column vectors}$$

$$= \binom{2}{-2} \qquad \text{Adding integers}$$

For (b), write

$$a - b = \begin{pmatrix} -2 \\ 1 \end{pmatrix} - \begin{pmatrix} 4 \\ -3 \end{pmatrix}$$

$$= \begin{pmatrix} -2 - 4 \\ 1 - (-3) \end{pmatrix} \qquad \text{Subtracting the numbers in the column vectors}$$

$$= \begin{pmatrix} -6 \\ 4 \end{pmatrix} \qquad \text{Subtracting integers}$$

For (c), write

$$2a + 3b = 2 \times \begin{pmatrix} -2 \\ 1 \end{pmatrix} + 3 \times \begin{pmatrix} 4 \\ -3 \end{pmatrix}$$

$$= \begin{pmatrix} -4 \\ 2 \end{pmatrix} + \begin{pmatrix} 12 \\ -9 \end{pmatrix} \qquad \text{Multiplying the first vector by 2 and the second by 3}$$

$$= \begin{pmatrix} 8 \\ -7 \end{pmatrix} \qquad \text{Adding the two vectors}$$

The magnitude, or size, of a vector is the length of the line segment that represents the vector. It is found using the theorem of Pythagoras.

Example 2. Find the magnitude of the vector $a = \begin{pmatrix} 4 \\ -3 \end{pmatrix}$.

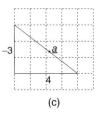

(c)

Solution. The vector is drawn in figure c. Write

$$a^2 = 4^2 + (-3)^2 \qquad \text{Pythagoras' Theorem}$$

$$a^2 = 16 + 9 \qquad \text{Squaring the numbers}$$

$$a^2 = 25$$

$$a = \sqrt{25} \qquad \text{Taking square roots}$$

$$a = 5$$

The magnitude of the vector is 5.

Vectors can also be added using a vector triangle, as explained in the following example.

Example 3. Pat loves swimming and decides to swim across a stream that flows uniformly at a speed of 2 km/h. In still water Pat can swim at 3 km/h. Her plan is to swim directly across the stream at right angles to the bank, but the water pulls her downstream. What is her resultant speed, and in what direction does she cross the river?

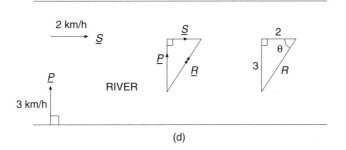

(d)

Solution. Let the velocity of the stream be $\underline{S}$ and the velocity of Pat be $\underline{P}$. These two vectors are added using a vector triangle in the following way. Draw the vector $\underline{P}$ (see figure d). Draw the vector $\underline{S}$ starting at the point where $\underline{P}$ ends. Then complete the triangle of vectors with the resultant vector $\underline{R}$ as the hypotenuse. Draw an arrow on the vector $\underline{R}$ in the direction from where $\underline{P}$ starts to where $\underline{S}$ ends, as shown in the figure. Write

$$R^2 = 3^2 + 2^2 \qquad \qquad \text{Theorem of Pythgoras}$$

$$R^2 = 9 + 4$$

$$R^2 = 13$$

$$R = 3.61 \quad \text{(to 2 dp)} \qquad \text{Taking the square root of 13}$$

$$\tan \theta = \tfrac{3}{2} \qquad \qquad \text{Tan} = \frac{\text{opposite}}{\text{adjacent}}, \text{using trigonometry}$$

$$\theta = \tan^{-1}\left(\frac{3}{2}\right) \qquad \text{If } \tan \theta = a, \text{ then } \theta = \tan^{-1}a$$

$$\theta = 56.3° \qquad \qquad \text{Using the calculator}$$

Pat's resultant speed is 3.61 km/h at an angle of 56.3° with the bank.

References: Components of a Vector, Line Segment, Pythagoras' Theorem, Translation, Trigonometry.

VECTOR TRIANGLE

References: Components of a Vector, Vector.

VELOCITY

Velocity is defined to be the rate at which the displacement of an object is changing as time changes. The basic unit of velocity, as of speed, is meters per second. Another common unit is kilometers per hour. If an object is traveling at a constant speed of v meters/second for t seconds and covers d meters, then the formulas connecting these quantities are

$$\text{distance} = \text{speed} \times \text{time}, \qquad \text{speed} = \frac{\text{distance}}{\text{time}}, \qquad \text{time} = \frac{\text{distance}}{\text{speed}}$$

Example. Amanda runs the 100 meters in 12.9 seconds. Find her speed in meters/second and in kilometers/hour, assuming she runs at a constant speed throughout.

Solution. Write

$$\text{Speed} = \frac{\text{distance}}{\text{time}}$$

$$= \frac{100}{12.9} \qquad \text{Substituting distance} = 100, \text{time} = 12.9$$

$$= 7.75 \quad \text{(to 2 dp)}$$

Amanda's speed is 7.75 meters/second.
   Now write

$$\text{Speed} = 7.75 \times 3.6 \qquad 1 \text{ meter/second} = 3.6 \text{ kilometers/hour}$$

$$= 27.9$$

Amanda's speed is 27.9 kilometers/hour.

Reference: Displacement.

VELOCITY–TIME GRAPHS

Reference: Acceleration.

VERTEX

Reference: Edge.

VERTICAL

Reference: Horizontal.

VERTICAL LINE TEST

Reference: Correspondence.

VERTICAL PLANE

Reference: Inclined Plane.

VERTICALLY OPPOSITE ANGLES

When two straight lines intersect at a vertex there are two pairs of congruent angles. Angles at the vertex that are opposite each other are called vertically opposite angles, and are equal in size. In this geometry theorem, "vertically" has no reference to the word vertical, but is derived from the word vertex (see figure a):

Angle a = angle b and   Angle c = angle d

(a)

Example. Figure b shows an open pair of scissors. If angle $x = 47°$, find the size of angle y.

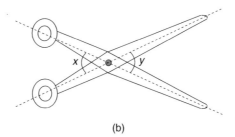

(b)

Solution. Write

$y = 47°$ Vertically opposite angles are equal

Reference: Geometry Theorems.

VOLUME

The volume of a solid shape is a measure of the three-dimensional space it occupies. It is measured in cubic units, which is written as units^3. We can find the volume of a solid shape, say a cuboid, by counting the number of cubes that its three-dimensional space occupies. The volume of a cuboid measuring 3 cm by 2 cm by 3 cm can be found by counting the number of cubic centimeters (abbreviated cm^3) it occupies. There are three layers of cubes, and in each layer there are six cubes, so the volume of the cuboid is $6 + 6 + 6 = 18$ cm^3.

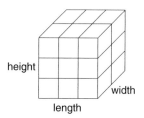

Alternatively, the volume of the cuboid can be found using the formula (see the figure)

Volume = length × width × height

$$= 3 \times 2 \times 3 \qquad \text{Substituting length} = 3, \text{width} = 2,$$
$$\text{and height} = 3$$

$$= 18 \text{ cm}^3$$

For examples of finding the volumes of well-known solids see the respective entries. The units commonly used for volume are

- Cubic millimeter, mm^3
- Cubic centimeter, cm^3
- Cubic meter, m^3

The relationships between these units are

$$1000 \text{ mm}^3 = 1 \text{ cm}^3$$
$$1,000,000 \text{ cm}^3 = 1 \text{ m}^3$$

When we are finding the volume of liquid that a vessel holds we say we are finding the capacity of the vessel.

References: Capacity, Cube, Cuboid, Metric Units.

W

WEST

Reference: East.

WHOLE NUMBERS

Reference: Integers.

WIDTH

Reference: Breadth.

X

X-AXIS

Reference: Cartesian Coordinates.

X COORDINATE

Reference: Cartesian Coordinates.

Y

$$Y = MX + C$$

References: Cartesian Coordinates, Gradient-Intercept Form, Graphs.

YARD

Reference: Imperial System of Units.

Z

ZENO'S PARADOX

Reference: Paradox.